Fair Science

WOMEN IN THE SCIENTIFIC COMMUNITY

Jonathan R. Cole

New York   Columbia University Press

Columbia University Press Morningside Edition 1987
Columbia University Press New York Guildford, Surrey

Printed in the United States of America

Library of Congress Cataloging-in-Publication Data

Cole, Jonathan R.
  Fair science.

  Reprint. Originally published: New York : Free
Press, © 1979.
  Bibliography: p.
  Includes index.
  1. Women in science. 2. Women in science—Social
aspects. I. Title.
Q130.C64 1987   508'.8042   87-13215
ISBN 0-231-06629-5 (pbk.)

  The extract on pp. 296–97 from "Gender and the Constitution" by Ruth Bader Ginsburg
[*University of Cincinnati Law Review* 44, no. 1 (1975): 1–42] has been reprinted here by permission of the *University of Cincinnati Law Review*.

This edition is reprinted by an arrangement with Macmillan Publishing Company, a division of
Macmillan, Inc.

For my brother,
Stephen

CONTENTS

TABLES

LEGAL CASES CITED

Albemarle Paper Co. et al. v. *Moody et al.*, 422 U.S. 405 (1975).
Arrington v. *Massachusetts Bay Transport Co.*, 306 F. Supp. 1355 (D. Mass. 1969).
Allan Bakke v. *The Regents of the University of California*, S.F. 23311, S. Ct. No. 31287, p. 2.
Brown v. *Board of Education*, 347 U.S. 483 (1954).
Bunting v. *Oregon*, 243 U.S. 426 (1917).
Byrne v. *Boadle*, Court of Exchequer, (1863), 2 H. & C. 722, 159 Eng. Rep. 299.
Califano v. *Goldfarb*, 45 U.S.L.W. 4237 (1977).
Carter v. *Gallagher*, 3 EPD Par. 8205 (D. Minn. 1971).
Chance v. *Board of Examiners*, 458 F. 2d 1167 (2d Cir. 1972).
Cheatwood v. *South Central Bell Telephone and Telegraph Company*, 303 F. Supp. 754 (1969).
Civil Rights Act of 1964—Title VII, Section 703, 42 U.S.C., 2000e-2 (1972).
Civil Rights Act of 1964—Title VII, Section 703(e) and (h).
Craig v. *Boren*, 45 U.S.L.W. 4057 (1976).
DeFunis v. *Odegaard*, 94 S. Ct. 1704 (1974).
Diaz v. *Pan American World Airways, Inc.*, 442 F. 2d 385 (5th Cir. 1971), cert. denied, 404 U.S. 950 (1971).
Doe v. *Bolton*, 410 U.S. 179 (1973).
Dothard v. *Rawlinson*, 97 S. Ct. 2720 (1977).
Fowler v. *Schwarzwalder*, 351 F. Supp. 721 (1972).
Furman v. *Georgia*, 408 U.S. 238 (1972).
Goesaert v. *Cleary*, 335 U.S. 464, 69 S. Ct. 198, 98 L. Ed. 163 (1948).
Griggs v. *Duke Power Co.*, 401 U.S. 424, 91 S. Ct. 849, 28 L. Ed. 2nd 158 (1971).
Hicks v. *Crown Zellerbach* Corp., 319 F. Supp. 314 (E. D. La. 1970).
Korematsu v. *United States*, 323 U.S. 214 (1944).
Lochner v. *New York*, 198 U.S. 45, 25 S. Ct. 539, 49 L. Ed. 937 (1905).
Muller v. *Oregon*, 208 U.S. 412, 28 S. Ct. 324, 52 L. Ed. 551 (1908).
Palsgraf v. *Long Island R. Co.*, 248 N.Y. 339, 162 N.E. 99 (1928).

PREFACE TO THE MORNINGSIDE EDITION

IN 1983 BARBARA MCCLINTOCK received a Nobel Prize for her work in biology. Some fifty years earlier, as a young and obviously precocious scientist, she had achieved worldwide recognition for her research on the cytogenetics of maize. Leading scientists of those years recognized that she was special. Marcus Rhoades, a prominent geneticist and close collaborator of hers at Cornell, put it forthrightly: "I've known a lot of famous scientists. But the only one I thought really was a genius was McClintock."[1] The irony about McClintock in those days at Cornell was that despite her reputation as a brilliant cytogeneticist, she did not have opportunities for regular tenure-track jobs at major research universities. Being a woman simply excluded her from such posts. Her situation was paradoxical: a "central woman" working at the forefront of her field, deeply respected by the most distinguished members of her field, and yet a "marginal woman" unable to gain full acceptance among the predominantly masculine community of scientists. She was not truly a full member of the club in those days; she was at once an insider and an outsider.

This paradox characterizes many of the women scientists who, through persistence and fortitude, were able to make it into science in the first several decades of the twentieth century.[2] In 1979, in *Fair Science,* I looked at those generations only tangentially; I focused on more recent generations of women in science. But even for these later generations, who came of age in the 1950s and 1960s, opportunities for full scientific careers were not equal for men and women. Much of the conspicuous discrimination that women confronted in the early part of this century had been eliminated, and, indeed, recognition for many aspects of work was based largely on performance criteria rather than gender. Still, opportunities for high-ranking positions at the most distinguished univer-

---

1. Evelyn Fox Keller, *A Feeling for the Organism: The Life and Work of Barbara McClintock* (New York: W.H. Freeman, 1983), p. 50.

2. Women were not alone, of course, in terms of exclusion from central positions in American science during the first three decades of the century. Anti-Semitism made it nearly impossible for Jewish scientists to hold tenure-track jobs at major universities.

sities, and being honored with science's most prestigious prizes remained simply more limited for women. With new empirical data in hand, I outlined in *Fair Science* a trend toward greater equality between the sexes. Has the movement toward gender equality been completed over the last eight years? Have the observations made about women in *Fair Science* been transformed into historical observations? The evidence indicates clearly that gender equality in science has not been achieved. Are the problems that existed for women in science, and the central questions raised in *Fair Science*, still of contemporary relevance? The central questions, in fact, remain much as they were in 1979.

The past decade has been marked by persistence and change in the position of women in science.[3] The sheer number of women entering scientific occupations has grown dramatically since 1970. The percentage of women Ph.D.s in all of the sciences has more than doubled, but the percent of Ph.D.s awarded to women remains low in the physical sciences and engineering. This growth in numbers results from a variety of factors: increased opportunities for women in all professional occupations, including science; slowly changing societal values that make scientific careers for women more acceptable both inside and outside of science; and a simultaneous decline in the proportion of extremely able young men interested in scientific careers rather than ones in business, law, and medicine.

Despite changing numbers, differences persist between men and women both in their promotion to high rank and the age at which they achieve those ranks. Women continue to be less apt than men to be recognized for their work by the highest forms of honors, although the numbers and proportion of women elected to the National Academy of Sciences has risen markedly over the past decade. As of 1986, there were still only 50 women members of the Academy out of almost 1,500 members, when it would take about 150 members to have women represented roughly in proportion to their numbers in science. Thus, they remain underrepresented in this most prestigious organization despite a fivefold increase in their numbers since 1972, a rate of increase far in excess of the 50 percent increase in the size of the Academy.[4] Women are less apt than men to direct major laboratories or to organize large research groups. And salary differentials still exist, although they have been reduced somewhat, especially in the lower ranks. Each of these features of scientific life were noted in *Fair Science* in 1979; each continues to need careful analysis.

Recent evidence also suggests some gains made by women in the past decade. There seems to be little evidence today of systematic discrimination against women in initial hiring to academic jobs, little evidence that gender plays a role

---

3. For a recent comprehensive review of the empirical studies documenting these gender patterns in science, see Harriet Zuckerman, "Persistence and Change in the Careers of American Men and Women Scientists and Engineers: A Review of Current Research," in *Women: Their Underrepresentation and Career Differentials in Science and Engineering*, Linda S. Dix, Ed. (Washington, D.C.: National Academy of Sciences Press, 1987), pp. 123–156.

4. *Ibid.*

in access to publication or to scarce resources in the form of research grants and contracts. This is particularly so among the younger entrants into academic science.

A major puzzle, identified in *Fair Science*, continues to elude solution: male scientists tend to publish more scientific papers than women, and this is so for scientists matched by field, professional age, source of Ph.D., and other attributes. This disparity turns up early in careers, before differential resources, such as laboratory directorships, could influence the rate of production. After a dozen years we find that men have published almost twice as many papers as women. Perhaps more intriguing is the empirical fact that this disparity increases over the course of careers. Furthermore, these patterns of scientific productivity have not changed markedly over the past fifty years. They were found for men and women receiving their Ph.D.s in the 1930s, and they were found to be of roughly the same magnitude for a generation of men and women receiving their doctorates in 1970. There are now some fifty studies that have confirmed the findings reported in *Fair Science*.

These productivity differences are not statistical artifacts nor are they spurious. Plausible explanations for the differences have not proven to be adequate. The suggestion that women are excluded from various forms of publication or are treated differently in the journal review process have not been borne out by evidence. Women are as apt to publish papers by themselves, show no difference in their propensity toward collaborative publication, are as likely as men to be first authors on published work, and do not have their papers turned down by scientific journals at a higher rate than do men.

Another plausible explanation for the "productivity puzzle" is that women's responsibilities as wives and mothers interfere with the time that they can devote to scientific research—reproductivity interferes with productivity. That explanation was questioned with data presented in *Fair Science*, in which I showed that married women published more scientific papers than those who were unmarried, and those with children actually published as much or more than those without them. That finding was questioned, but it has not proven wrong in subsequent studies. In a variety of works completed since 1979, including a recent study by Harriet Zuckerman and me, the evidence has been consistent: marriage and motherhood do not adversely affect the research productivity of women scientists.[5] Marriage and motherhood entail costs for women scientists, but they are not felt in terms of research productivity. The gender differences in scientific productivity remain a puzzle in need of solution.

Some inequalities between men and women scientists have been affected by affirmative action plans introduced in the 1970s. Affirmative action, at least at universities and colleges, has led to more clearly defined procedures in searching for qualified candidates for jobs and has led to closer scrutiny of sal-

---

5. Jonathan R. Cole and Harriet Zuckerman, "Marriage, Motherhood, and Research Performance in Science," *Scientific American* 255, no. 2 (February 1987): 119–125.

ary inequities. But these are formal changes which may not produce significant substantive changes. The more basic question is: Have attitudes, values, and presuppositions about women as scientists changed among faculty members so that gender is treated in a neutral or even a positive way in the search process? I believe that there are signs of some changes in values, especially among younger men who have entered science over the past decade. These signs are found in the increased proportion of women being promoted to tenured associate and full professorships, even after government pressure for affirmative action has diminished.

When *Fair Science* was published, passage of the Equal Rights Amendment remained a possibility. Since 1980, under the Reagan administration, little pressure has been placed on universities to alter the composition of their faculties. While cases continue to be litigated under the Civil Rights and Equal Pay Acts, as well as under state statutes, there remains no clear direction outlined,in recent court decisions. The Supreme Court has moved toward qualified support of affirmative action plans involving temporary quotas, but these cases have not involved academic hiring or promotion involving women.[6] Thus, the legal status of women in science remains cloudy and much as it was when the book was first published.

While *Fair Science* created a significant amount of discussion and was followed by substantial research on issues of formal types of recognition, I believe that key parts of the book remain relatively unexplored, despite the potential for elaboration. For one, scholars have not, by and large, attended to problems of reputation-building in science (chapter 4), and more specifically, to the differences in the reputations of men and women scientists. Although we should not undervalue the importance of formal types of recognition, that is, departmental affiliation, academic rank and title, and salary, many scientists place the highest value on reputation among peers. And the evidence presented in this book suggests that there are substantial differences in both the visibility and perceived quality of work produced by men and women.

While there have been reductions in inequalities of status among men and women scientists since the 1960s, even in the area of promotion to tenured ranks, there remain substantial gaps between men and women in the domain of the informal activities of science. Evidence from recent interviews conducted by Harriet Zuckerman and me, as well as by other scholars, suggest that it is from the tangible and intangible experiences associated with doing science from day

---

6. As I write, the Supreme Court has produced its first major affirmative action decision focusing on the right of employers to use a limited quota system for promoting women over men, even when the men have scored somewhat higher on an employer's application. This decision allows employers to give priority to female job candidates in order to redress past job segregation, even where there is no evidence of past gender discrimination by the particular employer. In effect, the decision protects employers who want to use limited quota systems from reverse discrimination suits. The Court was divided sharply in this case; it remains unclear what implications the decision will have for academic promotions—particularly for promotions to high rank in academic science departments. Paul E. Johnson v. Transportation Agency, Santa Clara County, Calif. 55 U.S.L.W. 4379 (1987).

to day that women rightly feel most excluded.[7] To say that women of science have now entered the central scientific community, and that they have achieved formal equality with men in many measurable ways, does not say that the opportunities for teenage boys and girls interested in science are equally open. Nor does it say that women who have chosen science have an equal chance of ending up in the inner circles of science, nor that they will be equal participants in the "invisible colleges" of the scientific establishment. Resistance to full participation, to the full citizenship of women in the scientific community, continues to exist.

Many women continue to be excluded from the very activities that allow for full participation and growth, for productivity and change. These are, by and large, the informal activities of science: the heated discussions and debates in the laboratory, inclusion in the inner core of the invisible colleges, and full participation in the social networks where scientists air current ideas and generate new ones. This is not invariably true, but it is a situation reported by many women, who say with unusual frequency that many male colleagues who only talk "shop" with men in their departments, invariably turn to talk about family, vacations, and other nonscientific topics in their conversations with women. Thus, women continue to be excluded frequently from the exchanges that help to shape scientific taste and sharpen the eye for a good research problem. Many women perceive that they remain excluded from those activities that define full membership in a community. In *Fair Science* I began to discuss these issues of citizenship. They remain central issues that still require substantial examination, particularly in terms of the way that organizational structures, such as laboratories, departments, and universities, influence the character of the informal interactions in the everyday life of scientists.

Fair Science contains an extensive discussion of aspects of justice within the scientific community. In terms used by sociologists, the question is raised: Do meritocratic, or universalistic, principles predominate over biased, or particularistic, principles in the distribution of scientific positions and recognition? A theoretical distinction of some importance was not sufficiently articulated in the book, however. In more recent work I have begun to examine the relationship between local injustices and systemic justices. There is an important distinction to be made here. In *Fair Science* I argue that rationally based universalism tends to predominate over particularism in science, although there are notable departures from this ideal. That conclusion applies to science as a social system, and is drawn therefore from the aggregative level of analysis. This does not mean, of course, that many of the specific decisions about hiring, promotion, salaries, awards, and other forms of recognition are meted out on purely meritocratic bases.[8] There are in fact many instances of local injustices. But, it

---

7. A more extended discussion of some of these ideas can be found in J. R. Cole, "Women in Science," *American Scientist* (July–August 1981):385–391.

8. I develop this idea further in J. R. Cole, "The Paradox of Individual Particularism and Institutional Universalism," in *Justice and the Lottery*, ed. Jon Elster (Cambridge, England: Cambridge University Press, 1987).

is possible to have biased decisions at a local level, which when aggregated suggest high levels of fairness. This seems paradoxical, of course. The critical idea, which needs development, lies in the dynamic social processes that sort candidates for jobs, promotions, salary increases, and awards. There may be many eligible candidates for scarce positions or awards. The determination of who should be hired, promoted, awarded, or given a raise among those "equally" qualified, may depend on particularistic criteria. Yet all or most of the relatively few candidates in the pool are considerably "better" than the many who are not. Thus if we compare those who receive awards with those who do not, quality tends to be correlated with recognition. At the aggregate level, merit turns out to be associated with recognition. This need not be the outcome of reward systems, but it does seem to operate in science. In sum, there may be a substantial degree of individual particularistic behavior in the decisions about rewards and yet at the level of the social system, the larger outcome can point to significant levels of universalism.

In *Fair Science,* I adopted a particular point of view about the nature of discrimination and fairness. My definition and perspective were predicated on the belief that in a just society, rewards were distributed on the basis of the quality of role performance, regardless of other attributes of individuals. The quintessentially "just" community would be one that disregarded "functionally irrelevant" characteristics such as race, religion, and sex, and rewarded excellence wherever it was found. Interesting questions have been debated recently, however, on the legitimate bases for distributing rewards and recognition.[9] Suppose that performance was not the sole criterion used; suppose that in order to further greater systemic equality, gender was used as one criterion in the distribution of rewards. Should research and other forms of scientific performance be the central or only value guiding the development of reward systems, or should larger questions of equity, into which gender or race is built, play a role in developing reward systems? One might argue that the secondary and tertiary consequences of the domination of systemic equity principles over individual equity principles would benefit science in the longer run. The perception, for example, that opportunities were open to them might lead more young women to choose careers in science. However, what cost might there be for the development of scientific knowledge and for the basic normative order of science if the bases of reward were to shift, in part, from performance to social attributes? These questions about the nature of justice and the consequences of accepting one or another theory of justice need to be debated so that our presuppositions about what constitutes just processes and outcomes are better understood.

---

9. For an extended discussion of procedural and substantive theories of justice, see papers by Stephen Cole and Owen Fiss in *Women and the Pursuit of Science,* ed. Harriet Zuckerman, Jonathan R. Cole, and John Bruer (New York: W.W. Norton, 1987, forthcoming).

Although scientific life has changed significantly for women since 1979, few of the major problems identified in *Fair Science* have disappeared. No self-correcting mechanisms within science have eliminated these problems, and no external forces have caused them to fade away. The story of women in American science, past and present, cannot be translated into a simple allegory; it is not a story of the forces of good versus the forces of evil. It is tempting to view history in terms of such polarities, but they are almost always false to the complexities of social life. The history of women in science is, in fact, a complex story—one which was, until quite recently, both about exclusion from principal avenues of reward and recognition, and about inclusion and recognition for certain forms of research efforts and collaborations. For too long women labored in marginal positions, and men, who predominated throughout the profession, often held false stereotypes about women's natural abilities and their abilities to combine domestic and scientific life. These beliefs were converted into self-fulfilling prophecies: men, who controlled resources, defined women as less able, and, by not allocating resources to them on an equal basis, found that their performances often confirmed their initial beliefs.

But there is another side to this story that is often omitted from histories of women in science. Many women did have substantial research opportunities, worked with major figures in their fields, collaborated with them, or worked on their own. They fought hard to retain their autonomy and to receive the credit due them for their research work. If they were passive in terms of some forms of scientific recognition, they would often fight to have their research contributions recognized. And the women almost always won these battles for credit. They were not invariably or even typically excluded from authorship; their collaborations were acknowledged. Furthermore, their exclusion from basic career opportunities, while hardly preferred by them, provided them with the chance to devote all of their energies to something that they loved doing—research science. Their love for research led many extraordinarily talented women to accept research jobs at major universities rather than accept tenured professorships at teaching-oriented colleges, particularly at women's colleges where they felt they would lose touch with the research frontier of their field. Moreover, since many of these women entered science well before the women's movement of the 1960s, they tended to take it for granted that some types of jobs were not open to them. These beliefs and attitudes about exclusion and outright discrimination were derived from the cultural fabric of the times. In short, many of the distinguished and rank-and-file women scientists that Harriet Zuckerman and I have interviewed express ambivalence about their positions in the days before opportunities expanded for them.

Today, a young Barbara McClintock would not be excluded from a tenure-track position at a major university. But there are still those scientists, some of whom are extraordinarily eminent, who argue that basic biological or physiological differences between men and women make it unlikely that women will achieve preeminence in science. The renowned I. I. Rabi, surely one of America's leading twentieth-century physicists, has been quoted recently as saying: "It's simply different. . . . It makes it impossible [i.e., women's nervous systems] . . . for them to stay with the thing. I'm afraid there's no use quarreling with it, that's the way it is. Women may go into science, and they will do well enough, but they will never do great science." [10] Indeed, Rabi, who sponsored the work of future Nobel laureates and scores of other physicists who went on to distinguished careers, is reported never to have had a woman postdoctoral or graduate student, and he apparently did not support women for faculty positions. [11] Rabi is not an isolated case, of course. Many other distinguished scientists, some far younger than Rabi, also refused to sponsor the scientific careers of women.

Women still face formidable barriers before entering science. Why so few able and talented young women elect science as a career remains a fundamental problem that we know little about. Those who reach the starting line represented by the Ph.D., do so today with fewer fetters than did women in the past: their opportunities are greater, and their achievements more likely to be recognized and rewarded. But the problems identified in *Fair Science* are still faced by women who enter science. The puzzles identified remain puzzles; the explanations remain understudied and largely untested. In short, I believe that the questions I raised in *Fair Science* in 1979 are as germane now as they were when they first appeared in print—perhaps more so.

10. Quoted in Vivian Gornick, *Women in Science: Portraits From a World in Transition* (New York: Simon and Schuster, 1983), p. 36.
11. John S. Rigden, *Rabi Scientist and Citizen* (New York: Basic Books, 1987), p. 116.

PREFACE AND
ACKNOWLEDGMENTS

It can hardly be attributed to accident, or to the use of the King's English, that until 1971 the largest compilation of biographical information about living American scientists was entitled *American Men of Science*. Until recently women in science was not a subject of much concern. For one, historically there has been a striking paucity of women entering the traditionally man's world of science. For another, of those women who did pass through the gates into the scientific community, few moved from the back benches to central stage; few became leaders in science; and few received substantial formal or informal recognition for their achievements. Even those women who made truly outstanding contributions to the fund of scientific knowledge remained largely invisible both to the vast majority of their scientist colleagues and to members of the larger society. Occasionally a biography or two appeared about exceptions—the Marie Curies. But by and large a woman in science was part of a hidden community.

Only within the past decade has detailed, systematic, empirical work been pursued on the place of women in science. However, there are still very few scholars who have completed extensive studies of the careers of women scientists. This book represents a beginning at understanding the social and cultural forces that influence those careers. I hope that it will succeed in raising new questions in need of further inquiry—that it will have potential for elaboration.

Since the basic problems and questions I consider are discussed at some length in the Introduction, I will only mention here one principle that has guided my research: To fully understand the position of women in contemporary American science one must compare their situations, their dilemmas, their careers, with those of male scientists. Thus, although this work focuses on problems faced by women, it also is about the forces that shape scientific careers of men.

A great number of social scientists and friends have contributed to my inquiry and have helped me produce this book. I want to acknowledge a few who have been particularly helpful. This book is dedicated to my brother, Stephen Cole, my closest intellectual colleague. Steve initiated my interest in sociology and has since provided me with intellectual stimulation, challenges, and fun.

I am deeply indebted to my colleague and friend Robert K. Merton, who along the way reviewed the manuscript as only he could. Only those scholars who have experienced Merton as a critic can know fully the extraordinary influence he has on evolving ideas. Harriet Zuckerman and I have spent many hours discussing and debating the problems faced by women in science. She has been a great help in defining questions and in speculating on answers.

Special thanks are due to Bernard Barber and Elinor Barber who have provided invaluable critiques of the research as it progressed. Other colleagues have provided useful critical comments on portions of the manuscript, notably Peter Blau, Paul Brest, Cynthia Fuchs Epstein, William J. Goode, Alvin Klevorick, Evelyn Satinoff, Burton Singer, Phillip Teitlebaum, Donald J. Treiman, and James S. Young. Several particularly able graduate research assistants at Columbia helped collect the data used in the book: Mark Beatt, Kenneth Dym, Naomi Gerstel, Mark Johnson, Pamela Summey, and Marlene Warshawski. I thank William Kelly, Lindsey Harmon, and Clarebeth Cunningham of the National Research Council, and Helen S. Astin for making available materials that proved essential for the book. Errors that remain are, of course, my own.

There are a substantial number of scientists and scholars who have not had direct contact with my research, but whose work on sex roles has influenced the development of my ideas. In particular, I would like to thank Jessie Bernard, Rose L. Coser, Alice Rossi, Betty Friedan, Barbara Reskin, Ruth Bader Ginsburg, Eleanor Maccoby, Valerie Oppenheimer, and William Chafe.

A number of organizations facilitated this work and deserve special mention. Much of the writing was done during a year as a Fellow at the Center for Advanced Study in the Behavioral Sciences, Stanford, California. That unique environment gave me the opportunity for extended informal discussions with other Fellows about the place of women in contemporary science. Particular thanks to Gardner Lindzey, the Director of the Center, and to his staff, including Preston Cutler, Jane Kielsmeier, Chris Hoth, Mariam Gallagher, and Heather MacLane.

I thank the John Simon Guggenheim Memorial Foundation for a Fellowship to finish the manuscript. The Ford Foundation supported several aspects of this project, and at Ford a special thanks to Mariam Chamberlain for the work that she has done to foster basic research on problems of sex roles in contemporary society. I have benefited from Ford's grant to Columbia's program on sex roles and social change.

This inquiry would not have been initiated without the long term commitment of the National Science Foundation to the Columbia University Program in the Sociology of Science. Throughout the book's gestation period, I have had generous support from the NSF.

The Center for the Social Sciences at Columbia has been a hospitable environment in which to complete the final phases of the book. I have benefited from exchanges with Harold W. Watts, and in technical aspects of my work I have been aided by members of the Center's staff, including Claudia Iredell, Pnina

Grinberg, Renee Parker, and by my administrative assistant Madeline Simonson, who has done her utmost to keep me going from day to day.

I am indebted to Gladys Topkis, Kitty Moore, Elly Dickason, and Charles Smith of The Free Press for looking after the manuscript through its production phase.

Thanks to my mother, Sylvia Cole, for locating in Gray's *Elegy* the reference to "fair science," and for making sure that her son was aware of its several entendres.

Finally, special thanks to my wife Joanna who, though herself a student of seventeenth-century English prose and poetry, has raised penetrating questions and made useful observations about twentieth-century American science. She has been a source of strength throughout this enterprise.

Chapter 1

INTRODUCTION

ALTHOUGH MOST AMERICANS say they believe in equality, they have always shown the most extraordinary capacity to tolerate its absence. Behind this lies the mythical belief that social inequalities are not the result of social and political favoritism, but that they are generated out of differences in personal capacity; they are expressions of individual ability or lack of it. Inequality, which might be conceived of as the product of social vice, has been transformed into a symbol of social virtue. Tolerance for inequality has taken a variety of forms, of course, including a willingness to accept extraordinary variations in personal wealth and income; unequal distributions of power, prestige, and esteem; and profound inequality between the races and the sexes. Although strict individualism in its nineteenth-century clothing is no longer in vogue, it continues to affect the scenario that explains unequal social outcomes in terms of individual and group deficiencies. This orientation has served to reinforce the conditions making for self-fulfilling prophecies—conditions that tend to secure the maintenance of these same social inequalities. This has been particularly so in the cases of inequality between men and women in American society, where until recently inequality has been explained as being consistent with the "natural" differences between the sexes.[1]

---

1. There are by now a large number of works that examine historical perspectives on the "natural" differences between the sexes. For two particularly useful examples, see Rosalind Rosenberg, "In Search of Woman's Nature, 1950-1920," *Feminist Studies* 1, no. 2 (fall 1975): 141–154; Carol Tavris and Carole Offir, *The Longest War: Sex Differences in Perspective* (New York: Harcourt Brace Jovanovich, 1977), esp. Chapters 1 and 2. For many other sources that touch on this theme, see Chapter 6.

Gunnar Myrdal makes this point in a brief discussion in *An American Dilemma* that draws a parallel between the problems faced by American Negroes and American women.

As in the Negro problem, most men have accepted as self-evident, until recently, the doctrine that women had inferior endowments in most of those respects which carry prestige, power, and advantages in society, but that they were, at the same time, superior in some other respects. The arguments, when arguments were used, have been about the same: smaller brains, scarcity of geniuses and so on. The study of women's intelligence and personality has had broadly the same history as the one we record for Negroes. As in the case of the Negro, women themselves have been brought to believe in their inferiority of endowment. As the Negro was awarded his "place" in society, so there was a "woman's place." In both cases the rationalization was strongly believed that men, in confining them to this place, did not act against the true interest of the subordinate groups. The myth of the "contented woman," who did not want suffrage or other civil rights and equal opportunities, had the same social function as the myth of the "contented Negro." In both cases there was probably—in a static sense—often some truth behind the myth.[2]

No one today would seriously argue that throughout its history American society has treated the "other half" of its population equitably. Even in the early twentieth century women were labeled, as Myrdal's remarks suggest, as "physiologically feeble-minded," as biologically weak, as psychologically frail, and as incapable of sustained hard work, much less of artistic or scientific creativity.[3] Indeed, in 1908 the United States Supreme Court in *Muller* v. *Oregon* accepted the principle of "protective legislation," which was brought to the Court somewhat ironically in historical retrospect by the liberals of the day, who wanted to protect women against the ill effects of long working hours. Defined as less able, women have been cut off from educational, business, and other occupational life chances open to men. Since the conditions for a self-fulfilling prophecy have been put firmly in place, women have been forced to defend themselves against and to overcome the consequences of the prophecy.[4] They have been forced frequently to live through their men or to carve out marginal positions for themselves within traditional American culture. There are of course exceptions: women who have made it in a male-dominated society. But if we take even a cursory glance at the rolls of leaders in American business, in American arts, in American science, and indeed in all American institutions (with the exception of the family), we find few women among those who have been most honored and esteemed. And this has resulted largely from structural impediments to pathways leading to achievement.

---

2. Gunnar Myrdal, *An American Dilemma*, vol. 2 (New York: Harper & Brothers, 1944), pp. 1073, 1077.
3. These perceptions of women are elaborated on in Chapter 6.
4. The "inventor" of the concept of the self-fulfilling prophecy is Robert K. Merton. Since Merton's initial formulation the concept has been developed widely in the social sciences, and is one of those social science ideas that has entered the parlance of general culture. For the original formulation, see "The Self-Fulfilling Prophecy," *Antioch Review* (summer 1948): 193–210; reprinted in *Social Theory and Social Structure*, enlarged ed. (New York: Free Press, 1968), pp. 475–492.

So we begin with a general and hardly surprising observation that there has been and continues to be widespread sex-based discrimination, without attending here to a causal explanation of it. To leap, however, from the general conclusion of the existence of such discrimination to the even more universal generalization that the level and intensity of sex-based discrimination has been uniform in all American institutions, and at all times in our history, would be grossly to over-simplify a complex phenomenon. There is of course an impulse to draw global inferences from limited knowledge about the status of women; it makes our studies seem that much more important. But this undercuts a fundamental role to be played by social scientists: to specify the conditions under which sex-based discrimination obtains more or less strongly.

Sex discrimination can be widespread without being spread uniformly throughout a society. There are many diverse questions about the phenomenon that require thorough investigation: What are the extent, the intensity, the shapes, and the forms of sex-based discrimination in the occupational structure of American society? How do these dimensions differ among several occupational categories: between the professions and semiskilled or clerical workers? Within the professions, to take only one classification, is sex discrimination as widespread among scientific occupations as among the legal or medical professions? At an even finer level of resolution, is the magnitude of sex discrimination as extensive and as intense in academic-science occupations as in industrial-science occupations? Specific answers to these types of questions are anything but self-evident.

Sex discrimination has been packaged in many forms, and frequently this packaging differs in different institutional spheres. In some occupations it is next to impossible for women to obtain jobs or the training required for jobs, making questions of salary differences and promotions within these jobs moot. In other occupations inequalities in salaries attributed to discrimination may be large and widespread, but opportunities for mobility may be relatively open. In still other occupations the barriers to promotion may be so high as to effectively preclude women from prestigious positions. In certain occupations the major form of sex discrimination may lie in unequal fringe benefits for men and women, or in the total absence of some benefits, such as sick leaves to cover maternity.[5]

To understand better the causes and consequences of sex-based discrimination, I believe that we need detailed quantitative and qualitative studies that describe the patterns and analyze the processes of social stratification within occupations and industries, within different geographical regions, within different organizational settings. This book attends only to several aspects of the problem of the social status of women in one set of occupations, American science.

Science is a particularly good setting to begin such a larger study. There is

5. See, for one example of differential benefits for men and women, the recent United States Supreme Court decision allowing employers to refuse pregnancy disability benefits for female workers.

now a growing historical and sociological literature on the patterns of stratification and rewards in American science.[6] And while most studies of stratification and inequality have examined patterns in the total occupational world, sociologists of science have been concerned particularly with examining inequality within the scientific community.

Several general findings have emerged from reports by these sociologists of science. Let us consider a number of the most germane. Of first importance, to a remarkable extent, perhaps unequaled in any other institution, the distribution of scientific rewards is based on the quality of scientific role performance in general, and on the quality of research publications in particular.[7] Forms of honorific recognition, including election to honorific societies and receipt of other prestigious prizes, seem to depend largely on the perceived merits of the research that scientists have produced.[8] On the less formal level of esteem and peer appraisals of scientists, the evaluations are also tied closely to the perceived quality of the research contributions of the scientist.[9] While appointments to prestigious academic departments are not as strongly linked to the quality of research performance as these other forms of recognition, they too are more heavily influenced by quality of research than by other social characteristics of scientists.[10] Finally, the receipt of research resources, which is both a means to recognition and a form of recognition in itself, is influenced primarily by the quality of proposals submitted for review and is not strongly influenced by the status characteristics of the applicants.[11]

The preponderance of evidence suggests, then, that science is a remarkably self-regulating social system, honoring great cognitive breakthroughs with extraordinary rewards and largely ignoring those scientists who make no mark with their research discoveries. Of course, there are notable cases of scientific fraud—plagiarism, falsifying data, and other forms of deviance. Science surely is not without its darker side, as incidents like the Piltdown Man forgery or the more recent Summarlain data fabrication clearly suggest.[12] But in the absence of

6. Extensive reference to works by sociologists of science on the reward system in science will be made throughout this book. Suffice it to note here only several of the basic works: Robert K. Merton, "Priorities in Scientific Discovery: A Chapter in the Sociology of Science,"*American Sociological Review* 22 (Dec. 1957): 635–659; Bernard B. Barber, *Science and the Social Order* (New York: Free Press, 1952); Warren Hagstrom, *The Scientific Community* (New York: Basic Books, 1965); Norman W. Storer, *The Social System of Science* (New York: Holt, Rinehart & Winston, 1966); Diana Crane, "Scientists at Major and Minor Universities: A Study of Productivity and Recognition," *American Sociological Review* 30 (Oct. 1965): 699–714; Jonathan R. Cole and Stephen Cole, *Social Stratification in Science* (Chicago: University of Chicago Press, 1973); Harriet Zuckerman, *Scientific Elite: Nobel Laureates in the United States* (New York: Free Press, 1977).

7. On the theory and supporting empirical evidence, see Cole and Cole, *op. cit.,* esp. Chapters 2–4.

8. See *ibid.,* Chapter 4, esp. pp. 93–95.

9. *Ibid.,* Chapter 4, pp. 99–122.

10. *Ibid.*

11. On the correlates of receipt of federal research funds, see Stephen Cole, Leonard Rubin, and Jonathan R. Cole, *Peer Review in the National Science Foundation* (Washington, D.C.: National Academy of Sciences, 1978).

12. Merton, "Priorities in Scientific Discovery" (*op. cit.*); Bernard Barber, "Resistance by Scientists to Scientific Discovery" *Science* 134 (1961): 596–602; Bernard Barber, John J. Lally, Julia Makarushka, and Daniel Sullivan, *Research on Human Subjects: Problems of Social Control in Medical Experimentation* (New York: Russell Sage Foundation, 1973); Hagstrom, *op. cit.;* Harriet Zuckerman, "Deviant Behavior and Social Control in Science," in *Deviance and Social Change,* vol. 1, ed. Edward Sagarin, Sage Annual Reviews of Studies in Deviance (Beverly Hills, Calif.: Sage Publications, 1977).

systematic contradicting data on the distribution of deviance of varying kinds, the historical record seems quite free of nefarious crimes and scandals; honor and recognition are rarely meted out to scientists who later turn out to be frauds. Indeed, throughout the hierarchy of recognition, there seems to be a high correspondence between the recognition received by scientists and the quality of the work that they have produced.

In short, although science is far from an unblemished meritocracy, there is a high degree of fairness in the distribution of scientific rewards, or to use the language of the sociologist of science, universalistic criteria of recognition (those which prescribe that scientists be judged solely on the merit of their performance rather than on the basis of characteristics such as race, sex, nationality, religion) predominate over particularistic criteria.[13]

These empirical findings are of singular importance because they focus on the tripartite relationship between rewards, false criteria of recognition, and scientific role performance. Few empirical studies of discrimination in other institutional spheres have been able to do this because they have lacked adequate empirical measures of the quality of role performance, despite the fact that almost all theoretical discussions of discrimination include productivity, performance, or creativity as key components in their models.[14] This is a major omission in most inquiries.

To be sure, these productivity concepts are not easily measured, and there are pitfalls to using simplistic measures. For example, to the extent that a woman's productivity is negatively affected by an accumulation of disadvantages in her background from T_1 to T_2, she is never truly on an equal footing with men at T_2 in her productivity potential.[15] While such possibilities must not be lost from view and should be taken into account in the formulation of models wherever possible, I will argue throughout this book that without some reasonable measure of productivity, it is virtually impossible to establish credible estimates of discrimination.

If productivity measures are essential, science becomes an even more strategic site for research on sex-based discrimination. As I suggest from the enumeration of several of the central findings in studies of social stratification in

---

13. There is now an extensive literature (and a lively continuing debate) in the sociology of science on the normative structure of science. For only a few highlights in this literature, see Robert K. Merton, "Science and the Democratic Social Structure," in *Social Theory and Social Structure* (*op. cit.*); Robert K. Merton, "The Ambivalence of Scientists," *Bulletin of the Johns Hopkins Hospital* 112 (1963): 77–97; Hagstrom, *op. cit.*; André Cournand and Harriet Zuckerman, "The Code of Science: Analysis and Some Reflections on Its Future," *Studium Generale* 23 (1970): 941–962; Ian Mitroff, *The Subjective Side of Science* (New York: American Elsevier Publishing Co., 1974); David O. Edge and Michael J. Mulkay, *Astronomy Transformed* (New York: John Wiley & Sons, 1976).

14. This point is expanded in Chapter 2, where appropriate references can be found.

15. Here we refer to the process of accumulating disadvantage, which is a theme in this book. For works in both sociology and economics that address processes of accumulating advantage and disadvantage, see Robert K. Merton, *The Sociology of Science: Theoretical and Empirical Investigations* (Chicago: University of Chicago Press, 1973), pp. 273, 416, 439–459; Cole and Cole, *op. cit.*; Zuckerman, *Scientific Elite* (*op. cit.*); Harriet Zuckerman, "Stratification in American Science,"*Sociological Inquiry* 40, No. 1 (spring 1970): 235–257; Paul D. Allison and John A. Stewart, "Productivity Differences Among Scientists: Evidence for Accumulative Advantage," *American Sociological Review* 39, No. 4 (Aug. 1974): 596–606; John Brittain, *The Inheritance of Socioeconomic Status* (Washington, D.C.: Brookings Institution, 1976).

science, work in the sociology of science has concentrated to a significant degree over the past decade on developing measures of the quality of scientific performance. While this was done initially with the intent to outline general patterns of rewards, it becomes an essential tool in evaluating the more focused questions about the level of sex-based discrimination in science. At the same time we should note that we still do not have as variegated a set of measures of performance as we would like to cover nonresearch roles such as teaching and administrative skills. But when it comes to measuring the research role of working scientists the indicators are fairly well articulated and tested.[16]

Productivity can be used not only as a key independent variable but also as an important dependent variable in the study of sex inequality. If research productivity differentials exist between male and female scientists, an explanation for these differences becomes an important part of analyzing the social processes that produce inequalities.

Although this book is about woman's place in the scientific community, it follows more general work on social stratification in science. A word about the shape of stratification in science and about one process that seems to strongly influence the contours of that shape is in order. If science is distinguished by its level of meritocracy, it is also unusual in the extent of its inequality. At first this combination may seem contradictory, or at least paradoxical. But it is not. A social system may produce extraordinary levels of inequality among its members and still be a meritocratic system if the distribution of rewards is based upon uniformly rational and universalistic criteria.[17] This seems to be so in science— or more precisely, it is so to the extent that we have studied the reward system.

The simple fact is that there are relatively few scientists who are talented enough to make extraordinary discoveries. It is these few, the elite, who receive the lion's share of rewards and honors in science; who produce the overwhelming majority of all papers, including, of course, the important ones. In fact, it has been reasonably estimated that roughly 10 percent of all scientists account for 50 percent of all scientific papers published.[18] In an earlier work, Stephen Cole and I found that the small elite of science, who are rewarded largely on the basis of the perceived quality of their work by peers, dominate receipt of the most coveted science prizes.[19] In the context of the history of science, it is the Nobel laureate, the National Academy of Sciences member, the occupant of "the Forty-First Chair," and the handful of other extraordinarily prolific scientists who are the

---

16. For a brief list of works on the relationship between citation counts and measures of quality of scientific research see footnote 35.

17. Plainly, universalistic criteria are not necessarily rational in terms of fulfilling the functions of an institution. If all scientists were rewarded on the basis of their hair or eye color, for instance, the application of uniform standards would be universalistic but not rational in terms of the goals of science. Rational criteria are those that are directly related to the basic goals of the institution or social system. Universalistic rewards for the quality of scientific discoveries would involve rational criteria, since the growth of knowledge through new scientific discoveries is the basic aim of the scientific enterprise.

18. Derek de Solla Price, *Little Science, Big Science* (New York: Columbia University Press, 1963).

19. Cole and Cole, *op. cit.*, Chapters 4 and 5.

bricklayers of science. They lay the foundation for the great historical transformations of scientific thought. Because of the skewed distribution of talent and the universalistic reward system, we get both an approximation of meritocracy and, simultaneously, extraordinary inequality in science.

This end product is facilitated by the process of accumulation of advantage and disadvantage. Social standing in the hierarchy of science is a result of a funneling process. Most young scientists, ironically, start at the top. A relatively small proportion of American graduate schools produce the majority of science Ph.D.'s. Between 1950 and 1966 approximately 5 percent of the doctorate-granting institutions supplied a quarter of all the doctorates, 11 percent supplied half; 25 percent supplied fully 75 percent of all the doctorates. And these doctoral suppliers are the most prestigious schools in the country. Furthermore, four universities—Harvard, Columbia, Berkeley, and Princeton—conferred degrees on about one-half of the American-trained Nobel laureates during the 1950–66 period. Finally, as of 1969 roughly three-quarters of National Academy members had received degrees from just ten universities.[20]

This suggests a strong relationship between early location in the stratification system and ultimate success. Only a small proportion of the students who are trained at centers of excellence stay there. Those who show a "spark," who are "comers," are kept on in teaching and research positions; those who appear less able (and errors in judgment are made) move to less distinguished academic settings or out of academic life. But more importantly, those who are labeled as potential stars have opportunities to interact, to exchange ideas, and to collaborate with exceptionally talented and established older scientists. Almost all autobiographies of famous scientists point to a special "master-apprentice" relationship. James D. Watson talks of Delbruck, of Luria, of life at the Cavendish; Alvarez, Libby, McMillan, and Seaborg were students of Ernest Orlando Lawrence; Chamberlain, Lee, Yang, Bethe, and Bloch all had studied with Fermi.[21] I could go on indefinitely. To summarize, by virtue of being in top graduate departments and "apprenticing" and interacting with influential scientists, by having superior resources and facilities to carry out research, some scientists have social advantages. Once located in a favorable position in this initial phase, the probabilities may no longer be the same for two scientists of equal abilities. The one who is strategically located in the stratification system may have a series of accumulating advantages over the one who is not. These advantages ultimately affect a scientist's achievement. This is not to say, incidentally, that the "advantaged" scientist does not also face a series of tough tests along the way.[22] As Harriet Zuckerman has observed:

> The accumulation of advantage in science involves both getting ahead initially and moving farther and farther in front. . . . [E]nlarged access to resources enables

---

20. Zuckerman, *Scientific Elite* (*op. cit.*), pp. 86–95.

21. *Ibid.*, Chapter 4.

22. Cole and Cole, *op. cit.*, p. 75.

talented individuals to perform more effectively and, in accord with the merit principle of science, to be rewarded more copiously. Since collegial recognition and esteem are the prime rewards for scientific achievement, honored standing must be converted into other assets more directly applicable to further occupational achievement: including assets such as influence in allocative decisions; access to gatekeeping positions... and, most important for the accumulation of advantage, new facilities for work. These resources can be used to improve subsequent role performance, thus renewing the cycle.[23]

The skewed distributions of scientific productivity and of subsequent rewards not only result from the "rich getting richer" but also from "the poor getting poorer." This is not the result of a zero-sum game; profits accruing to scientists for their discoveries rarely come out of the assets of others in the community. But there is an accumulation of disadvantage in science. Without resources and facilities, without daily or at least occasional interaction with other scientists with novel ideas, the probability of a scientist with a great deal of talent continuing to be a productive researcher is low. When subsets of scientists do not have access to opportunities for interaction and collaboration, they are apt to drop out of the small group of productive scientists. Thus the growing inequality between the "haves" and the "have-nots" of science results in part from a decline in productivity among those scientists who started their careers as moderately productive researchers, while the elite either remain moderately or highly prolific researchers.[24]

Potentially, this process can influence the careers of women scientists. If for one reason or another they do not attend superior training centers, do not apprentice for master scientists, do not have facilities to carry out their research ideas, their chances for recognition and esteem are diminished. If women are more apt to be in such disadvantaged positions than their male counterparts, then the career histories of men and women of science may be explained in part by processes of accumulation of advantage and disadvantage.

Few empirical studies of social stratification in science have focused on women.[25] And since many of the studies that have been completed have been based upon random samples of scientists—a decision that frequently turns up few women—or upon samples of scientists in fields that have few women to begin with, the meritocratic conclusions reported here stand more for patterns of re-

---

23. Zuckerman, *Scientific Elite* (*op. cit.*), pp. 61–62. The footnotes that appear in the original quote have been deleted here.

24. Stephen Cole has made me aware of this process of increasing inequality through attrition of publications among the lower producers. He has developed empirical materials to demonstrate this phenomenon that will be published shortly.

25. There are a growing number of autobiographical accounts of problems faced by women in science, but there are still only a few systematic, larger-scale, quantitative, empirical studies of women and men in science. Throughout this book I will refer to those studies that have been completed that do attempt empirical analysis. To note just four of the more important recent works, see Helen S. Astin, *The Woman Doctorate in America* (New York: Russell Sage Foundation, 1969); Alan E. Bayer and Helen S. Astin, "Sex Differentials in the Academic Reward System," *Science* 188 (1975): 796–802; Barbara F. Reskin, "Sex Differences in Status Attainment in Science: The Case of the Post-doctoral Fellowship," *American Sociological Review* 41, No. 4 (Aug. 1976): 597–612; Barbara F. Reskin, "Sex Differences in the Professional Life Chances of Chemists" (Ph.D. diss., University of Washington, Seattle, 1973).

wards for "men of science" than for all "scientists." I turned to the study of women in the scientific community because my previous work had just such limitations. I asked: Could an institution that so nearly approximates the ideals of a meritocracy fail to approximate such high standards in dealing with women of science? If so, this would be a strategic research site for examining conditions under which there is a notable deviation from an important general pattern.

This study turns, then, to the possibility of identifying systematic departures from the general pattern of predominant universalism. Are there sets of social conditions under which particularlistic standards predominate over universalistic ones in science? Are particularistic criteria applied more to some subsets of the scientific community than to others? Just as it was theoretically interesting and practically important to search out the general patterns in the distributions of rewards in science, so was it interesting and practical to search for conditions that make for marked departures from those patterns. The place of women in the scientific community provides an obvious case in which discrimination might be found to occur on a substantial scale.

There was some sound basis for believing that I would find notable differences when comparing the positions of male and female scientists. Consider the following ten widely held beliefs about the treatment of women in science.[26]

1. It is widely believed that women are neither fit for scientific careers nor interested in them. Furthermore, female students in universities and colleges believe that scientific careers are open only to women with the most exceptional talent—talent that must exceed that of men who enter the scientific community.

2. Many academics believe that women students are as able as men, but women are still uniformly discriminated against in admission to graduate school, and those that gain admission tend not to be granted financial assistance.

3. It is also widely believed by faculty members that more women are poor risks as graduate students. They are said to be less apt to complete their postgraduate degrees, and when they do, to take longer than men.

4. Many believe that women are less likely to have been trained at the most distinguished universities, since they are discriminated against in admissions and since their range of alternatives in choosing graduate schools is more limited than that of men.

5. It is believed not only that training women to be scientists is a poor investment, but also that if they get their degrees, they marry, have children, and stop working. Among those few who return to work, their skills and knowledge, which were once fully up-to-date, have become obsolete.

6. It is commonly believed that women scientists individually, and of course collectively, do not contribute as much to science as men; they are simply less productive scientists. Furthermore, it is believed that the obligations of family and children in addition to careers account for much of the observed difference.

7. There is a widespread belief that women scientists are consistently un-

26. The beliefs listed below are reviewed in terms of the existing literature of them in Harriet Zuckerman and Jonathan R. Cole, "Women in American Science," *Minerva* 13, No. 1 (spring 1975): 82–102.

derrewarded with regard to promotion and the class of university to which they are appointed, regardless of the quality of their work.

8. It is believed that women receive less informal recognition from colleagues for the work that they have produced, again regardless of its quality.

9. As a corollary to the above claims, it is believed that since women scientists have tended over the course of their careers to be deprived of resources needed to do good work—a condition that widens the gap between their scientific achievements and those of men—they suffer the consequences of accumulative disadvantage.

10. It is widely believed that the social conditions that have impeded access to and achievement in science are much the same for women today as they were fifty to seventy-five years ago.

In short, when it comes to the treatment of female scientists the basic equity in the distribution of scientific facilities, resources, rewards, and opportunities is being questioned. For many of the claimants, the efficient cause for these inequities is a network of "old boys," most of whom are powerful and know one another; as eminent scientists located in the most prestigious academic-science departments, they can effectively control through informal means appointments to the best departments, receipt of fellowships, and forms of honorific recognition, as well as publication outlets and funding sources. The belief goes beyond the matter of potential control, of course, to the actual and systematic use of this control to discriminate against women scientists. Furthermore, the claims suggest that discrimination exists in more subtle forms—at the level of informal exchanges of ideas and interaction. Women have limited access to the eminent "old boys" and are not as apt as their male counterparts to become their students, apprentices, or collaborators. These limited opportunities are reflected later on in the form of concrete rewards. These are bold claims, but they allow us to formulate a set of working hypotheses about the position and treatment of women in science.

A preliminary examination of the subject appears in my earlier book with Stephen Cole on social stratification in science, but the data then in hand could provide only limited estimates of the situation.[27] Also, Harriet Zuckerman and I have examined some of the above-listed beliefs on sex discrimination in science by reviewing the research evidence on the subject that could be pieced together from the existing literature.[28]

This literature, as fragmentary as it is, suggests that the observed distributions of rank and rewards in the domain of science can be thought of as the resultant of processes of "social and self-selection." Of these two generic stratification processes, much less seems to be known, through intensive empirical investigation, about the workings of self-selection. In describing systems of allocation of rewards and facilities, I proceed from the assumption that it is as important for understanding the observed outcomes to consider the factors in-

---

27. See Cole and Cole, *op. cit.*, Chapter 5.
28. Zuckerman and Cole, *op. cit.*

fluencing decisions of people to compete for rewards and facilities as it is to consider the factors determining the actual allocations among the competitors. Plainly, those who elect not to enter the competition—say, by not applying for jobs or fellowships, by not applying for admission to science departments— affect the actual distribution of men and women within jobs, within graduate science departments, or among fellowship recipients. Observed disporportions in the numbers of men and women scientists may be as much a result of (socially and culturally supported) self-selection as of social selection by colleges, universities, and other gate-keeping institutions.

In fact, the review of the literature by Zuckerman and myself indicated that processes of self-selection play a significant role in the final distribution of men and women scientists. We began, of course, with the striking fact that science has been cultivated almost exclusively by men. In 1967 roughly 11 percent of all Ph.D.'s in science were women, and although that figure has risen sharply in the last decade to 23 percent in 1976, when we look at the overall community of active scientists there is a strikingly small proportion of women, and an even smaller proportion in many of the physical and biological sciences.[29] Evidence in the extant literature indicates that smaller proportions of women than men are led to aspire to scientific careers; smaller proportions of women elect one or another science as an undergraduate major; smaller proportions apply for admission to graduate study in science, and those who do have generally made the decision to apply later than men. When combined with higher attrition rates among women graduate students in science (including those with financial support), these socially induced self-selective processes make for the comparatively small numbers of women holding the doctorate in science.[30]

In short, women scientists are a highly self-selected group—in a sense, a group of survivors. Like their male counterparts, they are far above average in terms of measured intelligence; but unlike men of science, they have gone against the grain of occupational stereotyping to enter a "man's world"; they have scaled, in all likelihood, a series of social and psychological high hurdles before reaching the take-off stage for a scientific career: earning the doctorate.

Then, more often than not, they have placed their careers on a roughly equal plane with other roles that have historically tended to dominate women's lives. Until the recent post-World War II growth of female participation in all sectors of the labor force, family and marital status tended to dominate a great range of

---

29. On sources of female doctorate production, see the annual summary reports produced by the National Research Council, National Academy of Sciences. These reports were prepared by the Office of Scientific Personnel. Data are available from this source until the early 1970s, when the annual reports were discontinued. See also Lindsey R. Harmon and Herbert Soldz, *Doctorate Production in United States Universities* (Washington, D. C.: National Academy of Sciences–National Research Council, 1963).

30. A review of the literature that deals with the process of self-selection out of science, with admission rates, and with attrition from graduate schools appears in Zuckerman and Cole, *op. cit.* For a discussion of some of the social and cultural barriers to scientific careers for women that frequently lead them to drop out of science, see, among others, the recent autobiographical sketches in Sara Ruddick and Pamela Daniels, eds., *Working It Out* (New York: Pantheon, 1977), particularly discussions by Evelyn Fox Keller, "The Anomaly of a Woman in Physics" (pp. 77–91), and Amelie Oksenberg Rorty, "Dependency, Individuality, and Work" (pp. 38–54).

women's activities. Not so for female scientists. Even among the average work-ingwomen of today, who often continue to work primarily to supplement family incomes, work is not given the same position of importance as are marriage and motherhood. But this has never been the case among academic female scientists. For those who passed through the gate-keeping processes and achieved doctor-ates in science, occupation always has had a position of primary importance—in many cases second to none. This fact is reflected in and partially explains the extraordinarily low rates of marriage among female scientists throughout most of this century. Today the primacy of the status of scientist is further reflected in the lengths to which women will go to pursue their careers—the varieties of child care that they use, and their growing willingness even to commute long distances for extended periods of time to hold down desirable jobs.

Nonetheless, on the whole, women scientists are not as mobile as men, more often feeling tied to a particular geographic location because of the work re-quirements of their husbands.[31] To what extent do women scientists refrain from applying for or accepting positions in outstanding departments located away from their husbands' place of work? How does this limit their bargaining position, as compared to men scientists, in the use of offers from competing colleges and universities to improve their salaries and other perquisites at their own institu-tions? The combined results of accommodative self-selection resulting in restric-tions on the actual mobility of married women scientists and the immobility imputed to women scientists by their colleagues obviously contribute to a process of accumulating disadvantage. But the extent of the effects of self-selection has yet to be adequately estimated.

Despite the lengths to which women scientists will go to maintain their careers, their participation in the marketplace of science still differs fundamen-tally from that of men scientists as a partial result of self-selection. For example, it seems that women who manage to get the doctorate tend to exhibit work patterns differing from those adopted by their male counterparts: In the aggre-gate, they do more teaching and less research, even when the type of work setting is controlled. Furthermore, compared with men scientists, women scientists re-port differentially less interest in research than in teaching, and to this extent tend to exclude themselves from access to rewards and facilities in a social system of science that puts a premium on research productivity rather than on teaching.[32]

---

31. Evidence on this point is fragmentary, but see Gerald Marwell, Rachel Rosenfeld, and Seymour Spilerman, "Residence Location, Geographic Mobility, and the Attainments of Women in Academia" (1976), preprint. Consider also the observation by Amelie Rorty in the essay cited in note 30:

The wives of intellectuals and professional men are often as highly trained as their husbands, often as intent on their work and scholarly projects. But what typically happens is that, at a time when both are just starting work, the man gets a better job offer than the woman; the woman follows him and takes her chances on finding something within the vicinity. There is rarely anything to match his working conditions, his stimula-tion; she is lucky if she finds anything at all. (p. 47)

32. On the differential interest of women and men in research, see John K. Folger, Helen S. Astin, and Alan E. Bayer, *Human Resources and Higher Education* (New York: Russell Sage Foundation, 1970); Alan E. Bayer, "Teaching Faculty in Academe: 1972–73," Research Report no. 8 (Washington, D.C.: American Council on Education, 1973).

Further outcomes of self-selective processes may be seen in the job location of scientists. Women less often than men find themselves working in industrial science, and they tend to remain longer in the jobs that they have. At the same time, contrary to widespread belief, women scientists who interrupt their careers for childbearing generally do so for only short periods of time, and most of them promptly return to the labor force. They do not self-select themselves out of the labor market.[33]

Thus there is some evidence that processes of self-selection are in part responsible for both the distribution of women in science and the recognition they receive once they enter the community. But the evidence is fragmentary.

In sum, the question of whether men and women scientists differentially self-select themselves out of competition for various kinds of rewards and facilities, or did so more in the past than today, remains largely unanswered. To what extent are the inequalities that can be identified in the status attainment of men and women scientists attributable to processes that exist within colleges and universities, including sex-based discrimination, and to what extent are such inequalities attributable to exogenous social facts related to self-selection?

The operation of processes of social selection, of which discrimination is one form, is no better articulated in the existing literature. Some prior studies of rewards meted out to men and women scientists suggest that discrimination exists—but the measurement of sex-based discrimination is usually equated in most of these studies to the observed inequalities between the sexes.[34] We will argue that such simple identities are misleading and tend to confuse the basic issues about sex-based discrimination in science. If the literature yields a paucity of sophisticated studies that attempt to account for observed inequalities by including a set of explanatory factors such as scientific role performance, it also consists of few adequately designed comparisons of the status of men and women scientists. The literature is of little help in focusing systematically on the extent of sex-based discrimination in hiring and promotion; on peer recognition of men and women scientists; on comparative career models for men and women scientists; and on long-term historical changes in the status of women scientists. This book tries to fill some of these gaps. Concretely, I want to address in some detail the following questions:

Are women scientists subject to sex-based discrimination in matters of hiring, promotion to high rank, and the receipt of honorific recognition? Under what conditions is sex-based discrimination more or less likely to obtain?

How do patterns of scientific productivity—both quality and quantity of research output—compare for men and women scientists?

---

33. On the labor force participation of academic women, see Astin, *op. cit.*

34. The recent works that most often draw the simplistic equation between inequality and discrimination are affirmative action reports. It is rare to find multivariate statistical analyses of hiring data, although some more sophisticated analysis has been done in treating salary differences. I have not encountered any affirmative action reports that include measures of performance in their analysis of inequality. For detailed discussion of this point, see Chapter 7.

What is the "cost" of being a woman in the academic scientific community? Do inequalities between the sexes persist after patterns of scientific role performance are taken into account?

How visible to others in the scientific community is women's research compared to that of men? How does the perceived quality of their work compare?

Do the determinants of peer recognition in science differ for men and women?

What is the distribution of measured intelligence among men and women scientists? How does IQ relate to the scientific recognition they receive?

How have observed patterns of discrimination on the basis of sex changed since 1900?

What processes of accumulating advantage or disadvantage, if any, affect the career mobility of men and women scientists?

What are the methodological problems of "measuring" the extent of discrimination and of "proving" its occurrence? How do the standards of proof on such matters in the social sciences differ from the standards in the law?

What are some of the problems with establishing affirmative-action programs as remedies for the low proportion of women in the scientific community?

Although a substantial part of this book focuses directly on the question of sex-based discrimination in science, that issue remains only one of my concerns. As the above list of questions suggests, I am also interested in describing general processes of social stratification in science and in comparing those factors that influence achievement for men and for women. Some of these processes I can describe and model with data in hand, while others, for which I have no direct evidence, I can only conjecture about. The influence of "masters" on the careers of "apprentices," the influence of subtle processes of self-selection or of motivation on the quantity and quality of scientific productivity, are important to estimate, but in the absence of data for some of the key concepts, thorough investigation must wait for the future.

LIMITATIONS TO THIS STUDY

The results reported here are based largely on the analysis of three data sets that are described fully in the substantive chapters of the book. Two of them are "contemporary," insofar as they deal with scientists who received their doctorates within the past twenty years, and who, by-and-large, are still active in the social system of science. A historical data set also was created from a variety of sources dealing with several samples of scientists who received their doctorates over a period from 1911 to 1950. Although these are substantial data sets, they are subject to several limitations. I will note several of the limits to the data reported in this book, especially since I am dealing here with complex and sensitive issues.

The book contains information on the social standing of women and men in the physical, biological, and social sciences. But it does not contain information on women in all of the sciences. For example, there is little information about women in physics. In part this simply reflects the paucity of women physicists, but it also reflects a conscious effort on my part to select fields that have a significant number of women and to obtain some representation from scientific fields that may have varying social and cognitive structures. Consequently, more data are drawn from chemistry, from the biological sciences, and from a variety of social sciences than from fields such as astronomy or physics. I attempted to generate samples of women scientists that were sufficiently large so that I could construct multivariate models. It is arguable, of course, that there are different patterns of treatment for women in fields in which they are at best token representatives of their sex. Several conflicting hypotheses are plausible in the absence of data. One could argue that where there are virtually no women in a discipline they do not represent a real threat to men in positions of power and consequently they are treated more equitably than in fields where women are found in greater numbers. Correlatively, one could argue that in fields with few women sex-discrimination is less visible, sanctions for discrimination are infrequent, and men are more apt to define women as incapable of doing the work involved in those fields; hence there are high levels of discrimination. In short, it is possible that systematic analysis of fields not covered in this study will turn up somewhat different results.

Because of the number of fields that are covered and the number of variables used in some of the statistical models, there are points in the book where the size of subsamples on which the analysis is based becomes quite small. If one examines subsets of the data—for example, women biologists who are married and have two or three children—one reduces the sample size to a point where the statistical results may be unstable. Consequently, the reader should approach those findings based upon small samples with caution.

Discussion of questions raised in this book is based almost entirely on the analysis of quantitative data gathered for men and women scientists. There is a conspicuous absence of qualitative interviews with female scientists. I have of course "interviewed" scores of female scientists since beginning this study. But systematic qualitative data were not collected. This represents a shortcoming in the book, since it is my belief that qualitative and quantitative materials almost invariably complement each other. There may be a variety of subtle forms of discrimination against women that cannot be easily measured and quantified. The consistency of subjective perceptions by women scientists about the types and intensity of sex-based discrimination cannot be examined in the light of quantitative data gathered to test perceptions, except as these perceptions have been articulated in other sources. Additional research is required on the relation between the results obtained through quantitative studies such as this one and the results of focused interviews with women who work in the scientific community.

In any quantitative empirical study involving concepts such as "discrimination," and "quality of research performance" there can be only a partially satisfactory relationship between the abstract concepts and the empirical indicators used to examine them. In this book I have made use of several indicators that surely will be subject to criticism by some readers who are unacquainted with the sociology-of-science literature. In anticipation, consider several important examples. I have measured the quality of scientific research by counts of citations to the work produced by men and women scientists. How can the quality of research be measured adequately by counting citations? This is now a familiar question to sociologists of science. The fact is that a great deal of work has been done on the use of citation counts as an indicator of the quality of research discoveries. While citations are hardly an ideal measure of quality, their frequency turns out to be strongly correlated with many other independent appraisals of the quality of research performance, such as the prestige of honorific awards held by scientists, and peer evaluations of the quality of research contributions.[35] In the absence of counts of citations, the number of publications does reasonably well as a measure of research performance. There is a strong although not perfect association between citation counts and number of publications. In sum, such counts have been found to be good rough indicators of the research performance of scientists when we deal with samples of substantial size. They are far less useful in making evaluations of the quality of work of individuals. Throughout this book I use citation and publication counts as indicators of research performance, but I never used them to assess the contribution of any particular man or woman scientist. In my view this would represent a reification of the measure.

At several points in this book I have used measured intelligence scores (IQ), as an indicator of the native ability of scientists. Plainly, using general test scores as an indicator of scientific ability has many drawbacks. Nonetheless, as with counts of citations and publications, IQ represented the best available indicator of the concept, and it was used because, regardless of the problems with it as a fine-grained measure, it allowed me to address several important substantive issues. I fully expect that in time there will be improvements in the measures of both the quality of scientific work and native scientific ability, but for now we have to work with less-than-perfect measures, recognizing the problems inherent in their use.

---

35. The following references represent only a small portion of the literature on the use of citations as a measure of quality of scientific research. A summary of the major argument appears in Cole and Cole, *op. cit.*, Chapter 2. Other sources include: Alan E. Bayer and John Folger, "Some Correlates of a Citation Measure of Productivity in Science," *Sociology of Education* 39 (1966): 381–390; Warren O. Hagstrom, "Inputs, Outputs, and Prestige of American University Science Departments," *Sociology of Education* 44 (fall 1971): 375–397; Derek de Solla Price, "Citation Measures of Hard Science and Soft Science, Technology and Non-science," in *Communication Among Scientists and Engineers*, ed. C. E. Nelson and D. K. Pollack (Lexington, Mass.: D. C. Heath & Co., Lexington Books, 1971), pp. 3–22; Daniel Sullivan, "Competition in Bio-medical Science: Extent, Structure, and Consequences," *Sociology of Education* 45 (spring 1975): 223–241; Jonathan R. Cole and James A. Lipton, "The Reputations of American Medical Schools," *Social Forces* 55, No. 3 (March 1977): 662–684.

There are three other notable limits to this study. One is that I am dealing only with women and men in American science. The findings reported here apply only for the American scientific community. I do not have any data on the comparative position of women scientists located in other nations. This comparative work remains to be done. A second limit is that I deal by and large only with the academic-science community. While it is certainly true that most of the important pure-science discoveries in the United States are produced at universities, colleges, and large scientific laboratories, there surely is more scientific work going on outside of academe than within it. I have not attended at all to the place of women in industrial or governmental science establishments. And in many ways it is far easier to deal with the position of women in academic than in industrial science. For one thing, the academic reward and evaluation system tends to be uniform across fields and across universities and colleges.[36] Furthermore, the norms of scientific recognition, of hiring, and of criteria for promotion tend also to be more uniform than in industry. Were I to deal with industrial science, I would face problems of measuring scientific performance, of establishing types of reward systems that operate in different industrial settings, and of establishing labor-supply patterns and market conditions for women in different geographical regions and in different types of industries. In short, a great deal of work remains to be done on the position of women scientists outside the walls of academia.

Finally, the material presented in this book covers only a segment of the scienfific career. I take up the discussion at a point when much of the race has been run: the postdoctoral careers of the samples of men and women. Moving forward from the doctorate to advanced states in the scientific career, I am concerned primarily with the patterns of rewards for men and women after they have finished their graduate training. Many of the factors that shape scientific careers are surely products of early childhood, adolescent, and undergraduate college experiences. The cultural forces, the social values, the stereotypes that influence occupational choice as well as performance in the occupation of choice, have made their impact by the time we meet our men and women. That is why I say that we deal with survivors. It would be particularly useful to have data that extend farther back in time than those available to us. But such longitudinal analysis will have to be done in the future.

PLAN OF THE BOOK

In concluding this introduction, an overview of the content of the book may help to place individual chapters in perspective. Before tackling the substantive issues that form the core of this work, we must attend to problems of defining and

---

36. On the similarity between the academic reward systems in several scientific fields, see Jonathan R. Cole and Stephen Cole, "The Reward System of the Social Sciences," in *Controversies and Decisions: The Social Sciences and Public Policy,* ed. Charles Frankel (New York: Russell Sage Foundation, 1976), pp. 55–88.

measuring discrimination. Definitions of discrimination are anything but standardized among social scientists. There have been· a variety of different definitions, and these variations often reflect the different angles of vision of the students of discriminatory behavior. Thus, the economist Gary Becker will define discrimination in economic terms, while the sociologist Robert Merton will define it sociologically.

Furthermore, a major problem in dealing with sex- or race-based discrimination is in establishing criteria for a judgment about its presence or absence, and its extent. A major issue in the study of sex- or race-based discrimination is the provision of proof of discrimination, and as we shall see, this is not an altogether straightforward matter, since discrimination is almost never measured directly. How does one go about "proving" or "disproving" the existence of sex-based discrimination? What techniques have social scientists used to empirically test claims of discrimination? What are some of the underlying assumptions in the techniques that are used to measure the level of sex discrimination? How do social science models of proof differ from those found in other scientific disciplines? These questions are explored in Chapter 2. At the close of the chapter, I compare briefly the standards of proof that we find in social science inquiries and in legal cases that involve charges of sex- or race-based discrimination, and I indicate how social science evidence and standards of proof are affecting court decisions in fair-employment cases with increasing frequency. This discussion addresses the problem of the disjunction between the producers of social science knowledge and the consumers of it among lawyers and judges.

After concluding the discussion of problems of measurement and proof, I turn in Chapter 3 to several of the central questions about the status of women in contemporary American science: Do women and men have equal access to jobs in high-prestige academic settings? Do women as often as men receive formal honorific recognition? Do women obtain promotions to high rank with the same probability and at the same speed as men? Do women produce as much science and as highly regarded science as men? Under what conditions do we find strong or weak evidence of sex-based discrimination? For example, in a meritocratic social system that rewards excellence regardless of gender, we should expect to find men and women who are equally highly productive as research scientists receiving roughly equal rewards. But in such a system, it is at least as important that there be an equality in rewards for equal performance throughout the range of performance. Thus, among men and women scientists who are equally unproductive there also should be equal recognition. But does such equality actually exist?

Empirical data gathered from samples of men and women scientists who have been "matched" on several social characteristics are used to answer these questions. The careers of these matched samples of scientists are traced for more than a dozen years immediately following receipt of the Ph.D. Patterns of scientific productivity are examined, and estimates of the effects of various status characteristics on the research performance of scientists are generated.

In Chapter 4, I turn to a form of recognition that is largely ignored in the literature of sex roles: the production of reputations among male and female workers. Through the use of peer appraisals of the scientific contributions of samples of men and women scientists, I begin to examine the "inequality of reputational standing" between the sexes. Do male and female scientists have an equal probability of being prominent among or esteemed by their scientist colleagues? What determines the reputations of scientists, and do these factors differ for men and women? I identify four basic types of reputations among scientists based upon the visibility of their work and its perceived quality among others in their field. After examining the distribution of males and females within these types, I turn to the influence of educational and other social background characteristics as determinants of reputations above and beyond the influence of research performance. I am interested in the probabilities of men and women being esteemed or prominent scientists given variations in their backgrounds and scientific role performance. Previous work in the sociology of science has shown that peer appraisals of contributions are strongly influenced by performance variables. But is this as much so for women as for men? What is the reputational cost, if any, of being female in the scientific community?

In Chapter 5 the focus of attention shifts somewhat to an examination of the parallels between processes of stratification in the scientific community and in the larger American occupational structure. As a point of departure, I briefly discuss studies of the larger occupational structure in terms of the relative influences of "ascribed" as opposed to "achieved" status in determining achievement. I then compare the influences of ascribed and achieved status in the process of social stratification in science with that found in American society as a whole. This chapter introduces into the analysis a measure of ability—IQ scores—which is used as a proxy for "scientific ability." This allows us to compare the initial aptitudes of men and women scientists and to estimate the effect of IQ scores on forms of scientific recognition. Furthermore, by adding the IQ data to indicators of educational background and to research performance, I can create simple models of the scientific career and compare the career histories of the male and female scientists. I also discuss the influence of IQ on achievement within academic science relative to its influence on achievement (to the extent that it is known) within other occupations.

Through Chapter 5, I focus on the current standing of women in the scientific community. The perspective shifts in Chapter 6 to the question of trends in the treatment of women scientists since the first decade of the twentieth century. Having identified what currently exists, I refocus on where we have come from. As noted above, there have been claims that women in the academic community are little better off today than they were in the early part of the century, that the level of sex-based discrimination has remained remarkably unchanged. But little data have been brought to bear on this belief. In general, there is a striking paucity of historical data on the place of women in science. Except for standard biographical accounts of a few notable women scientists, such as Marie Curie, almost no systematic data have been collected on the more general status of women in science. The fact is that some efforts at gathering data were made in the first quarter of the century, but little was done with those materials. This chapter focuses on both historical materials drawn from traditional sources and on limited quantitative data that I have collected for cohorts of female and male scientists who received doctorates at the beginning of each decade in the twentieth century. The data available for both male and female scientists in the early years are fragmentary. Accordingly, results reported have to be approached with caution. Fortunately, however, biographical compilations such as *American Men of Science* and abstracts of scientific publications do date back to the first decade of the century, and they provide a starting point for the collection of systematic historical data on men and women.

Specifically, Chapter 6 briefly looks at the changing cultural definitions of "woman's place" in the American labor force from the last quarter of the nineteenth century to the early 1950s. I then begin to examine trends in the position of women in science. What changes, if any, have taken place in the access to quality higher education for women interested in academic careers? Have there been significant changes in the patterns of hiring female scientists and promoting them to high rank once they have been hired? Has there been any shift in the extent to which women scientists are full participants in the scientific community in terms of the prestige of their jobs? What are the historical patterns of scientific productivity of men and women? What have been the trends in the salaries of male and female academic scientists? What changes have taken place in the nonprofessional statuses occupied by women scientists over time, such as the proportion that marries? At the turn of the century women scientists clearly held positions that were marginal to the scientific community. This chapter considers whether the extent of that marginality has changed.

If Chapter 6 focuses upon the social positions of women scientists in the past, Chapter 7 turns to the future. One notable attempt to remedy sex-based discrimination in colleges and universities has been the development of affirmative-action programs. Designed to ensure the adequate representation of women and minorities, and equal salaries for men and women in equivalent positions, affirmative action as a method for removing inequalities has been the subject of heated debate. In this final chapter I attend to several basic problems connected with affirmative-action policy. I am particularly concerned with problems of statistical definitions of compliance or noncompliance; with problems of measuring the quality of role performance; with the more general problem of the relationship between principles of individual and group justice that lies behind affirmative-action policy; and with examining the policy as part of larger social processes, such as the process of accumulation of advantage and disadvantage for male and female scientists.

Chapter 2

PROBLEMS OF PROVING AND MEASURING DISCRIMINATION

AUTHORS OF BOOKS and articles that discuss discrimination by sex or by race rarely attend to two fundamental problems—the problem of proving and the problem of measuring. Some pay homage to these matters and then proceed as if they do not require elaboration. Yet proof and measurement of discrimination are critically related to implicit or explicit theories of causation accepted by social scientists. Moreover, they are central to inquiries intended to bear on the formulation and the evaluation of social policies. They lie at the very heart of affirmative-action, antidiscrimination programs, for instance, yet they are never discussed in them. I will return to the linkage between problems of proof in the social sciences and in the law, but first I want to discuss briefly the general difficulties of proving scientific theories and hypotheses, and particularly of proving sex discrimination by social science methods. At the chapter's end, I will present the definition and measure of sex discrimination used in this book.

PROVING THEORIES IN THE SCIENCES

Until recently in the history of ideas, knowledge meant "proven" knowledge. The essence of science was not only that it minimized the number of unproven utterances or unprovable statements, but that it created a method of building systems of proven knowledge and of interrelating them. Today almost no philosophers, historians, or sociologists of science believe that scientific knowledge is proven knowledge.[1] The old notion that scientific theories or hypotheses could be proved was replaced, first by the idea that different theories had greater or lesser degrees of probability, and then by the idea of falsification. It is not difficult, with the advantages of hindsight, to see the fundamental problem in proving a theory. There simply is no way to demonstrate conclusively that successive tests of theories will have the same result. The idea that strictly logical deductions can produce proven knowledge no longer can find supporters.

A significant shift in orientation came with Karl Popper's early ideas on falsification.[2] His basic position was that although scientific knowledge cannot be proved, it can be disproved, or falsified. The scientist is expected to formulate theoretical statements that can be shown to be counterfactual. If experimentation or empirical research indicates that the data do not fit the theory or hypothesis, the theory must be discarded. In a more sophisticated form of falsification, the scientist who produces the theory must also state the conditions under which he or she will reject it. A tall order, but one in which Popper puts great stock.[3] But what are the problems in attempting to falsify a theory or hypothesis? Let me address just two, in order to suggest how difficult it is to disprove a theory.

The first is the age-old problem of relating theoretical statements to empirical or experimental statements. The instruments of experimentation, the methods of obtaining empirical evidence, can almost always be questioned. Unless there is extraordinary intellectual consensus within a discipline on the methods appropriate to test theories—an unlikely situation in most social science disciplines—a proponent of a particular theory can fall back on the argument that the evidence produced to falsify a theoretical prediction was generated from an inadequate or wholly inappropriate experimental instrument. Social scientists are acutely aware of this basis of attack, since there is substantial disagreement within the disciplines on the reliability of measuring instruments. Surveys, opinion polls, participant observation, and ethnomethodological techniques can all be attacked when necessary as failing to be appropriate methods of capturing the "subtle" relationships underpinning a theory. But the problem is also faced in the more mature sciences. Imré Lakatos suggests this in a recent essay dealing with the problematics of falsification:

> Galileo claimed that he could "observe" mountains on the moon and spots on the sun and that these "observations" refuted the time-honoured theory that celestial bodies are faultless crystal balls. But his "observations" were not "observational" in the sense of being observed by the-unaided-senses: their reliability depended on the reliability of his telescope—and of the optical theory of the

---

1. Imré Lakatos, "Falsification and the Methodology of Scientific Research Programmes," in *Criticism and the Growth of Knowledge,* ed. Imré Lakatos and Alan Musgrave (Cambridge, England: Cambridge University Press, 1970), pp. 91–195; also Imré Lakatos, "History of Science and Its Rational Reconstructions," in *Boston Studies in the Philosophy of Science,* vol. 8 (Dordrecht, Holland: D. Reidel Publishing Co., 1970), pp. 92–182.

2. An extended discussion of Popper's contribution appears in Lakatos and Musgrave, *op. cit.* For Popper's own work, see among others *The Logic of Scientific Discovery,* 1934 (reprinted, New York: Harper & Row, Torchbooks, 1965).

3. On this and other aspects of Popper's contributions to the philosophy of science, see Paul Arthur Schilpp, ed., *The Philosophy of Karl Popper* (LaSalle, Ill.: Open Court, 1974), 2 vols. See also Lakatos' essays cited above.

telescope—which was violently questioned by his contemporaries. It was not Galileo's-pure, untheoretical-*observations* that confronted Aristotelian *theory* but rather Galileo's "observations" in the light of their theory of the heavens.[4]

If this was true for Galileo's telescope, it is surely no less true for surveys, for polls, and for econometric techniques. They all involve implicit theories, and each can be and frequently is seriously questioned as a method for generating falsifying evidence.

The second problem is the infinite possibilities of using auxiliary hypothesis to buttress theories against counterfactual evidence. Again, Lakatos' ideas are instructive—particularly his discussion of why theories are not often rejected in the face of negative evidence. He constructs a story to make his point:

A physicist of the pre-Einsteinian era takes Newton's mechanics and his law of gravitation (N), the accepted initial conditions, I, and calculates, with their help, the path of a newly discovered small planet, *p*. But the planet deviates from the calculated path. Does our Newtonian physicist consider that the deviation was forbidden by Newton's theory and therefore that, once established, it refutes the theory N? He suggests that there must be a hitherto unknown planet *p'* which perturbs the path of *p*. He calculates the mass, orbit, etc., of this hypothetical planet and then asks an experimental astronomer to test his hypothesis. The planet *p'* is so small that even the biggest available telescopes cannot possibly observe it: the experimental astronomer applies for a research grant to build yet a bigger one. In three years' time the new telescope is ready. Were the unknown planet *p'* to be discovered, it would be hailed as a new victory of Newtonian science. But it is not. Does our scientist abandon Newton's theory and his idea of the perturbing planet? No. He suggests that a cloud of cosmic dust hides the planet from us. He calculates the location and properties of this cloud and asks for a research grant to send up a satellite to test his calculations. Were the satellite's instruments (possibly new ones, based on a little-tested theory) to record the existence of the conjectural cloud, the result would be hailed as an outstanding victory for Newtonian science. But the cloud is not found. Does our scientist abandon Newton's theory, together with the idea of the perturbing planet and the idea of the cloud which hides it? No. He suggests that there is some magnetic field in that region of the universe which disturbed the instruments of the satellite. A new satellite is sent up. Were the magnetic field to be found, Newtonians would celebrate a sensational victory. But it is not. Is this regarded as a refutation of Newtonian science? No. Either yet another ingenious auxiliary hypothesis is proposed or . . . the whole story is buried in the dusty volumes of periodicals and the story is never mentioned again.[5]

By implication, counterfactual evidence that cannot be explained by the dominant theory or theoretical orientation is usually forgotten or ignored. Lakatos is by no means alone in holding this view of the robustness of theory. Thomas Kuhn and Michael Polanyi both give multiple examples of theories that are maintained in the face of falsifying evidence.[6] To avoid misunderstandings,

---

4. Lakatos, "Falsification and the Methodology of Scientific Research Programmes," *op. cit.*, p. 98.

5. *Ibid.*, pp. 100–101. Lakatos' footnotes have been omitted from the quotation.

6. Thomas Kuhn, *The Structure of Scientific Revolutions* (Chicago: University of Chicago Press, 1962); Michael Polyani, *Personal Knowledge* (London: Routledge & Kegan Paul, 1958).

none of these historians and philosophers of science argues that theory will be maintained without any empirical support. But in the social sciences, especially, there is an abundance of data that may be variously interpreted and used to support competing theories. Perhaps it was this general condition that motivated Sir Arthur Eddington, the noted British astronomer, to observe: "It is also a good rule not to put overmuch confidence in the observational results that are put forward until they have been confirmed by theory." Critical questions arise: Under what conditions will one theory be discarded for another? How does science change and progress? Under what intellectual and social conditions will science witness fundamental transformations in thoeretical orientations, in themata, or in scientific perspectives? Although there have been several attempts to address such questions, no reasonably satisfactory "theories" have been developed. These questions remain a central concern of sociologists and historians of science.[7]

If theories and lower-level hypotheses cannot be either conclusively proved or falsified, how does science proceed? Most scientists work under helpful and heuristically valuable "false beliefs." They do not attend to the philosophy of proof. But they do believe that if proof is not possible, certainly disproof is. Furthermore, they believe that research results inform and affect theoretical developments and correlatively that theory informs research, and indeed this may be the case in many scientific specialities.[8] But this involves a necessary suspension of disbelief. We act *as if* theories can be falsified, and will be discarded if contrary evidence is uncovered; *as if* there are "crucial experiments"; *as if* there are definitive counterfactual examples. We search for evidence that either supports or refutes the theoretical predictions and propositions, or questions the axioms and underlying assumptions of the theory.[9]

It is probably functionally necessary for the growth of science to proceed by using certain false assumptions about proof. Problems arise primarily in periods of strong intellectual or social controversy—not in periods described by Kuhn as "normal science." But the possibility always exists for scientists to raise questions of "proof" when data or evidence do not conform to prevailing values or beliefs. This is particularly so when social scientists address such questions as:

---

7. Among the best treatments extant on these issues are: Stephen Cole, "The Growth of Scientific Knowledge: Theories of Deviance as a Case Study," in *The Idea of Social Structure: Papers in Honor of Robert K. Merton*, ed. Lewis A. Coser (New York: Harcourt Brace Jovanovich, 1975), pp. 175–220; Gerald Holton, *Thematic Origins of Scientific Thought: Kepler to Einstein* (Cambridge: Harvard University Press, 1973).

8. On the relations between the two, see Robert K. Merton, *Social Theory and Social Structure* (New York: Free Press, 1968), Part 1; Jonathan R. Cole and Harriet Zuckerman, "The Emergence of a Scientific Specialty: The Self-Exemplifying Case of the Sociology of Science," in Coser, *op. cit.*, pp. 139–174.

9. See the first essay on methodology in Milton Friedman, *Essays in Positive Economics* (Chicago: University of Chicago Press, 1953). Herbert Hyman tried to "test" Merton's idea about anomie by questioning the empirical validity of certain of its underlying assumptions. This is a typical method of criticism. Equally instructive is Merton's rejoinder to Hyman's empirical analysis, which turns the data into support for the anomie argument. See Herbert H. Hyman, "The Value Systems of Different Classes," in *Class, Status, and Power*, 2nd edition, ed. Rheinhard Bendix and Seymour Martin Lipset (New York: Free Press, 1966), pp. 488–499; Merton, *op. cit.*, pp. 224–230.

Does sex or race discrimination exist? What are its causes? What are the levels of its intensity? What are its consequences?

Textbooks authored by even the most sophisticated social scientists testify to this orientation toward testing theories. Arthur Stinchcombe's important text, *Constructing Social Theories,* for instance, lays out the logic of proof clearly.[10] His argument is essentially this: Theoretical statements imply empirical statements ($A \supseteqq B$), which are then examined. If the empirical statement is unsupported by data, the theoretical statement is false; if it is supported, the theoretical statement is made more "credible." If we can find several empirical implications of the theory or hypothesis that are supported by independent sets of data, the theory is even more credible. But there is no discussion by Stinchcombe of the resiliency of theories in the face of empirical contradictions, or of the conditions under which theories or hypotheses are actually rejected. Most students learn from such valuable texts and from courses in theory and method that it is ultimately possible to accept or reject theories and hypotheses on the basis of empirical evidence.

Other longstanding sociology-of-knowledge questions relate social attitudes, values, and other attributes of scientists to the acceptance or rejection of theories and hypotheses: How much do personal attributes affect the predispositions of members of a scientific audience to accept experimental results as supporting or undermining theoretical or empirical propositions? How do personal attributes of scientific investigators affect their choice of problems, their gathering and presentation of evidence, their acceptance or rejection of "proof"?

In fact, we know very little about the influence of empirical and experimental tests on developing scientific theories, or about how this influence differs among scientific fields and specialties. In some scientific specialties, there appears to be little evidence that the development of theory, and its acceptance or rejection, is dependent on empirical inquiry—on empirical proof. As a case in point, consider briefly some recent developments in economic theories of race discrimination.

Although it was adumbrated if not partially anticipated by F. Y. Edgeworth in 1922, the neoclassical economic theory of discrimination was first rigorously formulated in 1957 by Gary Becker and later modified by others, most significantly by Kenneth Arrow and Anne Kreuger.[11] The various theories focus on what Becker calls a "taste for discrimination"—defined as a willingness by members of one racial group to sacrifice income in order to work with individuals of the same racial group. Becker defines a "discrimination coefficient" to measure an employer's taste for discrimination. He builds a definition of "market

10. Arthur L. Stinchcombe, *Constructing Social Theories* (New York: Harcourt Brace Jovanovich, 1968), pp. 15–56.

11. Gary S. Becker, *The Economics of Discrimination,* 2nd ed. (Chicago: University of Chicago Press, 1971); Anne O. Kreuger, "The Economics of Discrimination," *Journal of Political Economy* 71 (Oct. 1963): 481–486; Kenneth J. Arrow, "Models of Job Discrimination," Chapter 2, in *Racial Discrimination in Economic Life,* ed. A. H. Pascal (Lexington, Mass.: D. C. Heath & Co., Lexington Books, 1972); F. Y. Edgeworth, "Equal Pay to Men and Women for Equal Work," *Economic Journal* 32 (1922): 431–457.

discrimination" by first assuming that persons in two social groups contribute to production in the same way. If members of one social group are perfect substitutes in production for another group, and the labor market is perfectly competitive, then the difference in the wage rates between the two groups is defined as market discrimination.[12]

Becker's model of discrimination has been challenged on several theoretical fronts, frequently by economists who sympathize with his neoclassical perspective. Barbara Bergmann, for example, questions Becker's assumption that discrimination is a taste or a distaste for physical proximity.[13] She argues that definitions and theories of discrimination must take into account occupational segregation, or "crowding." "The effect of this crowding is that the marginal productivity of Negroes is lowered in comparison with that of whites of equivalent education."[14] The Becker model, resting on an avoidance principle, cannot be extended easily, if at all, to sex-discrimination models, in which physical distance is not an important consideration.[15] Finis Welch argues that the Becker model fails to account for employee preferences as a cause for discrimination in employment.[16] Furthermore, it does not account for consumer preferences. Kenneth Arrow, attending to the accumulation of critiques of Becker's theory, has suggested that employers' discriminatory behavior may be based not so much on tastes as on perceptions of reality.[17] Joseph Stiglitz believes that the taste models, including Arrow's modifications, are inadequate explanations, because wage discrimination requires prejudice within a given factor of production, an element not built into neoclassical models.[18] Lester Thurow finds neoclassical theories problematic because "they cannot provide a persuasive explanation of how discrimination manages to perpetuate itself," when all "of the economic incentives and pressures are on the side of elimination of [it]." Thurow argues that discrimination persists regardless of personal tastes for it.[19]

A challenge from a different angle of vision is the "dual labor market" hypothesis. Economists and sociologists have studied the processes of sex and race segmentation of labor markets. The separation of markets is both a cause and a consequence of discrimination. It "perpetuates segmentation by restricting certain workers to secondary markets not because of their education and skills but

---

12. Becker, *op. cit.* Of course, the assumptions for the economist are often the critical points of inquiry for the sociologist.

13. Barbara Bergmann, "The Effect of White Incomes on Discrimination in Employment," *Journal of Political Economy* 79, No. 2 (March/April 1971): 294–313.

14. *Ibid.,* p. 310.

15. On this point, see Lester C. Thurow, *Generating Inequality* (New York: Basic Books, 1975).

16. Fin s Welch, "Labor Market Discrimination: An Interpretation of Income Differences in the Rural South," *Journal of Political Economy* 75 (June 1967): 225–240.

17. Arrow, *op. cit.* See as well the informative review of this literature by Ray Marshall, "The Economics of Racial Discrimination: A Survey," *Economic Literature* 12 (Sept., 1974): 849–871.

18. Joseph E. Stiglitz, "Approaches to the Economics of Discrimination," *American Economic Review* 63 (May 1973): 287–295.

19. Thurow, *op. cit.,* pp. 180–181.

because they have superficial characteristics resembling most workers in the secondary markets."[20]

Each of these critiques represents a certain advance beyond Becker's work. But there is widespread disagreement among many of the critics about the appropriateness of the alternative theories. What is striking about all of these discussions is the paucity of empirical evidence brought to bear on disputed points. The gap between theory and research is wide, with theoretical developments (or what is taken for advance) apparently moving forward without information generated by empirical or experimental inquiries. In fact, there have been almost no empirical "tests" of alternative models of discrimination—no attempt to disprove one or several competing hypotheses.[21]

In fact, it is unclear whether econometric techniques could provide tests of the theories. Can the assumptions, hypotheses, and predictions of the neoclassical, dual-labor, or radical theories of discrimination be translated into econometric equations?[22] When a theory includes statements about attitudes, beliefs, and perceptions, but these factors are not measured and used in experimental treatments of the subject, how close can the fit be between the theories and the research intended to examine them? To put the issue sharply: Is the development of theories of race or sex discrimination at all dependent on empirical inquiry, and if it is, in what ways? Can complex theoretical models of discrimination be transformed into "falsifiable" propositions without losing the complexities essential to them?

PROVING DISCRIMINATION IN THE SOCIAL SCIENCES

Definitional Problems

The previous discussion notwithstanding, all of us think we know discrimination when we see it or experience it. We have *verstehen* knowledge. But it is quite another matter to be able to prove the existence of discrimination. Why, beyond the philosophical reasons, is this so difficult to do? For one thing, the definition of discrimination varies greatly from one scholar to another, largely depending on his particular angle of vision. This becomes clear after only a few examples. F. H. Hankins defines discrimination as the "unequal treatment of

20. Marshall, *op. cit.* This approach is worked out most completely in Peter B. Doeringer and Michael J. Piore, *Internal Labor Markets and Manpower Analysis* (Lexington, Mass.: D. C. Heath & Co., Lexington Books, 1971).

21. Dale L. Heistand, "Discrimination in Employment: An Appraisal of the Research," Policy Papers in Human Resources and Industrial Relations 16. A joint publication of the Institute of Labor and Industrial Relations (The University of Michigan and Wayne State University) and the National Manpower Policy Task Force, Washington, D. C., 1970.

22. I have benefited here from discussions with my economist colleagues at the Center for Advanced Study in the Behavioral Sciences, Alvin Klevorick and Joseph Pechman, and with Harold Watts of Columbia University.

equals.''[23] But as Hubert Blalock points out, equals with respect to what? All social characteristics other than race? One surely could not ask for such a stringent requirement. Perhaps only equals with respect to all ''relevant'' characteristics. But how do we determine, and who is to determine, what are the relevant criteria?

Robin Williams says that discrimination is ''the differential treatment of individuals considered to belong to a particular group.''[24] Again Blalock puts his finger on the problem: ''We encounter the same difficulty, namely, that of determining whether or not a given group membership is in fact a major cause of the differential treatment. This requires that individuals be equated on all other relevant variables, which implies a theory of social causation.''[25]

Gary Becker, as I have noted, builds a definition of marketplace discrimination on the ''taste'' of employers for dissociating themselves from a social group. But beyond all the aforementioned difficulties with this definition, there are problems over what constitutes and determines equal productivity, and what factors other than the one obvious characteristic dividing the groups could explain observed differences in wage rates. By defining discrimination in terms of equivalent productive units, Becker forecloses consideration of discrimination as a temporal and cumulative process. He does not attempt to consider conditions that produce a low probability of two groups being equal in productivity. Finally, there is conflation between the attitudinal and behavioral dimensions. Tastes are presumably attitudes, but wage discrimination refers to behavior.[26]

Residual Analysis

These selective definitions suggest that the usual analysis of discrimination looks first at differences in social and economic rewards of members of different social groups. And this is what the empirical researchers do. Thus they show that male full professors have higher salaries on average than female full professors. They demonstrate that there are proportionately more males than females located at prestigious academic departments; or they provide evidence that blacks are less apt than whites to hold prestigious or high-paying jobs. They might report that women represent 20 percent of the current Ph.D.'s in sociology but only 10 percent of the new assistant professors at Ivy League colleges. Do such findings demonstrate discrimination? Emphatically no. And most sophisticated social scientists would not claim that they do. Such findings do represent an operationalization of the concept of discrimination as group differences, but in fact

---

23. As quoted in H. M. Blalock, *Toward a Theory of Minority Group Relations* (New York: John Wiley & Sons, 1967), pp. 15-18.

24. *Ibid.*, p. 16.

25. *Ibid.*, p. 16.

26. For a discussion of the distinction between cultural values, norms, and behavior in the study of discrimination, see Robert K. Merton, ''Discrimination and the American Creed,'' in *Discrimination and National Welfare*, ed. R. M. MacIver (New York: Harper & Brothers, 1949), pp. 99-126. Interestingly, Merton never attempts to define discrimination explicitly in the paper.

they go no further than establishing social or economic inequality. And inequality is not discrimination per se.

When we assume that all inequality between races, sexes, ethnic groups, or any other pair of social categories is equivalent to the level of discrimination, we have imported a causal theory into our analysis. We have said, in effect, that an unmeasured variable, called discrimination, is the sole intervening variable and one that completely interprets the inequality between the two social groups. The claim is that all of the differences between the comparison groups can be explained by discrimination. This is "naive residualism." And naive residualism is the least acceptable way of empirically defining and of measuring discrimination, because it embraces the weakest causal assumption. Without a multivariate view of the world, it is impossible to make discrimination arguments even plausible.

The concept of residual argumentation must be distinguished from the residual, or error term, that is obtained from a regression equation. I am referring here to the residual effect of a measured variable such as sex or race on a dependent variable, not to the unexplained variance in a regression, that is, $\sqrt{1 - R^2}$.

Although many casual observers of social systems take these simple inequalities to be exact estimates of discrimination, few social scientists today adopt this approach. The crux of the estimation problem lies in the absence of any direct measure of discrimination. I am unaware of any economic or sociological study of sex discrimination, for example, that attempts to measure the various dimensions of discrimination directly, although social psychologists have occasionally tried their hand at direct measurement. Concepts such as prejudice, militancy, and segregation have all been measured directly, for better or worse. But not discrimination. Discrimination is almost always operationally defined as the unexplained difference between two groups—that is, between blacks and whites, males and females, and so on.

Sophisticated social scientists are uneasy about such definitions and measures, because they are fully aware of potentials for inaccurate estimates confronting the analyst of complex social situations who attributes residual difference to any single cause. Nonetheless, let us proceed, placing cautionary signs along the way. The modus operandi of the "sophisticated residualist" is to start with a zero-order correlation between a "suspect" status characteristic and some form of social reward or recognition. Otis Dudley Duncan, in his informative essay on inheritance of poverty or inheritance of race, begins by establishing inequalities in educational, occupational, and income achievement between blacks and whites in American society.[27] He then asks: How much of this inequality results from the cycle of poverty and how much from race? Finally, Duncan does the interesting and crucial thought experiment of asking: Suppose we were to eliminate, in a series of steps, differences between the races in family

---

27. Otis Dudley Duncan, "Inheritance of Poverty or Inheritance of Race?" in *On Understanding Poverty*, ed. Daniel P. Moynihan (New York: Basic Books, 1968), pp. 85–110.

size and years of schooling; how much difference would still exist in occupational and income achievement? Still further, supposing blacks and whites were equated as well in occupational status, would we still find differences in average incomes? The inequality that remains after a set of controls has been established is defined as the effect of race or of discrimination.

But it is clear that social scientists as cautious as Duncan tread gingerly on the final concluding inference of discrimination, and are hesitant to claim that all of the residual variance results from discrimination. Thus, for example, after attributing about half of the occupational gap between blacks and whites to educational differences, family size, and family background, Duncan says:

> The remaining [difference], not otherwise "explained" by the model, one is tempted to label "occupational discrimination." It is due, literally, to the fact that Negroes equally well educated as whites (in terms of years of schooling) and originating in families of comparable size and socioeconomic level do not have access to employment of equal occupational status. This disadvantage—or form of discrimination, if you will—carries over into dollar amounts of income.[28]

Elsewhere he also places discrimination in quotation marks, suggesting that he is conscious of the limits inherent in defining discrimination as the unexplained differences between the races. The tentativeness of the estimate is made explicit by Duncan: "Unless and until we can find other explanations for it, this must stand as an estimate of income discrimination (or the increase in Negro income that would result from an elimination of such discrimination)."[29] Elsewhere he notes:

> The measurement of discrimination itself presupposes analysis of [a causal relationship among indicators], . . . for the empirical evidence of discrimination consists, I suppose, in the demonstration of inequality on one measure, after the contributions to that inequality made by variables other than discrimination have been evaluated.[30]

Kenneth Arrow, commenting on Duncan's attempts to account for income differentials between blacks and whites, succinctly states the problem with sophisticated residualism: "No doubt failure to explain is not the same as proof of nonexplanation. There may easily be other supply factors overlooked or not easily quantifiable; motivational differences due to cultural variation [etc.]."[31]

Other social scientists, working in much the same mode, refer to the residual differences simply as "inequality," or, as Paul Siegel does, "the cost of being Negro."[32] It makes no difference, incidentally, whether the methodological approach to estimating discrimination is correlational or regression analysis, covariance analysis, multivariate tabular analysis, or analysis of variance. So

28. *Ibid.*, pp. 99–100.

29. *Ibid.*, p. 100.

30. Otis Dudley Duncan, "Discrimination against Negroes," *Annals of the American Academy of Political and Social Science* 371 (May 1967): 87.

31. Arrow, *op. cit.*, p. 85.

32. Paul Siegel, "On the Cost of Being a Negro," *Sociological Inquiry* 35 (winter 1965): 41–57.

long as there is no direct measurement of discrimination and so long as sex or race differences, with or without controls for other explanatory variables, are used as the estimate of discrimination, a residual approach is being used. When a wide variety of explanatory variables are introduced into the analysis and the resultant difference is tentatively equated with discrimination, sophisticated residualism is being used. It may be the best that we can do until better measures of discrimination are developed, but let us not fool ourselves into believing that this number *is* discrimination.

Although economists and sociologists almost always equate group differences with discrimination, social psychologists have occasionally attempted to measure discrimination directly by using quasi-experimental designs. One study matched psychologists in several respects except sex (implied by first names) and asked 155 department heads to indicate the appropriate level of appointment for each candidate and to appraise the psychologist's desirability as a member of the department. No attempt was made to determine the reasons for decisions.[33] A second study asked chairman of 179 science departments to evaluate pairs of curricula vitae of "candidates" for appointment as associate professors. A further wrinkle in the design was to subdivide the pairs—the pairs of one set showing qualifications of two average candidates, differing only in sex, and those of the other set showing a male candidate with mediocre qualifications and a female candidate with superior qualifications.[34] Although both of these studies had serious methodological shortcomings, they do represent small attempts to move away from residual-type analysis to direct estimates of discriminatory behavior.[35]

One problem with residual analysis is that the residual differences can be so variously interpreted. For example, Christopher Jencks in his interesting book *Inequality* attributes the unexplained differences in economic success between black and white Americans to "varieties of luck and on-the-job competence."[36] But several equally plausible and more specific explanatory hypotheses could be argued.

Cumulative Effects

Another problem with forming estimates of discrimination is to identify and isolate the cumulative effect of past experience or behavior on current outcomes.

---

33. L. S. Fidell, "Empirical Verification of Sex Discrimination in Hiring Practices in Psychology," *American Psychologist* 25, No. 12 (Dec. 1970): 1094–1098.

34. Arie Y. Lewin and Linda Duchan, "Women in Academia," *Science* 173 (3 Sept. 1971): 892–895.

35. Fidell's study fails to take into account principles other than sex bias that could produce the weak results obtained. Lewin and Duchan do not even present the statistical findings on which they base their conclusions, so the results must be treated with skepticism. For a better example of a similar mode of inquiry, and of the application of analysis of variance techniques to estimate the effect of race and sex statuses on college admissions, see Elaine Walster, T. Anne Cleary, and Margaret M. Clifford, "Research Note: The Effect of Race and Sex on College Admissions," *Sociology of Education* 44, No. 2 (spring 1971): 237–244. The findings reported in the Walster study are discussed in Chapter 3.

36. Christopher Jencks, *Inequality* (New York: Basic Books, 1972), p. 8.

Comparisons of social groups within specific categories often lead to an underestimation of actual discrimination. Suppose we are interested in the income differences between men and women. We could argue that to obtain realistic estimates of this inequality comparisons are only relevant within the same occupation. So we compare male professors with female and find, hypothetically, little difference in income. Are we to infer that there is no sex discrimination? Not if we consider that many females do not become professors in the first place because of discrimination. In short, any estimate of discrimination that fails to consider the influence of past bias on current and future outcomes will necessarily be an incomplete estimate. Accurate estimates of discrimination require measurement not only of the direct effect of discrimination but also of the additive and interactive effect of events at Time A on events at Time B. I shall have more to say on this matter.

Quality of Role Performance

A related and equally important theme will be the necessity of accounting for the quality of role performance in order for discrimination to be adequately estimated. In meritocratic and universalistic social systems, individuals' rewards are based ideally on the quality of their work, not on their race, sex, ethnic origin, religion, and so on. The principal functionally relevant criterion for rewards is high-quality performance. This is particularly so in science, but ideally so in all institutions. The other, irrelevant criteria are, by untested assumption, intrinsically unrelated to ability and consequently should not enter the evaluation process. Individual scientists should become prominent, should receive prizes and honors, should obtain chairs in the best departments of science, because their scientific work is superior to that of others. Science perhaps more than other social institutions approximates in most respects this universalistic ideal. Clearly, a requirement for a sophisticated analysis of inequalities in social rewards should be measures of the quality of role performance. No one assumes, for example, that all scientists produce an equal amount of work that is also of equal quality. We readily accept the idea of wholesale inequalities of talents among individual scientists. We know that a small fraction of the scientific community is responsible for the vast majority of scientific discoveries. We should also be willing to entertain the possibility that in the aggregate, for whatever reasons, some social groups do not perform on the job as well as others. It may turn out, for instance, that women, on average, produce "better" science than men. If so, their rewards, on average, should reflect this inequality in research performance. We may wish to ask why one social group performs better than another, and that question may indeed be answered by discrimination, but then we move into a more complex understanding of discrimination as a social process. We then are examining the effects of discrimination both on role performance and on social outcomes. Unfortunately, very few sociological or economic studies of discrimination have taken into account measures of the quality of role performance. I do have measures of role performance of male and female scientists, and these measures will be used to estimate levels of sex discrimination in science.

Once we realize the significance of performance as an intervening variable, the logical failure of inferring discrimination from simple inequalities is immediately apparent. For even if we find absolutely no correlation between sex status and awards, no percentage difference between the sexes in appointments to prestigious positions, no difference in average salaries, it would be unduly hasty to conclude that this demonstrates the absence of discrimination. After all, if a woman performs better on the job, and if there are no demonstrable impediments to the man's better performance, then the woman should be more highly rewarded. For when differences in role performance have been established, the absence of inequality suggests discrimination.

Inferences from Single Cases

Another familiar problem is the all-too-easy movement from inferences about individual experiences to conclusions about the aggregate level of social systems. How often one hears about individuals who "clearly were denied appointments because of their sex." From knowledge of a single case or a few cases, which frequently means partial knowledge derived from the "victim," extrapolations are made about general practices in an institution. Inferences about individual decisions may properly be drawn from individual case histories, but inferences about the extent and magnitude of discrimination in a social system based upon one or two cases is hazardous indeed. There are nonetheless good sociological reasons why such extrapolations are often made, and I will discuss them in the next chapter.

Inferences from Group Correlations

The logical counterpart of inferences drawn from single cases is the problem of inferences about individuals drawn from group correlations. Statistical discrimination is a widespread method for demonstrating the existence of social inequalities generally, and inequalities between the sexes particularly. Lester C. Thurow points out that such discrimination

> occurs whenever an individual is judged on the basis of the average characteristics of the group, or groups, to which he or she belongs rather than upon his or her own personal characteristics. The judgments are correct, factual, and objective in the sense that the group actually has the characteristics that are ascribed to it, but the judgments are incorrect with respect to many individuals within the group.[37]

Groups are treated objectively; individuals are not.

The problem of inferring qualities of individuals from group correlations is

---

37. Thurow, *op. cit.*, p. 172.

particularly important in all-or-nothing decisions, such as whether or not to hire, to promote, to dismiss, to increase salaries. It is all the more interesting because the actions taken by employers that effectively discriminate against a set of individuals with "suspect" characteristics can be based on entirely rational, cost-reducing decisions that are defensible as unbiased on the basis of "empirical proof."

Consider as one hypothetical example the often-discussed problem of differential dropout rates of male and female employees. Suppose that in science women are more apt than men to have discontinuous careers; although at any given time the vast majority of women scientists may be working, there is a correlation of .20 between sex status and continuous labor-force participation. Furthermore, suppose that the data on which this finding is based are impeccable—totally without measurement error or other types of error. This group difference is a "fact." An employer who has the choice of selecting either a man or a woman for a position chooses a man because he wants to maximize the return on his investment in training this new recruit. The employer is acting rationally in the face of known probabilities. But to the extent that on the basis of this correlation women consistently are not hired, the vast majority of women— who are no more likely than men to leave the labor force for a time—are denied opportunity as individuals. Those who possess one group characteristic (female sex status) but not the other ("dropout" status) are subjected to what might be called "indiscriminate discrimination." The Kafka-like aspect of this situation emerges when a person who in fact has the desired characteristics for a job, but who also has a disvalued status, such as being female, cannot demonstrate his or her functionally relevant characteristics to employers who are basing their decisions on the disvalued status.[38] Again, the conditions are in place for the emergence of self-fulfilling prophecies. The way in which individuals are institutionally tied to others who are "like them" in only one salient respect becomes a question for extended analysis.

We can easily multiply the number of examples, but note only one: the "ecological fallacy"—the inference of individual relationships from those between groups. Statistical discrimination is a case in point. A correlation showing that women produce fewer scientific papers than men may be a completely accurate group datum, but individual hiring and promotion decisions based upon it increase the likelihood of discrimination. To summarily reject individuals on the basis of even moderate associations between one of their characteristics and a valid job performance criterion represents a hazardous use of statistical correlations. This is a basic problem in recent antidiscrimination legal disputes, to which I shall return below. The problem of proof here is double-edged, for it applies not only to the objects of discrimination but also to the group of "discriminators." For example, showing that science departments in American uni-

---

38. *Ibid.*, p. 174.

versities tend to prefer, in the aggregate, males over females on the basis of a real or imaginary correlation does not prove, of course, that any individual department does so.

Finally, statistical discrimination in hiring has indirect effects on related forms of discrimination. If individuals are denied opportunities in all-or-nothing hiring situations, and if skills necessary for career advancement are acquired on the job rather than in exogenous settings, the initial effect of such discrimination is multiplied by the lack of access to training needed for promotions, salary increases, and so on. The extent to which inferences of proof as to individual capabilities are drawn from aggregate data becomes an important focus of attention for studies of discrimination.[39]

Inferences from Marginals

Another problem in estimating levels of discrimination has to do with the strong inferences often drawn from simple marginals. Many observers of science, although only a few social scientists, believe that the proportion of professors, doctors, lawyers, or dentists who are women is an indication of the existence of sex discrimination. For instance, if only 2 percent of physics Ph.D.'s are women, or if only 18 percent of all Ph.D.'s are women, these figures supposedly reflect discrimination in the academic community. If only 20 percent of all fellowship recipients are female, there must be sex discrimination in awarding graduate fellowships. Such statistics, standing alone, fail in two critical ways to tell us much about discrimination. They fail to tell us what proportion of applicants for graduate school admissions and fellowships were female, and what the relative qualifications were of the males and females who did apply. Authors or readers who draw strong conclusions from such figures totally neglect the important social process of self-selection. Such simple marginals also fail to indicate the source of discrimination. Women may face discrimination and bias throughout the process of career choice, but where does the discrimination occur, at what points in the life cycle, and what form does it take? The marginal figures on female Ph.D.'s or fellowship holders represent an end point in a selective process that extends back in time beyond the point of entrance into the academic community. As noted in the introductory chapter, values, norms, and social constraints on women's behavior from preschool through college years contribute to the final low proportion of women in a job category. It is actually quite possible for universities to act meritocratically toward men and women and still be left with a disproportionately large number of male Ph.D.'s or fellowship holders.[40]

---

39. Discussions of the cumulative effects of past behavior on careers can be found in *Ibid.*; Janice Fanning Madden, *The Economics of Sex Discrimination* (Lexington, Mass.: D. C. Heath & Co., 1973); Solomon W. Polachek, "Discontinuous Labor Force Participation and Its Effects on Women's Market Earnings," in *Sex, Discrimination, and the Division of Labor,* ed. Cynthia B. Lloyd (New York: Columbia University Press, 1975), pp. 90–122.
40. Several of these problems involve the application of false criteria for establishing causality. On such problems, see Hanan Selvin and Travis Hirschi, *Delinquency Research* (New York: Free Press, 1967).

"Half of the Table"

Several other even more elementary evidence problems are encountered frequently in discussions of sex or race discrimination. One that should be immediately recognizable is the "half of the table" problem. The associated arguments typically go: Ninety-six percent of the women who applied for jobs at Harvard were turned down; or only 10 percent of all women applicants received fellowships; or a great many women are discouraged from writing doctoral dissertations on the subject of their choosing. Of course, for women we could substitute any number of labels: men, blacks, whites, Jews, Arabs, and so on. The logic is the same regardless of the social attribute selected. What is totally absent from such statements is, of course, the relevant comparative group. Thus, if 96 percent of all men who applied for jobs at Harvard were also turned down, we cannot conclude that there was sex discrimination, although we would clearly want to carry the analysis further. Failure to keep in mind what the record would be for the other relevant comparative group—that is, for the other half of the table—renders the initial assertion logically impotent.

The Unit of Analysis

An additional difficulty in constructing plausible arguments about discrimination is selecting the proper unit of analysis. Statistical analysis of inequalities between men and women can be performed at different levels of aggregation. What appear as minor differences between men and women for an entire university or set of universities may conceal far greater differences, for example, if we examine individual academic departments or disciplines. Correlatively, apparently substantial inequalities between the sexes at the aggregate level may result from one glaring inequality in a subunit that outweighs the equality found in other subunits. Consequently, the selection of appropriate units for analysis is a matter of importance, and in general, units should be selected that are subject to decisions that potentially could be discriminatory. When disaggregation is not possible, the potential for confounding results and for spurious interpretations increases.

PROVING DISCRIMINATION IN LEGAL DISPUTES

There are, then, multiple difficulties in demonstrating the existence and in measuring the extent of discrimination in social science inquiries. Yet, despite all these problems, social science inquiries generally have the fortunate property of dealing with large aggregates and of drawing conclusions on the basis of large samples. Moreover, since these studies are not typically intended as policy statements, and are frequently prefaced with cautionary notes about the tentativeness of the conclusions, errors of inference are unlikely to directly affect individuals. This cushion for error does not exist in a legal dispute, in which there is a winner and a loser, and in which the decision affects the lives not only of the immediate parties but also, indirectly, of individuals whose cases are influenced by legal precedents.

Recently the courts have become a major forum for settling anti-bias disputes. Compliance with affirmative action guidelines, with provisions of recent civil rights acts, and with the constitutional requirements of both the Fourteenth Amendment's Equal Protection Clause and the Fifth Amendment's Due Process Clause is being tested by individuals and classes who believe themselves to be victims of discrimination. Consequently, problems of "proving" discrimination under the law become a central concern for those interested in sex and race discrimination. Since I will be discussing the problems of affirmative action programs, and since social science evidence is increasingly being used to fortify positions and to justify opinions, I want to consider some standards of "proof" established in constitutional cases that focus on sex and race discrimination.

A large portion of antidiscrimination law in the last decade has been based either on the Equal Protection Clause of the Fourteenth Amendment or on Title VII of the 1964 Civil Rights Act. The intent of the Civil War amendments, especially the Fourteenth, was to prevent discrimination against blacks. The *Brown* school desegregation decision and the nineteenth-century *Slaughter-House, Strauder,* and *Plessy* decisions were all based upon various interpretations of the Equal Protection Clause.[41] The point is that the Fourteenth Amendment was intended to exempt Negroes "from unfriendly legislation against them distinctly as colored," and to exempt them "from legal discriminations, implying inferiority in civil society."[42] But are there any distinctions or classifications based on race that are not prohibited? Are there any conditions under which racial classifications will not be held discriminatory under the Equal Protection Clause? Given that any racial classification in law will be treated as "suspect," how does one provide evidence of or prove discrimination?[43]

Since the Fourteenth Amendment does not explicitly deal with sex status, and its intent had nothing to do with eliminating sex discrimination, use of the Equal Protection Clause to redress apparent sex discrimination in employment has had only limited success. For example, in *Goesaert* v. *Cleary* 335 U.S. 464 (1948)

---

41. We are all familiar with the *Plessy* v. *Ferguson* and the *Brown* decisions. In *Strauder* v. *West Virginia* (100 U.S. 303 [1880]), the Supreme Court held that exclusion of Negroes from juries not only injured criminal defendants but also discriminated against Negroes, because it denied them equal participation in the administration of justice. However, in the earlier *Slaughter-House* case (83 U.S. [16 Wall.] 36 [1873]), which was the first decision to interpret the Fourteenth Amendment, the Court refused to construe the amendment in terms of its antidiscrimination function. For an extended, interesting discussion of the history and multiple interpretations of the Fourteenth Amendment, see Paul Brest, *Processes of Constitutional Decision-Making: Cases and Materials* (Boston: Little, Brown & Co., 1975), Chapter 5.

42. As quoted in Brest, *op. cit.,* p. 477.

43. The last time the Court upheld overt racial discrimination was in *Korematsu* v. *United States* (323 U.S. 214 [1944]), which dealt with restrictions on the civil liberties of Japanese-Americans during World War II. Justice Black's opinion for the Court states:

　　All legal restrictions which curtail the civil rights of a single racial group are immediately suspect. That is not to say that all such restrictions are unconstitutional. It is to say that courts must subject them to the most rigid scrutiny. Pressing public necessity may sometimes justify the existence of such restrictions; racial antagonisms never can.

the Supreme Court upheld a Michigan law prohibiting a woman from being a bartender unless her father or husband owned the establishment. This was litigated as a violation of equal protection.[44] Justice Frankfurter, delivering the opinion of the Court, said:

> The fact that women may now have achieved the virtues that men have long claimed as their prerogatives and now indulge in vices that men have long practiced, does not preclude the States from drawing a sharp line between the sexes, certainly, in such matters as the regulation of the liquor traffic. . . . The Constitution does not require legislatures to reflect sociological insight, or shifting social standards, any more than it requires them to keep abreast of the latest scientific standards.[45]

In short, until the 1970s the major developments in sex-discrimination law are not derivative of interpretations of the Civil War amendments. Rather, they grow out of interpretations of recent congressional legislation.

Until the 1970s sex-based discrimination cases that dealt with the Equal Protection Clause were decided by the Supreme Court within the context of social values that held closely to traditional sex-role stereotyping: Women required special "preferential treatment" and "protection." Gender-discrimination cases that have been decided recently on constitutional rather than on statutory grounds have not, by and large, been employment cases. The Court has followed a winding path in its constitutionally based decisions. As Ruth Bader Ginsburg has said: "From 1971 through 1975, with one dramatic exception [the *Roe* v. *Wade,* 410 U.S. 113 (1973) and *Doe* v. *Bolton,* 410 U.S. 179 (1973) abortion decisions], the Supreme Court treated the sex-based discrimination cases brought to it as occasions for *ad hoc* rulings."[46]

The strength of the empirical evidence presented to the Supreme Court in discrimination cases involving gender or race differ because of the Court's historical interpretations of the antidiscrimination clauses of the Constitution. For example, the Equal Protection Clause has been interpreted very differently for different types of classifications. While race, religion, and nationality have been defined by the Court as "suspect classifications" requiring "strict scrutiny," gender has not been so defined. The unwillingness of a majority of the Court to classify gender as "suspect" in any of its gender-discrimination cases as of 1977 is of no small import. There has been movement by the Court toward placing gender discrimination on an equal footing constitutionally with race discrimination, but a majority of judges has not been put together in any single case. The closest that the Court has come has been the four votes in *Califano* v. *Goldfarb,* 45 U.S.L.W. 4237 (1977). In the 1976 *Craig* v. *Boren* 45 U.S.L.W. 4057 (1976) decision, sometimes referred to as the Oklahoma 3.2 beer case, the Court

---

44. *Goesaert* v. *Cleary,* 335 U.S. 464, 69 S. Ct. 198, 93 L. Ed. 163 (1948).

45. *Ibid.* For a discussion of patterns of occupational segregation fostered and impeded by law, see Margaret J. Gates, "Occupational Segregation and the Law," *Signs* 1, No. 3, part 2 (spring 1976): 61–74.

46. Ruth Bader Ginsburg, "Women, Men and the Constitution: Key Supreme Court Rulings." (1978) Preprint, Columbia University Law School, p. 14.

recognized, following a series of cases since *Reed* v. *Reed* 404 U.S. 71 (1971), that classifications by gender "must substantially further important government objectives." In short, the Court has defined a double standard. In race-based discrimination cases evidence for use of racial classifications must demonstrate both the existence of a "compelling state interest" and the absence of alternative means for achieving that state objective; in gender cases evidence need only establish an important government interest. Consequently, gender-discrimination cases do not require the same tightness of "fit" between means and ends that must be shown in race-based discrimination cases.

Because of the Court's reluctance to accept gender as a "suspect classification" within the framework of the Fourteenth Amendment, a substantial portion of the gender-discrimination litigation involving employment has been argued before the Court in terms of violations of Title VII of the 1964 Civil Rights Act.

Title VII of the 1964 Civil Rights Act was drafted to prohibit discrimination in employment, and the Equal Employment Opportunities Commission (EEOC) was established to enforce it. The act prohibits discrimination by employers of fifteen or more persons engaged in an industry affecting commerce. It was amended in 1972 to cover public employees. Given the Court's broad interpretation of the commerce clause, this 1972 provision provides for almost all business of even limited size. The key provisions are Sections 703(a), 703(e), and 703(h). The first section forbids an employer

> (1) to fail or refuse to hire or to discharge any individual, or otherwise to discrimination against any individual with respect to his [sic] compensation, terms, conditions, or priviledges of employment, because of such individual's race, color, religion, sex, or national origin; or

> (2) to limit, segregate, or classify his employees or applicants for employment in any ways which deprive or tend to deprive any individual of employment opportunities or otherwise adversely affect his status as an employee, because of such individual's race, color, religion, sex, or national origin.[47]

Section 703(e) of the statute contains an important exception to this provision, which has produced most of the legal battles, but only for matters relevant to sex, religion, and national origin, and not for race or color. It allows

> an employer to hire and employ employees . . . on the basis of . . . religion, sex, or national origin in those certain instances where religion, sex, or national origin is a *bona fide occupational qualification* reasonably necessary to the normal operation of that particular business or enterprise [emphasis added].[48]

Section 703(h) allows employers to use standardized qualifying examinations as a basis of distinguishing applicants, provided that the test is not designed to discriminate. The scope of this provision was significantly limited by the Su-

---

47. Section 703, 42 U.S.C. § 2000e-2 (1972).

48. *Ibid*. A useful and extensive discussion of Title VII appears in Barbara Allen Babcock, Ann E. Freedman, Eleanor Norton, and Susan C. Ross, *Sex Discrimination and the Law: Causes and Remedies* (Boston: Little, Brown & Co., 1975), pp. 229–439.

preme Court's ruling in *Griggs* v. *Duke Power Co.* (1971).[49] In *Griggs* the Court held that eligibility examinations for jobs were invalid if the employer could not show a direct relationship between the test scores and successful job performance. It did not matter whether or not the employer's intention was to discriminate against minorities; if the application of a "neutral" policy disparately affected a particular group, it was illegal. The Court held: "Under the Act, practices, procedures, or tests neutral on their face, and even neutral in terms of intent, cannot be maintained if they operate to 'freeze' the status quo of prior discriminatory employment practices."[50]

The requirements for establishing a prima facie case of discrimination under EEOC guidelines are minimal. It is necessary to show only a disparity between the proportion of a "protected" group in the employment pool qualified for a position and the representation of that group currently in that position.

Further, in *Griggs* the Court agreed that accumulative disadvantages of inferior schooling should not be allowed to forever haunt the members of disadvantaged groups:

> The Court of Appeals [found that] . . . on the record in the present case, "whites fare far better on the Company's alternative requirements" [a high school education or passing a standardized general intelligence test] than Negroes. This consequence would appear to be directly traceable to race. . . . Because they are Negroes, petitioners have long received inferior education in segregated schools and this Court expressly recognized these differences in *Gaston County* v. *United States*, 395 U.S. 285 (1969). There, because of the inferior education received by Negroes in North Carolina, this Court barred the institution of a literacy test for voter registration on the ground that the test would abridge the right to vote indirectly on account of race.[51]

An equally important component of *Griggs* was that if a prima facie case can be made that an employment procedure is in fact discriminatory, the burden of proof that it is not shifts to the employer. The plaintiff is not required to establish causation. The defendant must demonstrate, generally with empirical evidence, that the procedure is related to actual job performance.

A significant number of recent antidiscrimination cases before the federal courts have tested the "test-invalidation" rule in *Griggs* and the "bona fide occupational qualification" (hereafter BFOQ) exception in Title VII. The BFOQ has been the chief defense used by employers. The Court has held to a narrow interpretation of the exception.[52] Its standards have been set down for employers who argue that their practices involve sex discrimination based on BFOQ. In two such cases the question was whether the physical requirements of a job, such as

---

49. *Griggs* v. *Duke Power Co.*, 401 U.S. 424, 91 S. Ct. 849, 28 L. Ed. 2nd 158 (1971).

50. *Ibid.*

51. *Ibid.*

52. *Weeks* v. *Southern Bell Telephone and Telegraph Co.*, 408 F. 2d 228 (5th Cir. 1969); *Cheatwood* v. *South Central Bell Telephone & Telegraph Company*, 303 F. Supp. 754 (1969); *Diaz* v. *Pan American World Airways, Inc.*, 442 F. 2d 385 (5th Cir. 1971), cert. denied, 404 U.S. 950 (1971); *Rosenfeld* v. *Southern Pacific Company*, 444 F. 2d 1219 (9th Cir. 1971).

lifting coin boxes, represented a BFOQ. In *Weeks* v. *Southern Bell Telephone* the Fifth Circuit Court defined one standard:

> In order to rely on the bona fide occupational qualification exception, an employer has the burden of proving that he had reasonable cause to believe, that is, a factual basis for believing, that all or substantially all women would be unable to perform safely and efficiently the duties of the job involved.[53]

A still stricter standard regarding sex stereotyping was set down in *Rosenfeld* v. *Southern Pacific Company* by the Ninth Circuit Court:

> In the case before us, there is no contention that the sexual characteristics of the employee are crucial to the successful performance of the job, as they would be for the position of a wet-nurse.... Rather, on the basis of a general assumption regarding the physical capabilities of female employees, the company attempts to raise a commonly accepted characterization of women as the "weaker sex" to the level of a BFOQ.... The premise of Title VII ... is that women are now on equal footing with men.... The footing is not equal if a male employee may be appointed to a particular position on a showing that he is physically qualified, but a female employee is denied an opportunity to demonstrate personal physical qualification. Equality of footing is established only if employees otherwise entitled to the position, whether male or female, are excluded only upon a showing of individual incapacity.[54]

Finally, in *Diaz* v. *Pan American World Airways* the question was whether being female was a BFOQ for the position of flight attendant. Pan Am presented evidence at trial "that Pan Am's passengers overwhelmingly preferred to be served by female stewardesses," that the special psychological needs of its passengers were better attended to by females, and that the performance of females was "better" in the sense that they were "superior" in nonmechanical aspects of the job. But the Fifth Circuit Court found against Pam Am, reversing a lower court's decision. The court stated:

> We begin with the proposition that the use of the word "necessity" in Section 703(e) [of Title VII] requires that we apply a business *necessity* test, not a business *convenience* test. That is to say, discrimination based on sex is valid only when the *essence* of the business operation would be undermined by not hiring members of one sex exclusively.[55]

In each of these cases the court was trying, *inter alia,* to limit the possibilities for statistical discrimination.

In 1975 the Supreme Court reaffirmed the *Griggs* decision in *Albemarle Paper Co.* v. *Moody*.[56] A class-action suit for present and past Negro employees sought injunctive relief under Title VII against the hiring and promotion policy of the Albemarle Paper Company. The major issues were the company's program of employment testing, its seniority system, and its liability for back pay. A central

---

53. *Weeks, op. cit.*

54. *Rosenfeld, op. cit.*

55. *Diaz, op. cit.*

56. *Albermarlc Paper Co. et al.* v. *Moody et al.,* 422 U.S. 405 (1975).

issue was whether Albemarle had demonstrated that its tests were job-related as defined in *Griggs*. According to *Griggs,* once the plaintiff has demonstrated a prima facie case of discrimination—which the Albemarle employees did—the burden of proof shifts to the defendant. The Court held that although Albemarle had made an effort to "validate" its tests as being job-related, the empirical evidence it gathered was insufficient and unconvincing. Justice Stewart, writing for the Court, held that the validation study was "materially defective in several respects." In summary, *Albemarle* held that "discriminatory tests are impermissible unless shown, by professionally acceptable methods, to be 'predictive of or significantly correlated with important elements of work behavior which comprise or are relevant to the job or jobs for which the candidates are being evaluated.' "[57].

In *Washington* v. *Davis* (1976), a fair-employment case, the Supreme Court attempted to clarify the distinction between its standards of review in constitutional as opposed to statutory cases, and it clarified its position on the need for demonstrating discriminatory intent among employers. In the case, two applicants to become police officers, both Negroes, claimed that the District of Columbia's Police Department's recruiting procedures, which included a standardized civil service test, were racially discriminatory and violated the Due Process Clause of the Fifth Amendment.[58] The validity of the test was the sole issue before the district court, since there was no claim that there had been intentional discrimination or that the relationship between the number of blacks on the police force compared to relevant population statistics suggested a differential impact. In fact, "[s]ince August 1969, 44% of new police force recruits had been black; that figure also represented the proportion of blacks on the total force and was roughly equivalent to [the proportion of] 20–29-year-old blacks in the 50-mile radius in which the recruiting efforts of the Police Department had been concentrated."[59] *Washington* v. *Davis* was a constitutional case, yet the Supreme Court found that:

> Because the Court of Appeals erroneously applied the legal standards applicable to Title VII cases in resolving the constitutional issue before it, we reverse its judgment in respondent's favor.... As the Court of Appeals understood Title VII ... employees or applicants proceeding under it need not concern themselves with the employer's possibly discriminatory purpose but instead may focus solely on the racially differential impact of the challenged hiring or promotion practices. This is not the constitutional rule. We have never held that the constitutional standard for adjudicating claims of invidious race discrimination is identical to the standards applicable under Title VII, and we decline to do so today.[60]

57. *Ibid.*, 422 U.S., at 431, quoting 29 CFR § 1607.4(c) of EEOC guidelines. The acceptable methods referred to are those set down in 1966 by the American Psychological Association, in *Standards for Educational and Psychological Tests and Manuals*. They include three basic methods of statistical validation: "empirical" or "criterion" validity, "construct" validity, and "content" validity.

58. *Washington* v. *Davis*, 426 U.S. 229, 96 S. Ct. 2040 (1976).

59. *Ibid.*, p. 2.

60. *Ibid.*, p. 4.

In a 7-2 decision, the Court held that the use of a standardized test (Test 21, which measured basic verbal skills) to screen applicants not only did not violate the Due Process Clause, but did not violate the respondents' rights under the Equal Protection Clause of the Fourteenth Amendment.

In the first section of the opinion, Justice White, writing for the majority, disclaimed the widespread interpretation of *Griggs* that an employer's purpose or intent is of no significance if the impact of his action is discriminatory: "Our cases have not embraced the proposition that a law or other official act, without regard to whether it reflects a racially discriminatory purpose, is unconstitutional *solely* because it has a racially disproportionate impact."[61] The Court was now suggesting, in effect, that discrimination is not simply equatable to differential impact. The outcome of an employment procedure is not sufficient evidence for establishing proof of discriminatory intent.

A basic question in *Washington* v. *Davis* is what standards are required to make a test acceptable and what kind of proof is necessary for demonstrating its acceptability. In *Griggs* and *Albemarle* the Court seemed disposed to support EEOC guidelines. In *Washington* v. *Davis* the petitioners had social science experts testify that Test 21 was, in fact, correlated with success in the police training program. The evaluation by D. L. Futransky of the U.S. Civil Service Commission concluded that the data "support the conclusion that T[est] 21 is effective in selecting trainees who can learn the material that is taught at the Recruit School." The Court then cited affidavits submitted by two expert independent witnesses, personnel research psychologists: (1) "It is my opinion . . . that Test 21 has a significant positive correlation with success in MPD Recruit School for both blacks and whites and is therefore shown to be job-related." (2) "It is my opinion that there is a direct and rational relationship between the content and difficulty of Test 21 and successful completion of recruit school training."[62] Curiously, the expert testimony cited in the Court's opinion was produced by the employee of an agency that makes wide use of the test in question; also, the affirmations of the findings were made by two research psychologists who just happened to be employees of the Civil Service Commission—the government agency that produced the original evaluation study.

Besides the absence of any mention of the strength of the correlations on which these opinions were based, and of the nature of the samples used to produce the association, there was no empirical evidence presented on the relationship between success in training and success as a police officer. The majority of the Court held that this final link was not required for acceptance of the test as a valid instrument, and it distinguished *Washington* v. *Davis* from *Griggs* and *Albemarle* by suggesting that the opinion represented a specification and a "much more sensible construction of the job-relatedness requirement."

61. *Ibid.*
62. *Ibid.*, note 17.

Justice Brennan dissented, holding that the decision effectively undermined the earlier Title VII decisions, and was based upon incomplete, if not faulty, empirical evidence: "The Court's conclusion cannot be squared with the focus on job performance in *Griggs* and *Albemarle*." Justice Brennan held that if the Court had followed the prior standards of validation it would have found that the "petitioners' proof is deficient in a number of ways."[63] His most significant objection was that "[t]here has been no job analysis establishing the significance of scores on training examinations, nor is there any other type of evidence showing that these scores are of 'major or critical' importance . . . there is no proof of a correlation—either direct or indirect—between Test 21 and performance of the job of being a police officer."[64]

In several recent Title VII cases the legal standards set down in *Griggs* and in *Albemarle* have been reinforced by the Court. In *E. C. Dothard* v. *Rawlinson* (1977), which focused on female applicants for the position of correction officer in the Alabama state penitentiary system, Justice Stewart, writing for the majority, stated:

> The gist of the claim that the statutory height and weight requirements discriminate against women (33.29 percent of women in the U.S. between the ages of 18 and 79 would be excluded from employment under the height requirements, compared to 1.28 percent of men; and 22.29 percent of women compared with 2.35 percent of men would be excluded under the weight requirement) does not involve an assertion of purposeful discriminatory motive. It is asserted, rather, that these facially neutral qualification standards work in fact disproportionately to exclude women from eligibility for employment by the Alabama Board of Corrections. We dealt in *Griggs* v. *Duke Power Co.*, and in *Albemarle Paper Co.* v. *Moody*, . . . with similar allegations that facially neutral employment standards disproportionately excluded Negroes from employment, and those cases guide our approach here.[65]

In the same case the Court did hold that gender was a BFOQ for "contact positions" in the penitentiary system.

There is a curious feature to the standards set forth in these opinions. Putting aside for the moment those sections in the opinions that focus on the unconstitutional intent, purpose, or motivation of decision makers, and that which rest on the clearly discriminatory consequences of a suspect classification, what basic problems confront courts in reaching their decisions? The law does not say that any prima facie distribution of males and females in an occupation can be equated necessarily with discriminatory intent or practice. Nor does it hold that all standardized tests are invalid tools for distinguishing qualified from unqualified potential employees. Nor that there are not exceptional circumstances under which bona fide occupational qualifications exist. The result is that courts are being asked to determine: When is discrimination actually a consequence of an

63. *Ibid.*
64. *Ibid.*
65. *Dothard* v. *Rawlinson,* 97 S. Ct. 2720 (1977).

employment procedure? When are tests actually measuring or predicting successful job performance? When is a social characteristic of an employee significantly related to or necessary for successful job performance? When are employment decisions based upon universalistic or particularistic criteria? And each of these questions almost invaribly is an empirical one.

Of course, the normal discovery and evidence procedures are followed in fact-finding. Expert witnesses are called in for their opinions; documents and records are analyzed. But the point is that increasingly these experts will be psychologists familiar with testing procedures, as well as economists, sociologists, and statisticians familiar with data-gathering and analytic techniques. On appeal the courts may seek additional advice on the adequacy of the data presented in support of or against hiring or promotion practices. In short, the courts are becoming the forum for social science debate; and their decisions, in some measure, are turning on the adversary procedure involving different pieces of social science evidence. The courts have begun to adopt the standards of proof and hypothesis-testing that are used to measure or estimate discrimination in the social sciences. And they are increasingly relying on "naive" or "sophisticated" residualism—or on even less sophisticated forms of proof—as evidence of discrimination.[66]

---

66. Title VII discrimination cases are significant as they relate to another aspect of proof—the locus of the burden of proof. As I mentioned, under Title VII only a prima facie case for discrimination is required to shift the burden of proof from plaintiff to defendant. Of course, this burden is worked into the statute. But there is a long tradition in the common law for such shifts in responsibility. Under the nineteenth-century legal doctrine of *res ipsa loquitor*, a plaintiff had only to demonstrate a defendant's negligence on a prima facie basis to shift the burden of proof to him. After drawing this parallel in my own mind, I came across the same basic observation in the *Harvard Law Review* 89, no. 2 (Dec. 1975): 392, Note 18: "While *Griggs* relied on congressional intent for the proposition that proof required only a showing of discriminatory effect . . . a parallel justification for the inference of discrimination purely from number has been developed in both the prior and subsequent law along the lines of res ipsa loquitor." Thus, I have stumbled once again into a "multiple discovery." Other areas of law also allow this shift; e.g., in civil tax cases where fraud is not alleged, the defendant must offer proof. This raises a series of sociological rather than legal questions: Under what conditions are such shifts of burden of proof likely to be found in the law? Why are the shifts in presumption allowed? Location of the burden of proof is unlikely to be simply a matter of identifying the party that is most apt to hold the "facts" in question. The location probably is influenced strongly by cultural values and norms that may be violated.

A further issue is the degree of difficulty posed in establishing the prima facie case. In Title VII cases the requirements have been minimal—a plaintiff having to show only that the outcomes of employment procedures are at significant variance with the demographic characteristics of the applicant pool, or the local constituency.

Even cursory analysis suggests that there is a great variation in the standards of proof required in different branches of the law. In criminal law a defendant must be proved guilty "beyond a reasonable doubt," but in tort law a plaintiff is required only to show with a "preponderance of evidence" that a defendant is guilty of negligence. The history of common law in England and the United States since the seventeenth century is one of changing standards of proof. (See, among others, Roscoe Pound, "Causation," *The Yale Law Journal* 671 (Nov. 1957): 1–18; Marc A. Franklin, *Tort Law and Alternatives* [Mineola, N.Y.: Foundation Press, 1971]: 59–94.)

The law invariably works with incomplete evidence. In most personal injury cases, for example, there rarely is any documentary, or "real," evidence. But even without "real" evidence, or eyewitness testimony, cases can be joined and won on the basis of circumstantial evidence. (See Franklin, *op. cit.*, p. 60.) There are a variety of doctrines in tort law that require only prima facie evidence as proof: the "but for" test and the doctrine of proximate cause, to mention two. (See *Byrne* v. *Boadle* [1863] for a classic res ipsa argument and J. Cardozo's opinion in *Palsgraf* v. *Long Island R.R.*, 248 N.Y. 339, 162 N.E. 99 [1928], for a classic proximate cause argument.)

As Guido Calabresi has recently noted, criteria of proof in tort law follow function, intent, or purpose. Over time these functions have changed spectacularly, reflected most recently in "no fault" liability. The functions of compensation and deterrence are directly related to standards of proof in tort law. The criteria of proof are, of course, not uniform, varying frequently from one state jurisdiction to another. (Guido Calabresi, "Concerning Cause in the Law of Torts: An Essay for Harry Kalven, Jr.," *University of Chicago Law Review* 43, No. 1 [fall 1975].)

But are the courts capable, using traditional discovery and adversary procedures, of finding the best evidence about actual rather than apparent discrimination; and are judges, law clerks, lawyers, law professors, and jurors trained to evaluate the quality of that evidence? The courts have placed themselves, or have been placed, in the position of evaluating the adequacy of the data presented by the two parties, the methods of sampling and of analyzing the data, the internal and external validity of the data, the relationship between aggregate-level correlations or associations, and the application of the law in individual cases. Perhaps the most striking recent incident in which the Supreme Court became involved in evaluating such data and studies was in the capital punishment case *Furman* v. *Georgia* (408 U.S. 238 [1972]). A central question was whether the application of the death penalty was discriminatory against racial minorities and other socially disadvantaged groups.[67] Data were also presented on the deterrent effect of capital punishment. The Court in a 5–4 vote decided *Furman* with these data in hand, and Justice Marshall (in one of nine separate opinions) concluded: "In light of the massive amount of evidence before us, I see no alternative but to conclude that capital punishment cannot be justified on the basis of its deterrent effect."[68] But up to this day, there continues to be widespread disagreement about whether the data actually "prove" the absence of deterrence.[69]

The *Furman* case represents relatively sophisticated methods of hypothesis testing compared to those often used by courts to establish the "facts" of discrimination. In *Chance* v. *Board of Examiners,* for example, the use of a standardized qualifying test was ordered discontinued after a court-ordered study found that in the aggregate "31.4% of minority candidates and 44.3% of nonminority candidates achieved passing scores."[70] A 13-point percentage difference, without controls for a broad set of possible explanatory variables, was taken by the Court as sufficient to establish a prima facie case of discrimination and to warrant discontinuance of the test. In *Fowler* v. *Schwarzwalder* (351 F. Supp. 721 [1972]), to cite only one other among many possible examples, the federal district court, in a Title VII case, "held that the fact that the city had a minority population of between 4.6% and 8% of the total population [depending on whether Mexican-Americans are included in the calculations], with a fire department in which between 1.3% and 2.2% were members of minority groups [depending again on inclusion or exclusion of the Mexican-Americans] established a prima facie case of discrimination."[71] A difference of 3 or 6 percentage points between the proportion of minorities in the "pool" of applicants and the

67. An interesting discussion of *Furman* appears in Brest. *op. cit.,* Chapter 8.

68. *Ibid.* Justice Douglas, in his opinion, also made extensive use of social science data purporting to show that the death penalty discriminates against blacks.

69. A great deal of recent empirical social science research has gone into the reargument of the death penalty case before the Supreme Court (1976). The data do not lead to a straightforward conclusion. See Isaac Ehrlich, "Capital Punishment and Deterrence: Some Further Thoughts and Additional Evidence," *Journal of Political Economy* 85, No. 4 (Aug. 1977): 741–788.

70. As quoted in Babcock et al., *op. cit.,* p. 396.

71. 351 F. Supp. 721 (1972) at 722

proportion actually holding positions in the job was sufficient here to shift the burden of proof, and ultimately to gain injunctive relief.

In other cases, such as *Griggs,* far greater differences have been found in the performance of whites and minorities; yet even in *Griggs* the test data were based not upon Duke Power Company employees but on employees in other places that used the same test. The point is that courts are now consistently drawing inferences about discrimination on the basis, *inter alia,* of simple zero-order statistical relationships.[72]

Not only are there few members of the legal profession who are familiar with the methods and assumptions of empirical social science research (although the number is clearly increasing in the nation's better law schools), but there are no institutionalized mechanisms for courts or lawyers to obtain the best evidence on empirical questions of discrimination, or to evaluate the results of social science inquiries that have the aura of scientism. In most cases, the procedures actually followed can only be described as haphazard.

Some may argue that the adversary process should take care of poorly presented evidence. But this is problematic if both parties are committed to their clients, if most lawyers do not understand the methods of social science, and if the courts are in no better position than the lawyers to evaluate empirical findings or to draw causal inferences from imperfect data and analysis. In fact, it seems likely that advocacy is a procedure peculiarly unsuitable for arriving at the best social science evidence. As Henry M. Levin has recently said: "There is always some social science evidence on virtually any phenomenon, so one must ask what types of evidence are likely to be drawn into the courts."[73] For one thing, the empirical evidence is likely to be either extraordinarily simplistic—e.g., marginal differences or simple associations—or highly complex, based upon methodologies that are far beyond the experience or evaluatory competence of the courts. For another, the evidence is apt to be presented as unambiguous and without qualification, since the "experts" presenting it are likely to dismiss those data that do not support their case, justifying this practice as part of the adversary procedure.[74]

To establish this point more forcefully, consider several cases, only one of which relates to questions of discrimination. Take the *Brown* desegregation decision first. Although it is generally held that the social-psychological evidence referenced and used in *Brown* was not necessary for the decision, a significant portion of Chief Justice Warren's opinion is devoted to the psychological and social effects of segregated schooling. Even at the time of *Brown* there was considerable skepticism about the adequacy of the empirical data cited. Today it

---

72. See as examples *Hicks* v. *Crown Zellerbach Corp.,* 319 F. Supp. 314 (E.D. La. 1970); *Arrington* v. *Mass. Bay Transport Co.,* 306 F. Supp. 1355 (D. Mass. 1969); *Carter* v. *Gallagher,* 3 EPD Par. 8205 (D. Minn. 1971).

73. See the provocative paper by Henry M. Levin, "Education, Life Chances, and the Courts: The Role of Social Science Evidence," *Law and Contemporary Problems* 39, No. 2 (spring 1975): 217–240.

74. See Levin's discussion of these matters, *ibid.*

is evident that many of the empirical questions raised in *Brown* are far from resolved.[75]

In a 1974 paper, Hans Zeisel and Shari Seidman Diamond discuss the Supreme Court's use of empirical data in determining whether in certain types of cases jury size can be reduced from twelve to six without any effect on trial results.[76] "In *Williams* v. *Florida,* which upheld the use of the six-member jury in criminal cases in state courts, the Court . . . was misled in believing that there was such evidence.' "[77] In *Colgrove* v. *Battin* the Court asserted that "four very recent studies have provided *convincing empirical evidence* of the correctness of the *Williams* conclusion that 'there is no discernible difference between the results reached by the two different-sized juries.' ' "[78] These four empirical studies were cited by the Court as proof. The Zeisel and Diamond paper returns to the original studies cited by the Court and evaluates the methodology used to reach the conclusions about the effect of jury size. In each case it finds sufficient methodological inadequacies to seriously question the studies' conclusions. The Court seemed to rely on the summarized conclusions of the investigators rather than on the adequacy of the study designs. Consider two of the studies. In Washington, civil trials "are held before six-member juries, unless one of the litigating parties requests a twelve-member jury.' "[79] Since this particular empirical study was not conducted in a laboratory, the investigators who studied the outcomes of 128 workmen's compensation trials had to take the juries as they were determined by the litigants rather than by the social scientists. Of course, the investigators were aware that there might be differences in the types of cases brought before juries of different sizes. But they slough this off by saying: "If we may properly assume that the assignment of jury size was essentially random . . . then we may conclude that the use of the smaller jury introduced no systematic bias into trial outcomes.' "[80] But in fact, as Zeisel and Diamond point out, "There is good evidence that lawyers are more likely to opt for the larger jury if the amount in controversy is larger.' "[81]

In the New Jersey study of jury size, it becomes clear upon reanalysis that (1) cases before twelve-member juries tended to be more complex; (2) "settlements and verdicts of twelve-member juries were, on the average, three times as great

---

75. See, among others, Nancy St. John, "Desegregation and Minority Group Performance," *Review of Educational Research* 40 (1970): 111 ff.; Frank Goodman, "De Facto School Segregation: A Constitutional and Empirical Analysis," *California Law Review* 60 (1972); David Armor, "The Evidence on Busing," *Public Interest* 28 (1972); Thomas Pettigrew et al., "Busing: A Review of 'the Evidence,' " *Public Interest* 30 (1973); Brest, *op. cit.,* pp. 454–475; Edmond Cahn, "Jurisprudence," *New York University Law Review* 30 (1955); Kenneth Clark, "The Desegregation Cases: Criticism of the Social Scientists' Role," *Villanova Law Review* 5 (1960).

76. Hans Zeisel and Shari Seidman Diamond, " 'Convincing Empirical Evidence' on the Six-Member Jury," *University of Chicago Law Review* 41, No. 2 (winter 1974): 281–295; see also Alvin Klevorick, "Jury Size and Composition: An Economic Approach," 1975 preprint obtained from the author.

77. Zeisel and Diamond, *op. cit.*

78. As quoted in *Ibid.,* p. 282.

79. *Ibid.,* p. 283.

80. *Ibid.,* p. 283.

81. *Ibid.,* p. 284.

[in dollar amounts] as for six-member jury cases"; (3) "the average deliberation time of six-member juries was 1.2 hours, compared to 1.8 hours for the twelve-member juries"; but (4) there was no difference between the juries when the trial results were for damages under $10,000.[82] The investigators were aware that these differences posed problems of inference about the effect of jury size, but they did not attend to them. The significant point is not so much that the social science studies used inadequate methods and analysis to reach their conclusions—this is all too often the case—but that the Court "swallowed" these conclusions and apparently did not have any mechanism for evaluating them before using them.

Few courts are prepared to evaluate the internal and external validity of empirical social science research. How many judges, law clerks, or attorneys have even superficial knowledge of problems of causal inference resulting from history, maturation, testing, instrumentation, selection, statistical regression, residual arguments, a variety of interactions, and so on?[83] Courts are in no better position to evaluate the adequacy of sampling procedures. It is not only that courts are not competent to handle these problems, and thus are forced to fall back on traditional modes of analysis to reach a conclusion as to the presence or absence of discrimination, but also that there are no established bureaus or organizations to provide the courts with either the necessary data or evaluations of the "best evidence."

In sum, the courts are increasingly using techniques of "proof" developed by social scientists in sex-discrimination cases. But the courts labor under a double disadvantage. Not being social scientists, few judges are capable of adequately evaluating the new evidence at their disposal, and they must make decisions that affect the lives of a significantly large number of individuals and institutions on

---

82. *Ibid.*, pp. 285–286.

83. Donald Campbell and Julian Stanley, "Experimental and Quasi-Experimental Designs for Research on Teaching," in *Handbook of Research on Teaching*, ed. N. L. Gage (Chicago: Rand McNally & Co., 1963), pp. 171–246.

A recent note in the *Harvard Law Review* discusses at considerable length the advantages to be gained by plaintiffs, defendants, and "the truth-finding process" by using multivariate statistical techniques, particularly analysis of variance and multiple regression. See "Beyond the Prima Facie Case in Employment Discrimination Law: Statistical Proof and Rebuttal," *Harvard Law Review* 89, No. 2 (Dec. 1975): 387–422. This is an informative article on the ways the use of simple percentage differences to judge discrimination can be improved upon by use of more advanced statistical techniques. It is instructive by its very appearance, since it signals the perception within the legal community of the need to know about social science statistical techniques and how they can be used in discrimination cases. The review article notes a variety of ways that statistics have been and are currently being used in discrimination cases. Yet the note reflects a surprising naiveté about the use of statistics. It tends to reify statistical tests of significance, suggesting for example that cases will depend on the ability of plaintiffs or defendants to demonstrate that race or sex does or does not have a statistically significant effect on a dependent variable (generally at the .05 level or higher). But clearly there are theories and assumptions adopted in using these statistics, and pitfalls in their blind application to many specific situations. Furthermore, there is difficulty in assessing the *substantive* meaning of associations that may be very small and yet statistically significant because they are based on very large samples. Correlatively, associations that are strong may not be statistically significant because they are based on very small samples. Even more interesting, the authors of the note totally miss the ways that zero correlations can mask discriminatory behavior. A total reliance on "significant correlations," rather than on substantive explanations, can lead to false inferences and incomplete tests. Nonetheless, the note's conclusion is essentially correct: "A large amount of current jurisprudence in employment discrimination law could still . . . have benefited from the application of multivariate statistical techniques, . . . and might have turned out differently." (p. 421)

the basis of incomplete and often poorly analyzed data. One can argue that the availability and use of these new techniques represents an advance over the status quo ante, but one should also recognize the need to address the emergent problems resulting from the use of these new forms of evidence.

ESTIMATING DISCRIMINATION IN SCIENCE

The term "discrimination" as it is used in this book is defined as the importation into a social situation of characteristics that are functionally irrelevant—specifically here to the quality of scientific role-performance. It is a behavioral and not an attitudinal concept. Given the basic values of science, sex is a functionally irrelevant status, unless it can be forcefully argued with reliable evidence that it is related independently of social and cultural factors to the ability to perform *qua* scientist. In some social settings, it is particularly difficult to define the boundaries of functional relevance. In the marketplace, we might be forced to discuss how factors that are irrelevant to production might be quite relevant to consumption. But in the marketplace of ideas, where science fundamentally is located, the problem of distinguishing spheres of relevance hardly applies. It is hard to view gender as a relevant characteristic in the production of scientific ideas. Perhaps in areas of consumption of ideas it can become relevant, but most probably because of bias and expectations about the capabilities of women rather than anything biologically intrinsic to gender. This does not mean, of course, that gender may not be discovered to be a functionally relevant characteristic even today because of the pervasive social structural impediments to female achievement from childhood on, which limits skills and abilities later on in their life cycle.

When analyzing empirical data, I use a sophisticated residualist approach. The effects of gender in producing inequalities are initially examined, and subsequently a set of factors that potentially can explain differences are analyzed. The remaining effect of sex status is tentatively used as an estimate of sex discrimination, although I often will speculate on additional possible explanations for which I have no data.

Sex discrimination here is more precisely residual sex inequality. In fact, when the word "discrimination" is used in this book, the reader might appropriately think of it in terms of residual inequality. As I have noted, this empirical method has its distinct limitations, and consequently the results presented should be treated with caution. So much for problems of proof. Let us move directly to the data that locate the place of women in contemporary American science.

Chapter 3

WOMAN'S PLACE IN THE SCIENTIFIC COMMUNITY

UNTIL THE TWENTIETH CENTURY, science was populated almost exclusively by men, and so the phrase "men of science" was almost equivalent to the non-sex-linked tag "scientists." Even today the situation has not drastically changed. There are relatively few women in science, and even fewer women among the scientific elite. As of 1970 in the United States, for example, only slightly more than one-tenth of science Ph.D.'s were women. Women comprised only 3 percent of 1970 Ph.D.'s in physics and astronomy; 8 percent in chemistry; about 15 percent in the biological sciences; 18 percent in sociology; and 24 percent in psychology.[1]

The proportion of women to the total number of doctorates awarded remained remarkably constant over the fifty-year period from 1920 to 1970. In the period from 1970 to 1977, there have been sharp increases in the proportion of female doctorates. I will discuss these shifts later in this chapter. A 1966 National Academy of Sciences report notes:

> Women receive 40 percent of the baccalaureate degrees granted in the United States, and the percentage is increasing. They receive 32 percent of the master's

---

NOTE: Parts of this chapter appeared in slightly different form in Jonathan R. Cole and Stephen Cole, *Social Stratification in Science* (Chicago: University of Chicago Press, 1973).

1. Source: *Summary Report, 1967: Doctorate Recipients from United States Universities,* prepared in the Research Division of the Office of Scientific Personnel, OSP-RD-1, (Washington, D.C.: National Research Council, National Academy of Sciences, May 1968); *Summary Report, 1968: Doctorate Recipients from United States Universities,* prepared in the Education and Employment Section, Manpower Studies Branch, Office of Scientific Personnel, OSP-MS-2, (Washington, D.C.: National Research Council, April 1969).

degrees, but this proportion has remained constant for many years. The percentage of women among United States doctorate recipients dropped from 15 to 9 percent between 1920 and 1950, but a gradual increase restored the value to 11 percent by 1960. Since 1960 [to 1966] the proportion of women among all doctorate recipients has remained constant, not only for the total, but for each summary field.[2]

The rolls of the National Academy of Sciences and the list of Nobel laureates register very few women. Only 8 of the 866 members of the National Academy are women. Only five of the 281 scientists who have received the Nobel Prize have been women; and of these five, Marie Curie, Irene Joliot Curie, and Gerty Cori each shared her prize with her husband.[3] How can we account for these facts? Has American science systematically excluded women from its ranks?

As I noted earlier, many scientists and others believe that women are discriminated against when they seek appointment to high-prestige science departments; that they are less likely than men to receive promotions to tenured positions; that they must wait longer for their promotions; that they are less often the recipients of honorific forms of recognition; and that their salaries are lower at every academic level.[4] Only recently, however, have women scientists begun to protest against these apparent injustices.

To establish the existence of discrimination against women, it is necessary to compare the careers of women and men scientists. We must estimate the influence of sex status on scientific recognition while controlling for the type of training the scientists have received and the quality and quantity of their scientific research.

To this end, I collected data from several sources on a sample of 565 academically employed men and women scientists in four fields: 298 biologists, 62 chemists, 159 psychologists, and 46 sociologists.[5] The sample was generated by selecting men matched to a sample of women who received their doctorates from American universities in 1957 and 1958. Data on female doctorate recipients

---

2. Source: *Doctorate Recipients from United States Universities, 1958-1966,* Publication 1489, (Washington, D. C.: National Academy of Sciences, 1967), p. 107.

3. Harriet A. Zuckerman, "Women and Blacks in American Science: The Principle of the Triple Penalty." Paper presented at the Symposium on Women and Minority Groups in American Science and Engineering, California Institute of Technology, 8 Dec. 1971, pp. 34-35.

4. See recent issues of *Science* which have had letters and papers about the place of women in science: 16 April 1971, vol. 172, no. 3980; 7 May 1971, vol. 172, no. 3983 (see a particularly strong protest, which is found increasingly in scientific journals, on p. 514); 16 July 1971, vol. 173.

5. Physicists and mathematicians are not included in this analysis because there are so few women Ph.D.'s in these fields that any separate analysis by field would involve a minute number of women scientists. Other social sciences and humanities are not included because I could not obtain reliable data on the publications and citations of academics in these fields. Finally, I have excluded nonacademic scientists, because the reward system in industry and government science is far more varied than in academic science. Since I wanted to consider a more or less uniform reward structure, I limited analysis to academics. In chemistry 55 percent of the women Ph.D.'s held academic positions, compared with 31 percent of the men Ph.D.'s. Women in chemistry were also slightly more likely than men to hold positions in universities. The same pattern was observed for the biological sciences: 74 percent of the women and 68 percent of the men were located in academic positions. In psychology the proportion of men in academic life was greater than the proportion of women: 52 percent to 38 percent. The number of sociologists is too small to make a meaningful comparison here. While women are slightly more likely than men to hold positions at colleges, they are not significantly underrepresented compared to men in university settings.

were collected in 1965 by Helen Astin through a mail survey.[6] Eighty-three percent of all 1957 and 1958 women Ph.D.'s, or 1,547, responded to the survey. I selected male matches for only the 749 women who received doctorates in the physical, biological, and social sciences. These male matches were drawn from the doctorate record file of the Office of Scientific Personnel (OSP), which includes annual data on 99 percent of all doctorate recipients, or about 17,000 cases.[7] Four matching criteria were used: year of doctorate, university where the Ph.D. was earned, field, and specialty.[8] For the most part, I report data on only the 565 men and women scientists who in 1965 were academically employed, because of the uniformity of the reward structure in the academic science community.

Career data for the women were obtained from Astin, and data on selected social characteristics and on social mobility were collected for the entire sample from *American Men of Science*.[9] Publication counts for the years 1957 through 1969 were gathered from the appropriate abstracting journals.[10] Citation data were collected from the *Science Citation Index*.[11] Finally, the measured intelligence and high school academic records were obtained from the Office of Scientific Personnel. Although I used the matching procedure to generate the sample, after the matches were obtained the men and women scientists were treated as an

---

6. I thank Dr. Helen S. Astin for the help she has given me in this study, without which it would not have been attempted. She has provided me with the basic data-set and codebook for the women scientists. Her results for women scientists are reported fully in *The Woman Doctorate in America* (New York: Russell Sage Foundation, 1969).

7. Astin's study also made extensive use of the doctoral record file of the Office of Scientific Personnel. I thank Dr. William Kelley and Ms. Clarebeth Maquire Cunningham of the OSP for their help in obtaining the basic matches used in this study.

8. Drawing the sample involved a number of contingency operations. First priority was given to matching on university department, second to matching on field, third on specialty, and fourth on year. If, for example, I was able to generate a match for a woman in biochemistry at the University of Missouri, but there was no Ph.D. in 1957, I searched for a match in 1958. If a match was found in 1958, I took this "adjusted" match. Moreover, I generated two male matches for every woman in the sample. This was done to ensure that I got a high proportion of the men with complete data comparable to that available for the women. Since my primary source for background information on the male scientists was *American Men of Science,* if the first match was not found in *AMS,* I turned to the second match. If neither man was found in the compilation, I selected the first match. Both male matches were, where possible, randomly selected from the available matches. In about 19 percent of the cases, I used the second match. There are at least two reasons why this procedure is unlikely to distort the results presented here. First, most of the male physical and biological scientists were found in *AMS;* a much smaller proportion of the social scientists could be found in the *Behavioral and Social Science* volumes of *AMS.* Wherever possible, however, missing information for social scientists was filled in by consulting the *Directory* of the American Psychological Association and the *Directory* of the American Sociological Association. Certainly the results for chemistry and biology are unlikely to be distorted at all; psychology and sociology could have some minimal bias, if the men in *AMS* are more eminent than the women to whom they are matched. Second, in many cases the second male match was linked to a woman who was in *AMS*. Consequently, by selecting the second match in this limited number of cases, I actually controlled somewhat for "minimal" eminence.

9. I used the 11th and 12th editions of *American Men of Science,* now *American Men and Women of Science.*

10. I used the following sets of abstracts for the publication counts: *Science Abstracts; Chemical Abstracts; Biological Abstracts;* and *Psychology Abstracts.* Both single-authored and multi-authored papers were counted. For sociologists, papers were counted which appeared in any one of twenty-nine leading sociological journals, based on the Glenn-Villemaz ratings.

11. Citation data were collected for six years: 1961, 1964, 1965, 1967, 1969, 1970. If I had wanted only a single measure of "quality" of scientific work, I undoubtedly could have taken counts for only one year. Extensive citation data were collected in order to examine the differences in the impact of early and late work for men and women scientists. In all of these counts, I have excluded all self-citations.

aggregate. This aggregation did not produce significant bias. For example, for the entire sample of academic scientists the zero-order correlation between sex status and prestige of Ph.D. department, which on the basis of my sampling frame was zero, was in fact only −.03.[12]

To ascertain the extent of recognition received by men and women scientists, I consider two forms of recognition: positional and reputational. As an indicator of positional recognition, I use the prestige rank of the scientist's academic department as measured by the Cartter and the Roose-Andersen studies sponsored by the American Council on Education. Universities not included in these studies were rated lower in prestige than those that were ranked, and all colleges were scored lower than universities. While this may involve occasional inaccuracies, in general universities are the locus of research activities and a substantial proportion of the eminent scientists in any era can be found at the better large universities. I considered academic rank an additional indicator of positional recognition, although rank obviously must be coupled with prestige of the academic department before it is a meaningful indicator of recognition. I use the number of honorific awards that the scientist has received as a rough estimate of reputation. Past studies have shown that number of awards is strongly correlated both with the prestige of awards held ($r = .70$) and with the reputations of scientists ($r = .63$).[13] Direct testimony about the subject of this study from a sample of their colleagues in science on how well known their work is within the scientific community or on the perceived quality of that work was not elicited.

SEX STATUS AND PRESTIGE OF ACADEMIC AFFILIATION

The few inquiries into the comparative positions held by men and women scientists report that women tend to occupy lower-prestige positions than do their male peers. For example, Budner and Meyer found among a sample of social scientists at American universities that women were more likely than men to be at lower-prestige institutions: 22 percent of the women were at schools of "high" quality compared with 38 percent of the men.[14] Correlatively, 55 percent of the women and 30 percent of the men were located at institutions of "medium-low" or "low" quality.[15] Rossi, studying the distribution of men and women sociologists, also reported that women were underrepresented in prestigious de-

---

12. Some bias may result from lack of complete data on the careers of all 565 academic men and women. Since the correlations are based on pair-wise comparisons, the data tend to overweigh slightly the men in the sample.

13. Jonathan R. Cole and Stephen Cole, *Social Stratification in Science* (Chicago: University of Chicago Press, 1973).

14. Stanley Budner and John Meyer, "Women Professors," as reported in J. Bernard, *Academic Women* (University Park, Pa.: Pennsylvania State University Press, 1964). This study discusses data drawn from Paul F. Lazarsfeld and Wagner Thielens, Jr., *The Academic Mind* (Glencoe, Ill.: Free Press, 1958). Budner and Meyer used the Berelson ratings of the quality of a university.

15. *Ibid.*, p. 93.

partments.[16] Studies commissioned by professional societies suggested that women were underrepresented in top departments. For example, in physics 0.9 percent of the faculty were women in the "top ten" departments; 1.7 percent of the physics faculty were women at 158 other universities; 2.5 percent were women at four-year colleges with M.A. or M.Sc. programs; and 6 percent were women at colleges with undergraduate programs only. These studies and reports by professional societies[17] rarely related the proportion of women faculty to the percentage of female Ph.D.'s in the hiring pool, or controlled for the quality of scientific role-performance as indicated, for example, by the quality and quantity of publications. Furthermore, none of these studies controlled for professional age or the type of training received by the scientists. My data, which do match men and women scientists both for professional age and where they received their training, and include extensive information on research output, thus allow me to determine more adequately the relationship between positional recognition and sex status.

I begin with an apparently striking datum: The zero-order correlation between sex status and prestige of academic department in 1965 is virtually zero ($r = -.02$), suggesting that men and women scientists with similar training tend to be located in academic departments of equal prestige seven years after receiving their Ph.D.'s. (For this analysis sex was coded: 1 = men, 2 = women.)[18] This zero-order correlation coefficient, however, conceals some variation among the four academic fields making up the sample. In chemistry and psychology, women are somewhat less likely than men to be found in top departments. The zero-order correlations for the two fields are $r = -.23$ and $r = -.17$, respectively. In biology, where there are more women in academic positions than in the other two fields, women do slightly better than men ($r = .06$), and the differential in favor of women is even greater in sociology ($r = .30$).[19]

It could be argued that as a result of discriminatory practices women are less likely than men to remain at top-ranked departments as their careers progress. In order to test this possibility I collected data on the location and rank in 1970— five years after the Astin survey—for a subsample of the 565 academic men and women. Data were obtained for those academicians who were listed in *American Men and Women of Science* (12th edition) or who were listed in the directories of the American Sociological or American Psychological Association and whose

16. Alice A. Rossi, "Status of Women in Graduate Departments of Sociology 1968–1969," *American Sociologist* 5, No. 1 (1970): 1–12; see also Alice A. Rossi, "Equality Between the Sexes: An Immodest Proposal," *Daedalus*, 1964: 98; Alice A. Rossi, "Women in Science: Why So Few?" *Science* 148, no. 3674, (28 May 1965).

17. See, among others, "Women in Physics," Report of the Committee in Physics Submitted to the Council of the American Physical Society, New York, N.Y., 30 Jan. 1972, p. 26 (mimeo.); *The Status of Women in Sociology, 1968–1972* (American Sociological Association, Washington, D.C., 1973).

18. Thus a negative sign before a correlation indicates that women are less likely than men to be in high-ranked departments, or less likely to hold high academic rank.

19. These zero-order correlations for the four fields are based on very different sample sizes. Psychology and biology produce many more women Ph.D.'s than do chemistry and sociology. Therefore, the size of these correlations should be examined in light of the numbers on which they are based. The correlations in sociology and chemistry are less stable than those in the other two fields.

last names began with A through K. I had data for 262 cases.[20] The zero-order correlation between prestige rank of department in 1970 and sex status does not significantly change: Men are no more likely than women to be in top-ranked departments some thirteen years after the Ph.D. ($r = .01$). The pattern of findings within the several fields is similar to that in 1965.

ACADEMIC RANK

Intense competition for promotion to the higher ranks of associate and full professor exists among American academicians, especially in prestigious departments. Scientists are concerned not only with achieving high rank, but with doing so as quickly as possible and in high-quality departments. It is some measure of distinction to be a "young" associate or a "young" full professor.

Apparently conflicting results are reported in work dealing with the academic rank of men and women scientists. Some studies, such as those by Perrucci,[21] Rossi,[22] Simon et al.,[23] and those administered by a number of professional associations,[24] conclude that women tend to hold lower ranks, and are particularly absent from tenured positions. Simon goes on to specify the conditions under which women do achieve equal academic rank. She found for her sample of doctorates in the natural and social sciences, as well as in the humanities and education, that unmarried women were just as likely as men to hold high-ranked positions. A more recent study by Bayer and Astin[25] was able to eliminate apparent sex differences in academic rank by controlling for the time women did not spend in the labor force after receiving their Ph.D.'s: They compared the ranks of men's doctorates in 1958 with women's doctorates in 1957 and found only minor differences in academic rank. However, each of these studies presents problems in interpretation, since once again professional age, educational background, types of institutional affiliation, and scientific role-performance variables have not always been adequately controlled.

Since all the scientists in my sample received their Ph.D.'s at the same time (1957–58), and therefore have roughly equal professional ages, the correlation between sex status and academic rank can be used as an indicator of the extent of academic recognition of the men and women scientists. The zero-order correla-

---

20. For these 262 cases the correlations for their 1965 positions and ranks were almost identical to the correlations for the larger sample. The sample size was limited because only volumes A–K of *AMWS* were available as of the summer of 1972, when these data were collected.

21. Carol C. Perrucci, "Minority Status and the Pursuit of Professional Careers: Women in Science and Engineering," *Social Forces* 49, (1970): 245–259.

22. Rossi, "Status of Women in Graduate Departments of Sociology (*op. cit.*).

23. Rita J. Simon, S. M. Clark, and K. Galway, "The Woman Ph.D.: A Recent Profile," *Social Problems* 15, (1967): 221–236, esp. 228–229.

24. *Op. cit.*, note 22.

25. Alan E. Bayer and Helen S. Astin, "Sex Differences in Academic Rank and Salary among Science Doctorates in Teaching," *Journal of Human Resources* 3, (1968): 191–200.

tion between sex status and academic rank for the entire sample is $-.34$; that is, women hold lower academic ranks than men. There are differences between fields, ranging from an insignificant difference in the ranks of men and women chemists ($r = -.08$) to substantial differences among biologists ($r = -.38$), psychologists ($r = -.44$), and sociologists ($r = -.44$). The lower correlation for chemistry may simply indicate more rapid mobility of all physical scientists to the higher academic ranks. Men may still be promoted sooner than women, but by seven years into their careers a high proportion of both sexes have reached the level of associate professor but are not yet full professors. A slightly higher proportion of academic chemists than social scientists did, in fact, hold the rank of associate professor by 1965. In the social sciences, where mobility takes somewhat longer on average, seven years may be a critical time to catch sex-related differences in academic rank. I also collected data on academic rank for the subsample of 262 scientists for whom I had 1970 data. The zero-order correlation between sex status and academic rank in 1970 declines somewhat from the 1965 level but remains significant ($r = -.28$), suggesting the possibility that discriminatory practices persist in the promotion of men and women scientists.[26]

As noted above, academic rank is an adequate indicator of academic recognition only when it is coupled with the prestige of the department in which high rank is achieved. The facts of American academic life are clear: If one stays in the business long enough, one is almost bound to be promoted to the rank of associate or full professor. Greatest recognition through rank is registered by high rank in a high-prestige department at a relatively early age.[27] In fact, Hargens has suggested that one mechanism by which the social system of science handles failure is by promoting scientists in rank while demoting them in terms of the prestige of the department with which they are affiliated.[28]

If seniority principles are used in promotion, then especially at less-distinguished departments there will not be a strong positive association between academic rank and measures of research performance. This should be less true of the top departments, in which research performance criteria are employed more frequently in filling scarce positions. This is supported by the data. In departments rated distinguished, strong, or good in the Cartter study, scientists who produce more ($r = .38$) and better work ($r = .37$) are more likely to be found in high-ranked positions, whereas in lesser departments quantity ($r = .02$) and quality ($r = .07$) of research output are virtually uncorrelated with academic rank. In both types of academic departments, however, women are not as likely to hold high ranks as men. Women in better departments are far less likely to hold top positions ($r = -.37$); at all other universities and colleges, they still are

26. The correlation between sex status and academic rank in 1965 for the subsample of 262 was slightly lower ($r = -.26$) than that for the entire sample, probably as a result of there being a higher proportion of women in the subsample who were consistently employed since receiving the Ph.D.

27. David Caplovitz, "Student-Faculty Relations in Medical School," Ph.D. diss., Columbia University, 1960.

28. Lowell Hargens, "The Social Context of Scientific Research," Ph.D. diss., University of Wisconsin, 1971.

less likely to be found in the ranks of associate and full professor ($r = -.19$) but the discrepancy is not as great as in the better departments.

SEX STATUS AND SALARIES

Various studies have found that women scientists' salaries are significantly lower than men's, even in similar institutions, within the same academic rank, and after accounting for length of tenure.[29] These studies, using different sampling techniques and focusing on different scientific disciplines, arrived at slightly different estimates of the differentials between the salaries of men and women scientists, but the pattern of the findings was consistent. Bayer and Astin, for instance, found that after five or six years of employment women in colleges and universities earn about 92 percent of the salaries of men in both the natural and social sciences.[30] Salary differentials by sex are greater among senior-rank than among junior-rank scientists, as one might expect from the greater variance in salaries at the higher ranks. None of these studies controlled for research performance. Since I did not have the salaries of the male scientists, I cannot estimate the differentials in salaries for the matched sample. Any adequate test for salary discrimination must take into account differences in the role performance of men and women. Without adequate controls for the actual performance of scientific roles, differences in salaries are not subject to easy interpretation. I will return to a discussion of salary differences in the chapter that considers historical trends.

REPUTATIONAL RECOGNITION: HONORIFIC AWARDS

Only limited data are available on the honorific recognition granted to women scientists. I have already noted their slight representation in the most elite academies of science and in the ranks of Nobelists. Simon reports that women are more likely than men to receive postdoctoral fellowships and memberships in honorary societies. Reskin, in a 1976 study of 450 doctoral chemists, also suggests that women are slightly more apt than men to receive a postdoctoral fellowship ($r = .20$). Her data indicated that several measures of the eductional background of the chemists, such as the quality of their undergraduate colleges and of their Ph.D. departments, were significantly related to the prestige of postdoctoral fellowships for men but not for women. Moreover, the prestige of

29. See, among others, Sylvia F. Fava, "The Status of Women in Professional Sociology," *American Sociological Review* 25, (April 1960): 271–276; Bayer and Astin, *op. cit.;* M. A. LaSorte, "Sex Differences in Salary among Academic Sociology Teachers," *American Sociologist* 6, (1971): 265–278; D. G. Lubkin, "Women in Physics," *Physics Today* 24 (1971); D. A. H. Roethal, "Starting Salaries—1970," *Clinical and Engineering News* (23 Nov. 1970).

30. Bayer and Astin, *op. cit.,* p. 196.

fellowships received was a significant predictor of future career success for men, but it did not have any predictive ability for women. The prestige of fellowships did, however, predict future scientific productivity reasonably well for both men and women.[31] An analysis of the composition of fellows of the Center for Advanced Study in the Behavioral Sciences between 1954 and 1971 showed that only 3 percent of the alumni were women.

For my sample of 565 scientists, I counted the total number of honorific awards and postdoctoral fellowships listed after their names in *American Men of Science*. For the entire sample, the correlation between sex status and number of honorific awards was $-.05$; that is, men were slightly more likely to have received awards. Once again, field differences exist: The zero-order correlation for chemistry was $-.19$; for biology, $r = .09$; for psychology, $r = -.07$; sociology, $r = .28$. In summary, there are only small differences in the level of this form of reputational recognition of men and women scientists in my matched sample. Sociology represents the most consistent divergence from the general pattern of results within the several fields. But since the number of cases upon which the sociology correlations are based is extremely small, one should be cautious in drawing strong inferences from these correlations; moreover, the result for sociology may simply reflect a heightened sense of social injustice among sociologists. As a field, sociology may have reacted more than other disciplines to the perception of sexual discrimination. This, of course. remains speculation.

The zero-order correlations from my data are summarized in Table 3-1. The overall pattern indicates that men do only slightly better than women in terms of the prestige of their academic departments and the receipt of honorific awards, and significantly better in terms of academic rank. Evidence from other studies suggests that women receive slightly lower academic salaries. Observed zero-order correlations such as these are often the basis for the statement that sex discrimination is operating in the scientific community. Among Astin's sample of women scientists, 25 percent cited experiences of discrimination in hiring practices; 40 percent experienced differentials in salaries; and 33 percent cited differentials in policies regarding tenure and promotion.[32] A recent survey of political scientists by Mitchell and Starr found that incidents of sex discrimination are just as likely to be reported by men as by women.[33]

But the gross data I have presented thus far, involving only zero-order correlations, are not enough to answer the question of whether academic women scientists are being discriminated against. However, unless it can be shown that

31. Simon, et al., *op. cit.*, pp. 232–233; Reskin, "Sex Differences in Status Attainment in Science: The Case of the Post-Doctoral Fellowship," *American Sociological Review* 41, no. 4 (Aug. 1976): 597–612.

32. Astin, *op. cit.*, p. 106.

33. J. M. Mitchell and R. R. Starr, "Aspirations, Achievementand Professional Advancement in Political Science: The Prospect for Women in the West," in *Women in Political Science: Studies and Reports of the APSA Committee on the Status of Women in the Profession, 1969–71* (Washington, D.C.: American Political Science Association, 1973).

TABLE 3–1. Zero-Order Correlations between Sex Status and Three Forms of Scientific Recognition for Four Scientific Disciplines

	ALL FIELDS	N	CHEMISTRY	N	BIOLOGY	N	PSYCHOLOGY	N	SOCIOLOGY	N
Rank of Department										
1965	−.02	423	−.23	42	.06	215	−.17	131	.30	35
1970	.01	262								
Academic Rank										
1965	−.34	469	−.08	54	−.38	244	−.36	130	−.44	41
1970	−.28	255								
Number of Honorific Awards	−.05	438	−.19	41	−.09	239	−.07	122	.28	36

NOTES:

There were missing data for this sample. The correlations are based upon pairwise computations. The "N" in the right-hand column gives the number of cases on which the Pearson correlation is based.

Sex status was coded: 1 = male; 2 = female.

Difference in N's between rank of department and academic rank is due to the fact that some scientists were located at medical schools which were not ranked by the ACE.

the differences in recognition received by men and women scientists are a result of differential role-performance, there is evidence of discrimination, and the social system is not operating in accordance with its universalistic norms. While it is tempting to draw inferences about the presence or absence of discrimination on the basis of simple bivariate relationships, they tell us little about equality or inequality of opportunity. For example, a correlation of zero or near-zero between sex status and prestige rank of department does not necessarily indicate that women are not discriminated against in terms of academic positions. For if the system is adhering to universalistic principles and if women as a group are publishing research superior to that of men, they should be rewarded in proportion to the quality of their work—and indeed should be overrepresented in superior departments. Correlatively, if men "outperform" women, a universalistic system will reflect this in the distribution of rewards.

It becomes all-important, then, to consider the influence of differential role-performance on the distribution of rewards. Previous studies addressing the sex-discrimination problem have failed to do this. But first we must entertain the possibility that sex status is, in fact, not a functionally irrelevant status for the performance of scientific roles. Perhaps female scientists have more native ability than male scientists. IQ scores cannot, of course, be used as an indicator of scientific ability. However, if we want a measure of native ability which is perhaps correlated with scientific ability, IQ is probably the best available indicator. The question arises: Is there any difference in the native ability of men and women Ph.D.'s?

The data do not support the hypothesis that differences in the level of recognition result from differences in the native ability of the men and women Ph.D.'s. Data presented by Harmon have shown that the measured intelligence of women Ph.D.'s is on the average slightly, if not significantly, higher than that of male Ph.D.'s.[34] My analysis of a subset of Harmon's data goes on to find that for the fields studied, at every level of doctoral-department prestige, women Ph.D.'s have on the average slightly higher IQ's than their male colleagues (see Table 3–2). While these differences are not always statistically significant, they are strikingly uniform.[35] A comprehensive empirical explanation for the pattern of IQ scores goes beyond my data. Speculatively, it may result from women being systematically denied admission to top graduate departments (although this seems unsupported by data to be discussed below), or it may result from women modifying their career patterns to conform with their husband's employment opportunities. Whatever the actual explanation, if we are willing to accept IQ as an indicator of native ability that is correlated with scientific aptitude, women and men start off on equal footing. On this basis, without intervening factors we would expect women scientists to fare at least as well as men.

---

34. Lindsey R. Harmon, *High School Ability Patterns: A Backward Look from the Doctorate*, Scientific Manpower Report no. 6 (Washington: National Research Council, 1965).

35. For a detailed discussion of the relationship between IQ and sex status, see Chapter 5.

TABLE 3-2.   IQ of Men and Women Scientists Receiving Doctorates at Science Departments of Varying Prestige

RANK OF PH.D. DEPT.*	MEAN INTELLIGENCE QUOTIENT**			
	Men	N	Women	N
Distinguished	68	(46)	72	(48)
Strong	65	(90)	59	(90)
Good	64	(61)	68	(55)
All Others	61	(76)	66	(55)
Totals	64	(274)	69	(249)
Group Statistics				
Mean	66.5			
Std. Dev.	8.8			
Variance	77.8			
N	(523)			

*The Cartter ratings of quality of graduate departments were used to obtain rankings for the graduate schools attended by scientists in this sample. These data include all of the recipients of doctorates in mathematics, physics, chemistry, biology, psychology, and sociology, whether they went on to academic careers or not. The pattern of findings for academics only, and for the fields examined separately, is the same as that reported in this table, although there were differences in means and variances.

**The IQ data, which were originally collected by Lindsey Harmon for a larger sample of doctorates, had all test scores standardized with a mean of 50 and a standard deviation of 10. These scores were based upon data obtained from high school records of the scientists. Clearly, the group statistics for our sample differ from those obtained by Harmon.

Since previous studies have demonstrated that rank in the stratification system of science depends heavily on the quality of published research[36]—and indeed on its production at a reasonable rate—we must examine the productivity patterns of men and women scientists. How, if at all, do the publication patterns of men and women scientists differ?

PUBLICATION PATTERNS OF MEN AND WOMEN SCIENTISTS

As a cohort, my sample of men and women scientists were neither extraordinarily prolific nor below average in terms of research output. The mean prod-

---

36. Cole and Cole, *op. cit.*

uctivity for the 565 academic scientists was nine papers over a span of twelve years. Productivity is very unevenly distributed among the scientists: Few published many papers, and a small number published a high proportion of the group's total. For example, in 1959, one or two years after receipt of the doctorate—prolific years for young scientists—53 percent of the sample failed to publish a single paper, and 34 percent published just one. In most years, about 70 percent of the scientists published nothing. In short, my data on scientific productivity corroborate studies by Price and others which show sharply skewed distributions of publications in science.[37] What has not been identified before is whether this skewed pattern holds equally for both male and female scientists. Gini concentration ratios of inequality[38] in scientific productivity over the span of twelve years are almost identical: $G_i = .502$ for men; $G_i = .504$ for women. For this sample, these statistics indicate that about 15 percent of scientists, male and female alike, account in each case for about 50 percent of the total papers produced by scientists of the same sex.

Our concern, of course, is not so much with productivity patterns within groups of male and female scientists as with the possible correlation between sex status and published productivity.[39] As a first step to analyzing more complex relationships, we note that male scientists in the matched sample are, on average, more productive than women as shown by the zero-order correlation for the entire sample between sex status and productivity during the twelve years of the scientists' careers ($r = -.30$)[40] (see Table 3-3). For these years, the median publication rate was eight papers for academic men and three for academic women. The differential in publication rates remains fairly constant over the first

---

37. Derek J. de Solla Price, *Little Science, Big Science*. (New York: Columbia University Press, 1963). The distribution of published scientific papers for men and women follows:

TOTAL NUMBER OF SCIENTIFIC PAPERS PUBLISHED BETWEEN 1958–1969	CUMULATIVE PERCENTAGES	
	Women	*Men*
0–1	34.4	17.1
2–3	53.4	27.2
4–6	68.7	41.6
7–10	85.3	58.7
11–15	92.6	75.5
16–20	96.7	84.2
20–24	98.8	87.2
25+	100.0	100.0

38. Otis Dudley Duncan, "The Measurement of Population Distribution," *Population Studies* 11, no. 1 (1957):40.

39. The total number of published papers, both single-authored and collaborative, was used as a measure of productivity. Total productivity here refers to the total number of published papers between 1958 and 1969; "early" productivity includes just the years 1959–65; "later" productivity refers to the years 1966–69.

40. The zero-order correlation between total published productivity and total productivity prior to 1965 is extremely high ($r = .95$). In order to estimate the reliability of publications and citation counts, I had two sets of coders independently code the data. The reliability coefficient for total publications (1958–70) was .97; for productivity from 1958 to 1965, it was .96. Citation counts had equally high coefficients: for example, total citations for six years yielded a reliability coefficient of .96. In this book I have not corrected the correlation coefficients for attenuation.

TABLE 3–3. The Relationship between Sex Status and Various Indicators of Scientific Output: Zero-Order Correlations

	ENTIRE SAMPLE OF SCIENTISTS*	N	CHEMISTRY	N	BIOLOGY	N	PSYCHOLOGY	N	SOCIOLOGY	N
Productivity (*Total Number of Papers*)										
First 8 Years (1958–65)	–.30	561	–.37	61	–.37	297	–.25	159	.04	44
First 12 Years (1958–69)	–.30	561	–.40	61	–.37	297	–.26	159	.05	44
Quality of Work First 8 Years (Total Citations: 1961, 1964, 1965)	–.18	516	–.32	61	–.18	297	–.25	158	*	
First 12 Years (Total Citations: 1961, 1964, 1965, 1967, 1970)	–.19	550	–.33	61	–.21	297	–.22	159	.21**	44**

*The correlation coefficients for the entire sample are based upon standardized variables. It was necessary to perform Z-score transformations on the data since the means and standard deviations differed from one field to the next.

**Citation data for most psychology and sociology journals were not abstracted by *SCI* during the early 1960s. Thus I collected citation data for only 1969 and 1970 for the psychologists. However, since the correlation between the total number of citations scientists receive in the 1961, 1964, 1965 index is highly correlated (*r* = .93) with the total number of citations, the use of later citations as an indicator of earlier use of work will not involve gross error.

twelve years of the careers of men and women scientists,[41] although the initial differences in productivity two years after receiving the doctorate are insignificant ($r = -.04$).[42]

There are numerous alternative explanations for the lower scientific productivity of women. Consider only two which are frequently offered. First, it is suggested that the traditional family obligations of women prevent them from spending as much time working on research as do men. This hypothesis can be tested by data on the marital status and number of children of both the men and women scientists. Productivity was regressed on three independent variables: sex status, marital status, and number of children. Sex differences in scientific productivity persist. The net effect of sex status on productivity after controlling for these family statuses is actually stronger than the zero-order correlation $b^*_{ps.mf}$ $= -.46$); that is, regardless of controls for marital status or number of children, the scientific productivity of women is considerably lower than that of men.[43]

---

41. The following correlations between sex status and research performance measures were obtained for individual years after receipt of the doctorate:

YEAR	PRODUCTIVITY (NO. OF PAPERS)	QUALITY (NO. OF CITATIONS)
	r (Sex and Productivity) (Males = 1; Females = 2)	r (Sex and Quality)
1957	−.02	
1958	−.04	
1959	−.12	
1960	−.17	
1961	−.19	−.10
1962	−.24	
1963	−.23	
1964	−.23	−.14
1965	−.25	−.16
1966	−.21	
1967	−.19	−.19
1968	−.22	
1969	−.22	−.21
	(N = 561)	(N = 561)

All productivity correlations from 1959 through 1969 are statistically significant at the .001 level. When women who had been unemployed at any time prior to 1965 are excluded from the computations, the correlations remain almost exactly the same.

42. Publication rate in the years immediately following the doctorate is a better predictor of later productivity for men than for women. In all fields the correlation between early (1957–61) and later (1966–69) productivity is higher for men than for women. For example, for the three fields for which I have very reliable productivity data, the following pattern obtained:

The Relationship between Early and Later Scientific Productivity for Men and Women Scientists: Zero-Order Correlation Coefficients

FIELD	MEN	WOMEN
Chemistry	.30	.19
Biological Science	.44	.32
Psychology	.43	.06

43. In presenting the partial regression coefficients I use the following subscripts: s = sex status; p = productivity; q = quality of research; m = marital status; f = family size; r = rank of current department (1965); a = number of honorific awards and postdoctoral fellowships; b = academic rank; c = college affiliation; u = university

However, marital status and family size would certainly seem to have a qualitatively different influence on the professional lives of male and female scientists. Linear regression results may conceal interaction between sex status and family statuses. To test for interaction, the sample was first divided into men and women and then further divided into those who were unmarried; married without children; those with one or two children; and those with three or more children. The mean number of publications for the twelve-year span was computed for each of the eight subgroups. The data in Table 3–4 testify that sex status has a much greater influence on publication patterns than do family statuses. Consider the following striking fact: Unmarried women scientists publish far less than men scientists in all family categories; family obligations negatively influence the productivity rates only of women having three or more children. Women with smaller families are actually more likely to publish than unmarried women. Although the means and variances are dissimilar, the pattern reported in Table 3–4 for both male and female scientists lends itself to Durkheimian interpretation. For both sexes, the stability and routinization of work patterns associated with marriage are actually positively correlated with a higher publication rate. The addition of children, and increases in family size, however, are negatively correlated with published productivity. As Durkheim suggests, there clearly seems to be an optimal level of social integration. In conclusion, family status per se cannot account for the differential rates of productivity of men and women scientists.

Another explanation offered for the lower scientific productivity of women is their occupational location. Academic women faculty are found more frequently than men in colleges, where research is not the norm or a prerequisite for promotion. Astin summarizes the hypothesis:

> Studies indicate that institutional affiliation (college or university) accounts for most of the differences observed [in productivity]. Essentially, academic women have a greater tendency to work in colleges, while academic men are employed by universities. Since university-employed persons publish more than college-employed persons, regardless of sex, the greater overall productivity of academic men is self-explanatory.[44]

To test this hypothesis, I regressed scientific productivity on institutional affiliation (i.e., college or university), and sex status. Although scientists at universities are in general more prolific ($r = .30$) and produce work that is more frequently cited ($r = .20$) than do those located at colleges, the data indicate that institutional affiliation does not modify the relationship between sex status and scientific output (which, of course, is logically necessary given the absence of any correlation between sex status and institutional affiliation). The partial re-

---

affiliation. The failure of family status to reduce the correlation between sex status and productivity results from the small zero-order correlation between productivity and marital status ($r = .13$, i.e., married scientists publishing slightly more than unmarried ones), and very low correlation between productivity and family size ($r = -.06$).

44. Astin, *op. cit.,* p. 85. Although Astin contends that academic affiliation can account for differentials in productivity, she does not present any data to support this hypothesis.

TABLE 3–4. The Relationship between Marital Status and Family Size on Mean Number of Scientific Publications* (1958–1969)

SEX	TOTAL	UNMARRIED	MARRIED WITH NO CHILDREN	ONE OR TWO CHILDREN	THREE OR MORE CHILDREN
Men	12 (298)	9 (32)	15 (33)	12 (124)	11 (109)
Women	5 (263)	5 (67)	8 (38)	6 (91)	4 (67)

Total Usable Cases 561
Total Missing Cases   4
Total Sample   565

*These means have been rounded to the nearest integer.

gression coefficient between sex status and productivity ($b^*_{ps.cu} = -.27$) is virtually equal to the zero-order correlation between the two variables ($r = -.30$). As a further test of this hypothesis, I computed the correlation between sex status and productivity only for scientists working at universities. The correlation was $-.35$, slightly greater than for the entire sample. In short, academic men scientists are more productive than their female colleagues in universities just as they are in colleges.

We might expect that the correlation between sex status and published productivity would be attenuated in better science departments. This is not the case: At the better academic departments ("distinguished," "strong," or "good," according to the Cartter ratings), I found a zero-order correlation of $-.34$ between sex and scientific productivity, and I found a correlation of $-.21$ at comparatively poorer departments. No matter how the sample of academic scientists was divided, published productivity differentials between men and women scientists persisted.

Other variables such as differences in teaching responsibilities, access to research funds, and opportunities to collaborate with other outstanding scientists might account for the differences in published productivity of men and women, but this must remain untested here for lack of data. It is possible that in order to account for the correlation between sex status and the rate of scientific output, it may be necessary to look outside the institutional structure of science, to examine carefully the prior experiences and socialization processes affecting women in the larger society—which may dampen motivation to succeed and influence their publication performance after they enter science.[45]

I have now discussed quantity of output and will shift to the question of quality. Although women publish fewer scientific papers than men, it is possible that they are of higher quality. Perhaps because women in general are under less pressure to "succeed" conspicuously, they may be more likely to be "perfectionists." The data show that the publications of the women scientists are actually of lower quality, in the sense that they are less frequently cited than are those of the men scientists ($r = -.24$). In the six years of the *SCI* for which I collected data, papers published by men received a grand mean of fifty citations and those by women a grand mean of seventeen.[46]

Let me recapitulate the findings and analysis to this point. I have found that the IQ's of women scientists are the same as, or higher than, those of men, but that the women produce fewer and less frequently cited papers, and that they receive slightly less academic recognition than men. It is not warranted to infer from simple zero-order correlations between sex status and recognition that women are being discriminated against. Such inferences are often misleading,

---

45. Rossi, "Women in Science: Why So Few?" *op. cit.;* M. S. Horner, "Femininity and Successful Achievement: A Basic Inconsistency?" in *Feminine Personality and Conflict,* ed. J. Bardwick et al. (Belmont, Calif.: Brooks/ Cole, 1970), pp. 45–77.

46. The correlation between early (1961–64) and later (1967–70) citations is higher for men than for women in both chemistry and biology, the two fields for which I have extensive citation data over time.

for they are predicated upon a two-dimensional model of social behavior. The task at hand is to bring into clear perspective the multivariate aspects of social reality affecting men and women in the social system of science. We must therefore examine the relationships between sex status and forms of academic recognition, controlling for differential role-performance.

The central issue is whether differential rewards can be explained by differential role-performance. Although differences in productivity of men and women scientists cannot be explained with the data that are available, substantial variability exists, of course, in the published productivity of women scientists. When both quality and quantity of research output are controlled, do men still receive greater recognition than women?

SEX STATUS AND RECOGNITION CONTROLLING FOR SCIENTIFIC ROLE-PERFORMANCE

Recall that for the entire sample of academic men and women in four fields, sex status was correlated with rank of department in 1965 ($r = -.02$) and in 1970 ($r = .01$). When productivity of the scientists is controlled, there is, if anything, a slight tendency for women to be overrepresented in higher-quality departments ($b^*_{ds.pq} = .06$).

Published productivity, as we would expect, had a greater independent effect than sex on rank of department ($b^*_{dp.sq} = .26$). This was true for all four disciplines. The zero-order correlations in 1965 between sex and department prestige within the four disciplines varied ($r = -.27$ in chemistry; $r = -.23$ in psychology; $r = .06$ in biology; $r = .30$ in sociology). The different zero-order correlations in the four fields do not imply that there is more or less "discrimination" in one field or another. For example, although women biologists in the matched sample are more likely to be found at high-prestige departments than their male peers, they could still be the objects of discrimination. Again, if women biologists are on average more productive and produce higher-quality work than their male peers, then it is possible that they are still not being rewarded in direct relation to their performance.

When department prestige is regressed on sex status, productivity, and quality of research, the association between sex status and rank of department is reduced significantly in chemistry, psychology, and sociology. In chemistry, for example, the partial regression coefficient for sex status, after controlling for quantity and quality of research, is $b^*_{rs.qp} = -.06$. As expected, scientific output is a more influential factor in determining a scientist's rank of department than is sex status. The net effect of research quality is $b^*_{rq.sp} = .23$; of research quantity, $b^*_{rp.qs} = .43$. For psychology, the zero-order correlation is reduced substantially after controlling for scientific output ($b^*_{rs.qp} = -.10$). Here, too, quality of research is a more important influence on rank of department ($b^*_{rq.sp} = .36$), and quantity has only a moderate independent effect ($b^*_{rp.sq} = .24$). In the biological sciences, the pattern is different. When I control for the amount and quality of work, women are still slightly more likely to be found at higher-ranked departments. Indeed, the partial regression coefficient increased slightly: $b*_{rs.qp} = .15$. These data for different fields must be considered with utmost caution, since they are based upon a very small sample of men and women. Nonetheless, the pattern in sociology offers an instructive comparison; here the Pearsonian correlation suggests that women are more likely than men to be located at better departments. We might infer incorrectly from this datum that reverse discrimination exists against men. However, controlling for scientific output reduces the size of zero-order correlation ($b*_{rp.sq} = .17$). This result suggests that regardless of the simple relationship between sex status and prestige of department—that is, whether the data show men or women more likely to be located at prestigious departments—quality and quantity of output can account for most of the correlation. The results are summarized in part A of Table 3-5.

When I examined the influence of research performance on the association between sex status and prestige of department in 1970, I again found little evidence for sex-related bias. The partial regression coefficient for sex status, after controlling for quality and quantity of research, was insignificant ($b*_{rs.qp} = .08$). The belief that women are slowly weeded out of top departments as their careers progress is not supported by these data.

Although women are as likely as men to hold positions at good universities, they might more often be found in nonacademic line-job classifications, such as research associate or research director. This is not the case. As of 1965, approximately 29 percent of the male as compared to 32 percent of the female scientists were either research associates or directors. There were virtually no differences when departments of differing prestige were considered separately.

In summary, the data from this sample clearly do not support the belief that women are discriminated against in terms of their access to and retention in top science departments. When their educational origins and the quality and quantity of their research output is taken into account, women are just as likely as men to be found in major science departments. Of course, most scientists are not at distinguished departments, and this apparently is just as true for male as for female scientists. When sex differences are found, the data suggest that the influence of gender on hiring is not substantial.

The story becomes more complex, however, when we turn to the issue of advancement to higher academic ranks. The zero-order correlations and standardized partial regression coefficients for the entire sample and for each of the four fields are presented in part B of Table 3-5. Earlier I noted the substantial association between sex status and academic rank ($r = -.34$). Controlling for quantity and quality of scientific output does not reduce the relationship between sex status and academic rank in 1965 ($b*_{bs.qp} = -.32$). At each level of productivity, women are less likely to receive promotions than men. Again, this result suggests that productivity ($r = .13$) and quality of output ($r = .15$) are not strongly correlated with academic rank. These are the first data in this study

TABLE 3–5. The Relationship between Sex Status and Three Forms of Scientific Recognition for Four Scientific Fields: Zero-Order Correlations and Regression Coefficients in Standard Form

FORMS OF RECOGNITION	CORRELATION BETWEEN SEX STATUS AND FORM OF RECOGNITION (1965 SAMPLE)	BETA COEFFICIENTS FOR SEX STATUS CONTROLLING FOR QUALITY AND QUANTITY OF RESEARCH (1965 SAMPLE)	
	r	$b*$	N
A. Rank of Current Dept.			
Total Sample	−.02	.05	423
Chemistry	−.23	−.06	42
Biology	.06	.15	215
Psychology	−.17	−.10	131
Sociology	.30	.17	35
B. Academic Rank			
Total Sample	−.34	−.32	469
Chemistry	−.08	−.13	54
Biology	−.38	−.39	244
Psychology	−.36	−.31	130
Sociology	−.44	−.50	41
C. Number of Honorific Awards			
Total Sample	−.05	.00	438
Chemistry	−.19	−.06	41
Biology	−.09	−.05	239
Psychology	−.07	.00	122
Sociology	.28	.23	36

*Since these results are based upon pairwise correlations, the total numbers do not add up to 565. The difference between the numbers on which the correlations are based and the total sample of academic scientists represents missing data.

suggesting that women may be treated differently from men in the reward process of science.

What, then, is the relationship between sex status and promotion in specific types of departments? Is sex status likely to have a greater independent influence on promotion in the better or in the lesser departments? In top departments, where both men and women scientists are in general found to be more productive, sex status may not be as important a factor in promotion decisions. To answer this question, I divided the sample by the same criteria as before into those men and women in the more and less prestigious departments. For each group separately, I regressed academic rank on sex status, productivity, and quality of scientific output. In both types of departments, sex status is still significantly related to academic rank.[47] In the better departments, the zero-order correlation between sex status and academic rank is $r = -.47$; controlling for the quality and quantity of output does not significantly reduce this association $b^*_{bs.qp} = -.38$). Although the correlation in less distinguished departments is lower ($r = -.19$), it remains unchanged when quality and quantity of output are accounted for ($b^*_{bs.qp} = -.19$). Thus, sex status definitely influences promotion within the scientific community. These findings are not statistical artifacts; there apparently are marked differences in the ranks of men and women scientists even after controlling for their scientific role-performance.

But before we can estimate the level of this discriminatory treatment of women scientists, it is necessary to examine influences other than discrimination that could, at least in part, explain these patterned differences in academic rank.

We have already seen that scientific productivity was not a strong determinant of academic rank in 1965, when, on the average, the scientists were seven or eight years past their doctorates. Although quality and quantity of work are more influential in promotion in the better departments than in lesser ones, they explain little variance in promotion. One factor alluded to above which has a strong effect on promotion is professional seniority. Most scientists, male and female, with notable exceptions, of course, must put in their years at a given rank before they become candidates for promotion. Exceptions to this pattern often result from competition between academic departments for exceptionally gifted young scientists. If seniority is an important ingredient in promotion decisions, it probably has more negative consequences for female scientists than for males, since women are more likely to interrupt their professional careers for some period of time closely following receipt of the Ph.D.

Astin reports: "Of the 1,214 women doctorates who were fully employed at the time of the survey, 957 (79 percent) had never interrupted their careers; 18 percent reported career interruptions lasting from 11 to 15 months, with a median period of 14 months."[48] The influence of such interruption on academic rank can

---

47. Within distinguished, strong, and good departments, the zero-order correlation between sex status and academic rank is $r = -.47$; the partial regression coefficient between sex status and academic rank, controlling for quantity and quality of output, is $b^*_{bs.qp} = -.38$. For all other departments the correlation between sex status and academic rank is $r = -.19$; the partial regression coefficient is identical to the zero-order correlation.

48. Astin, op. cit., p. 58.

be estimated. When Bayer and Astin took this employment pattern of women into account, they found only minor differences in the rate of promotion of men and women in the sciences.[49] In order to correct for different lengths of time in the labor force, they compared women who earned their Ph.D.'s in 1957–58 with men who earned their doctorates in 1958–59. Comparisons between these groups indicated that sex had no influence on academic rank. Since my sample of women is a subsample of Astin's and since I am considering promotions in 1965, seven or eight years after the doctorate—a particularly important period for mobility of academics—the sample affords the opportunity to investigate whether career interruptions of these women might have influenced their rate of promotion.

For the males among the 565 scientists, data on employment patterns were not available. I made the assumption that all the men scientists in the sample had been continuously in the labor force, a reasonable assumption for the period 1958–65, one of expanding opportunities for recent Ph.D.'s. In order to estimate the effects of female career interruptions, I excluded from this analysis all women who had interrupted their professional careers for any time from the receipt of their degrees to the time of the Astin survey. If seniority affects promotion, we would expect a reduction in the correlation between sex status and academic rank when women with career interruptions are excluded. This is precisely the pattern that is found: The zero-order correlation between sex status and rank is $r = -.24$, reduced from $r = -.34$, when employment history is taken into account. Controlling for scientific output does not further reduce the correlation.[50]

Since seniority is an element in determining rank, further reductions in the correlation between sex and academic rank should be observed when we examine later points in the careers of men and women scientists. I had academic ranks in 1970 for the subsample of 262 scientists. The zero-order correlation between sex status and rank in 1970 was $-.24$; when I eliminated those women who had career interruptions prior to 1965, the association was reduced to $-.18$. Finally, when I controlled for the influence of scientific productivity on academic rank in 1970, which is more highly correlated with rank in 1970 ($r = .27$) than in 1965 ($r = .13$), the effect of sex status is reduced to approximately one-third of the

---

49. Bayer and Astin, *op. cit.*, pp. 191–200.

50. The correlation between length of unemployment in months and academic rank is $-.28$ in 1965 and $-.39$ in 1970. One group of women with a high probability of taking some time off from their careers are those who have young children. Accordingly, women with preschool-age children and pre-teenage children would be less likely to receive promotions than women who have continuously been in the labor force, since they either have had less time to spend on their careers or have actually taken leaves of absence. This is the case. In 1965 the correlation between number of preschool children and rank was $-.18$; in 1970 it was $-.36$. I divided the sample into women who had appointments at more prestigious universities and those in less prestigious universities and colleges. At the high-prestige institutions, the correlation for women between number of preschool children and academic rank was $r = -.36$; it was even stronger between number of pre-teenage children and academic rank ($r = -.49$). These data point to the difficulty in locating the reasons for the correlation between sex and academic rank. Is it primarily due to particularism operating within science or to factors external to the scientific community? At lower-ranked departments, the zero-order correlation between number of preschool children and academic rank is somewhat lower ($r = -.18$). There is no correlation between number of pre-teenage children and academic rank at the less distinguished settings ($r = -.01$).

original zero-order correlation, $b^*_{bs.p} = -.10$. While sex status continues to play some role in determining academic rank in 1970, its independent effect is sharply reduced. In short, these data suggest that seniority weighs heavily in the process of academic promotion, and that women scientists more than men are adversely affected by the application of this principle.

Finally, consider the influence of a scientist's academic affiliation on his promotion chances. For the entire sample, department prestige does have a significant influence on rank in 1965, independent of productivity ($b^*_{br.sp} = -.13$), suggesting, as we would expect, that it is more difficult to achieve high rank in top departments. However, these results conceal the real story: Prestige of department has a strong influence in determining rank for women ($b^*_{br.p} = -.37$) and none at all for men ($b^*_{br.p} = -.03$). Women in prestigious departments are much less likely to be promoted to tenure ranks than are women in lower-ranked departments. Men are just as likely to be promoted in both the better and lesser departments.

Let us return now to the indicator of reputation for which I have data: the receipt of honorific awards and postdoctoral fellowships. (In Chapter 4, I will discuss other indicators of reputation collected for a different sample.) For the sample of 565 men and women scientists in four fields, I found a very slight correlation between sex status and number of honorific awards ($r = -.05$). When I control for the quality and quantity of published research, the influence of sex status on this form of recognition completely disappears ($b^*_{as.qp} = .00$). Sex status has no influence in determining receipt of awards and postdoctoral fellowships in any of the fields other than sociology, where women are more likely than men to receive honorific recognition. The partial regression coefficients are presented in part C of Table 3-5. Again, high productivity and quality of research are the primary determinants of honorific recognition. For the entire sample, the independent effect of quality on reputational success was $b^*_{aq.sp} = .20$. In terms of honorific recognition, there is little evidence that women scientists are discriminated against in the scientific community.

The data suggest that there is differential treatment of women scientists in terms of academic promotions, although a substantial part of this difference results from patterns of labor-force participation. However, women are not significantly underrewarded in terms of honorific recognition or in terms of their academic affiliation, once their achievements as scientific researchers are also considered.

A CONDITIONAL HYPOTHESIS OF DISCRIMINATION

I have been trying to specify the conditions under which sex status influences recognition granted to scientists, focusing primarily on whether differences in productivity and quality of published research explain the slightly unequal treatment women appear to receive in science. Unless interaction terms are used, regression procedures average differences in subpopulations, and often mask important substantive results which would emerge from consideration of distinct subgroups of scientists.

A truly equitable social system, in which people receive equal rewards for equal role-performance and unequal rewards for unequal performance, not only should reward the very productive equally, but should also reward the less productive equally. I hypothesize that functionally irrelevant characteristics, such as sex, will be more quickly activated when there are no or few functionally relevant criteria on which to judge individual performance. They will also be imported in situations where there is only limited agreement on the criteria relevant for judgment. A basic criterion for judging role performance is the quality, and to a lesser extent quantity, of research produced by a scientist.[51] Of course, if neither of two research scientists being considered for an academic job has contributed to the advance of scientific knowledge through their work, then the criterion of research quality cannot be applied. In the best science departments virtually every member publishes scientific papers. But at less prestigious colleges, far-removed from the center of scientific activity, a high proportion of the faculty often publish no research papers. I hypothesized that sex status would more likely be imported into a social situation in which scientists published little or nothing than in one containing very productive scientists. Totally unproductive women might be treated differently than totally unproductive men. In fact, in the absence of published papers to testify to the seriousness of the scientists, some scientists in positions to judge others may well contend that "sex" is, in fact, a relevant basis for awarding positions. These judges might argue that women are much greater risks than men, that they are more likely to withdraw from science for a variety of reasons, and that therefore they should not be hired in the first place.

A critical test of this hypothesis calls for the location of science departments in which no one publishes research. The data in their present form could not be arranged for such a test. Therefore, I decided to examine the career patterns of one extreme group: those truly "silent" scientists who had not published a single scientific paper during the first seven or eight years in their careers, yet had not dropped out of science.

Among this group, does "Susan the silent" suffer in comparison with "Stanley the silent"? Evidence substantiating such a pattern would suggest, although hardly prove, that sex status has indeed been activated, that sex differentials obtain. The data allow us to make a first test of this idea. Consider appointment to departments of varying prestige. Are silent women scientists less likely than silent male scientists to have jobs in distinguished, strong, good, or even adequate departments? Fifty-five of the 565 men and women scientists had not published any scientific papers between 1958 and 1965. However, in 1965 sex

---

51. Cole and Cole, *op. cit.*, Chapter 4.

status for this silent group made no real difference in terms of job location: The correlation between prestige of department and sex status was .07, a statistically insignificant result, which implies, if anything, that silent women do slightly better than silent men. The data seem to show that sex status is not activated in terms of appointments to more prestigious departments.

Before rejecting the idea that sex status will become a criterion of judgment in the absence of performance criteria, I decided to subject it to one further and more exacting test. Seven years after receipt of the Ph.D. might not be a sufficient period, since many decisions about retaining and promoting faculty are made after that interim. I therefore looked at those academic scientists who had been silent for a longer period of time; who, as far as I could tell, had not up to 1969 published a single paper from the point of receiving their degrees in 1957 or 1958. This turned out to be a subsample of only twenty-seven scientists, roughly 5 percent of the total. The correlation between sex status and prestige of department for this group was $-.28$ (statistically significant at the .08 level), compared to .01 for the subsample of 262 scientists—suggesting that among long-term silent scientists, sex status may be used to a limited extent in allocating rewards. The theory should be further tested with more extensive data before its validity is judged.

Examining the academic rank of silent scientists also produces only partial support for the theory. On the one hand, among the group from 1958 to 1965 there is a zero-order correlation of $-.29$ between sex status and academic rank, which is slightly weaker than the association for the entire sample, including productive scholars. This does not mean that sex status has been more influential in evaluating silent women than it has been in evaluating productive ones. On the other hand, if we examine the slightly longer time period, we do find some tentative support for the theory. In 1970 the association between sex status and academic rank was $-.41$ for silent scientists, a significant increase above the $-.28$ correlation for the sample of 262. This result suggests once again that sex status is more likely to be a criterion of evaluation for the extremely small subpopulation of men and women who were totally unproductive for twelve years following receipt of their doctorates than for more productive men and women scientists. Again, these results are based upon extremely small samples, and we must be cautious about the support they lend to the theory.

I have not located any other empirical sociological studies that discuss the activation of functionally irrelevant statuses. However, Walster et al., in a study of college admissions, present data that support my hypothesis.[52] They completed admissions applications and sent them to 240 randomly selected colleges. By design, "applicants" were identical, with variations only in their race, sex and ability (lower grades and lower SAT scores). Walster et al. did not find statistically significant effects of race or sex on admissions. However, interaction

---

52. E. Walster, T. A. Cleary, and M. M. Clifford, "The Effect of Race and Sex on College Admissions," *Sociology of Education* 44, (1971):237–244.

between sex and ability did influence decisions. "At the low ability level, males were preferred over females. At the higher ability level, this difference disappeared."[53] The lack of ability among college applicants is conceptually close to the lack of publications among scientists seeking jobs or promotions. Walster's finding is similar to mine. When a rational criterion—e.g., ability, as measured by SAT scores or publications—is lacking, sex may be used in evaluations.

Silent scientists represent one end of the publication continuum. At the other extreme are the "prolific" scientists who publish an unusually large number of papers. Recall that some 15 percent of the sample produced roughly 50 percent of all its scientific papers. Perhaps among these extremely active scientists women are treated differently than men. Superproductive women may be perceived as "pushy," overly aggressive, and threatening to the males within their departments.[54] Thus, high productivity is a second condition that might activate sex status as a factor in academic recognition. To test this conditional proposition, I examined the correlation between sex status and positional recognition for the most productive men and women scientists—the top 15 to 20 percent of the sample. There were 102 prolific scientists in 1965 and 63 in 1970. Very productive women were just as likely as very productive men to be in highly prestigious departments; the correlations ($r = .04$ in 1965; $r = .08$ in 1970) were not very different from those found for the entire sample. There is no evidence, therefore, that prolific women are systematically excluded from or pushed out of good science departments. But they are not nearly as likely to hold high-ranking positions within those departments as are their equally prolific male colleagues. The association between academic rank and sex status among prolific scientists was $-.58$ in 1965 and $-.44$ in 1970, both figures being significantly greater than the zero-order correlations for the entire samples.

An interpretative problem arises in dealing with the prolific scientists that is not encountered when dealing with silent scientists. Silence, after all, is simply defined as zero publications. Men and women in this category are "equally" silent. But within a very highly productive group there may, in fact, be systematic differences in publication rates. The case is this: Men within the prolific group published an average of 24 papers and received an average of 116 citations during the years 1958–69; women averaged 15 papers and 71 citations. Over these twelve years, 10 percent of the male scientists produced more scientific papers than did any of the women scientists. Thus, it is impossible to control precisely for quantity and quality of output of men and women scientists in this small sample. The empirical data discussed here represent only a first test of these theoretical ideas; further tests are necessary.

Much of this evidence, then, converges to form the following pattern: Women can arrive and remain at top locations within the stratification system,

53. *Ibid.*, p. 243.

54. In the 1930s and 1940s, some Jewish scholars who were highly prolific and who thought they were consequently perceived by their colleagues as aggressive often contended that they were especially subject to religious discrimination and were underrewarded given the level of their scientific productivity.

but they are less likely to be upwardly mobile within those contexts despite roughly equal scientific output. The general patterns obtained when the entire sample is analyzed are mirrored and slightly highlighted when we examine extreme subpopulations of men and women scientists.

ACCUMULATIVE DISADVANTAGE

A number of papers and monographs have suggested that scientists who are placed in structurally advantageous positions as a result of outstanding role-performance are given certain advantages by virtue of attaining these positions. There are accumulating effects of being in an outstanding science department—for example, future recognition. There is, however, a second and darker side of the process of accumulation of rewards in science. It is the accumulation of failures: the process of "accumulative disadvantage."[55]

A social system can operate universalistically with respect to a group of scientists at one point in time and still be discriminating against that group. Specifically, suppose that the social system of science is less likely to support the education of a woman by means of a graduate fellowship, and is less likely once she has received her degree to give her financial support in the form of resources and facilities necessary to carry out research. If women scientists receive less support, it should not be surprising if they are less productive than men. Their futures become predictable. When they come up for hiring and promotion decisions, their publication records are carefully reviewed and found inferior to those of men with the same type of background, and they lose out in the academic marketplace. The self-fulfilling prophecy, which is based on the assumption that women are less motivated, less productive, and less reliable scientific risks than men, now is strikingly supported by data. Gate-keepers for resources can now present data to justify not giving equal financial support to women scientists, since they are less likely to produce significant research with the funds. And so the conditions for the self-fulfilling prophecy are reinforced. Plainly, if the conditions for a self-fulfilling prophecy exist at an early point in time, a universalistic judgment later on will ultimately produce inferior status for the judged group.[56]

Several investigators who have spoken to the issue of sex discrimination in science contend that women scientists who have as much ability as male scientists suffer from compounding career disadvantages. Once denied access to resources and facilities, women must struggle simply to reach parity with their male colleagues. Harriet Zuckerman and I have referred to this process as the "principle of the triple penalty." The principle states that groups such as blacks and women suffer not only from direct discrimination and from cultural defi-

---

55. For a recent discussion of how these processes affect the careers of scientific elites, see Harriet Zuckerman, *Scientific Elite (op. cit.)*.

56. For a full discussion of the structural conditions making for self-fulfilling prophecies, see Robert K. Merton, "The Self-Fulfilling Prophecy," in *Social Theory and Social Structure (op. cit.)*.

nitions that define certain careers as inappropriate but also from being placed initially into second-rate structural positions, which makes it difficult or impossible for them to produce the outstanding work that is necessary for moving out of such positions.[57] This principle is predicated on the assumption that women scientists are not treated in the same way as men in the initial phases of their careers. We must ask whether this is empirically correct. Are women less likely to be admitted to top graduate departments of science, regardless of their ability? Do women receive less financial support when in graduate school than men? Are women's efforts at research given less support during their training period? Are women working on research projects given the less intellectually demanding jobs to perform? Must women more often than men choose dissertation topics that are less compatible with their interests? More research is needed on such matters involving the initial definition of talent and potential of men and women scientists.

Some scanty evidence is available, however, on the extent to which women and men are admitted to, and given financial support for, graduate training. Women appear to receive their doctorates from top-ranked departments in the same proportion as men. Folger, Astin, and Bayer examined the proportion of men and women Ph.D.'s from departments rated as distinguished or strong in Cartter's ACE study of the quality of graduate education.[58] They found that about 50 percent of the male and the female Ph.D.'s were being trained at these top departments. Berelson, in an earlier study of graduate education, also found that the probability of women receiving degrees from one of these "top twelve" schools was the same as it was for men.[59] Data reported by Astin on admission to medical schools show similar results: In 1964–65, "47.6 percent of the women applicants were accepted, as compared with 47.1 percent of the male applicants."[60] Recent studies of admissions practices at Stanford, Berkeley, and UCLA show that women graduate school applicants in the physical and social sciences are admitted in almost the exact proportion to which they apply.[61]

The distribution of fellowship support for men and women graduate students offers no prima facie evidence of discriminatory practices. Women are just as likely as men to receive financial support. Astin, comparing the proportion of women and men Ph.D.'s receiving either fellowship or graduate assistantships, found that 57 percent of the women had received aid compared with 58 percent of the men.[62] Jessie Bernard also reported: "The National Science Foundation

57. Harriet Zuckerman and Jonathan R. Cole, *op. cit.*, p. 84.

58. John K. Folger, H. S. Astin, and A. E. Bayer, *Human Resources and Higher Education* (New York: Russell Sage Foundation, 1970).

59. Cited in J. Bernard, *op. cit.*, p. 89.

60. Astin, *op. cit.*, p. 103.

61. R. H. Pearce et al , *Women in the Graduate Academic Sector of the University of California: Report on an Ad Hoc Committee of the Coordinating Committee on Graduate Affairs* (Los Angeles: University of California, June 1972); *The Study of Graduate Education at Stanford: Report of the Task Force on Women* (Stanford: Stanford University, June 1972).

62. *Ibid.*

awards in 1959 were given to women in about the same ratio as to men; 12 percent of the applicants were women, and 12 percent of the awards went to women.''[63] Bernard further noted that women receive fellowships to the Center for Advanced Study in the Behavioral Sciences at Palo Alto in proportion to the numbers of females recommended for fellowships. She concluded:

> However convincing individual cases of prejudiced discrimination are, it is difficult to prove its existence on a large or mass scale. The most talented women may be and, indeed, are victimized by it, but apparently not academic women en masse. At least the evidence from awards and from the number of academic women in proportion to the qualified pool available is far from convincing.[64]

To add a few more pieces of evidence to that already presented, consider the conclusion reached by James A. Davis in his study of some 34,000 college graduates in 1961. He found that "women . . . have no disadvantage or advantage in offerings" of fellowships for graduate education.[65] Both Simon and Reskin, as already noted, report that women are slightly more likely than men to receive some form of postdoctoral fellowship support. Cynthia Atwood found that women during the years 1968–73 received graduate fellowship support in greater proportion than that in which they applied: Women represented about 4.6 percent of the fellowship applicants in the physical sciences, and 7.6 percent of the recipients; similarly, they constituted 9.7 percent of the applicants in the social sciences, and 17.7 percent of the recipients.[66] A report on the status of women in sociology indicated that as a group women sociology graduate students received financial support in 1971–72 in direct proportion to the percentage of applicants who were female: 34 percent of the new applicants were women; of those accepted, 34 percent were women.[67] And finally, my own data showed no significant differences between the number of postdoctoral honorific awards received by men and women in the matched sample.

Clearly, these data only suggest that women are not treated differently than men in terms of support necessary to do full-time graduate work. It is possible, of course, that the average woman who applies for admission to graduate school and for financial assistance is significantly more able than the average male applicant. If this were demonstrated, we would expect that women would be both admitted and awarded fellowships at a disproportionately high rate. Available data from Graduate Record Examination scores of men and women applicants for graduate admission do not support this hypothesis. Men do better than women on

---

63. Bernard, *op. cit.*, p. 50.

64. *Ibid.*

65. Quoted in Bernard, *op. cit.*, p. 51.

66. Source: Cynthia Atwood, *Women in Fellowship and Training Programs*, Association of American Colleges, Nov. 1972, as cited in Lewis C. Solmon, "Women in Graduate Education: Clues and Puzzles Regarding Institutional Discrimination" (mimeo.).

67. Source: *The Status of Women in Sociology 1968–1972*, Report to the American Sociological Association of the Ad Hoc Committee on the Status of Women in the Profession (Washington, D.C.: American Sociological Assn., 1973), p. 22.

the quantitative part of the test, and women do slightly better than men on the verbal part.[68]

In summary, the data that do exist offer little support for the hypothesis that women are the victims of a self-fulfilling phophecy and suffer from accumulative disadvantage. However, these observations focus only on a limited period in the career histories of women scientists. The processes of accumulative disadvantage may well begin at a far earlier age for women, and there may be processes that impede the progress of women which are not easily quantifiable.

"THE HAUNTING PRESENCE OF FUNCTIONALLY IRRELEVANT STATUSES"

When the findings of this chapter have been presented to interested scholars, I have often received looks of incredulity. While the design, methods, and mode of analysis seem adequate for testing the hypotheses considered, somehow the results run contrary to "common sense" understanding of the place of women in science, and more importantly, contrary to individuals' own unsystematic impressions. I often hear comments along the following line: "The study does seem to show little discrimination, but I can't believe the data because a friend of mine—a woman in Department X—was clearly the object of discrimination, and moreover, I have heard of many such cases in academic science." When so many scientists view these results with skepticism, it is cause for pause and reflection. Of course, I first checked the empirical findings; then I attempted to account for these reactions. Are there possibly sociological reasons for such resistance?[69]

Consider two possible reasons for this skepticism. Part of it emerges because there are, in absolute terms, so few women in academia generally and in science particularly. Individuals react to these absolute numbers without realizing that few women start the academic race to the Ph.D. and to careers in science. Furthermore, individuals assume that a "foul" has been committed after the race has begun rather than before. This spurious interpretation of the low number of women scientists may account for part of the skepticism, but I believe that a larger part results from a social structural phenomenon described by Robert K. Merton as "the haunting presence of functionally irrelevant statuses."[70] Merton employs this concept to deal with ambiguities built into social structures that lead to individual suffering and victimization. The pattern begins with individuals

---

68. Source: *GRE: Guide to the Use of GRE Scores in Graduate Admissions, 1970–71* (Princeton, N.J.: Educational Testing Service, 1971), p. 19. For the period 1968–71 the mean verbal score for men was 499; for women, 511. The mean quantitative score for men was 549; for women, 472. A more reliable comparison would, of course, show the comparative scores only of those men and women applying to graduate schools in scientific fields. These data are not currently available in a published source.

69. Bernard B. Barber, "Resistance by Scientists to Scientific Discovery," *Science* 134, no. 3479 (1 Sept. 1961): 596–602.

70. I have benefited here from extensive discussion with Robert K. Merton.

who occupy, in a statistical sense, unusual combinations of social statuses. Examples of rare combinations in American society are the black physician, the forty-year-old professional athlete, the famous man's son, the male nurse, the female engineer, the female scientist. When examining the occupational careers of people who occupy such status combinations, it is often difficult for either the "victim" or the observer to determine whether these people are being judged in terms of their performance within an occupation or in terms of the esteemed or disesteemed status. Thus, the famous man's son must decide whether he received rewards because of his own capacity or because of his father's position of eminence and power. The observers of these rewards also must judge the basis on which they are being given. Merton stresses that the ambiguity over the basis of social judgment is a consequence of unusual status combinations.

It may help us to understand the skepticism of some scientists toward this study's findings if we consider Merton's concept. When a woman scientist either is or is not rewarded—by receiving or not receiving an appointment to a prestigious department, by being granted or not being granted tenure, by receiving or not receiving a postdoctoral fellowship or research grant—the functionally irrelevant status, sex, is activated and imported into the social situation as an explanatory variable. Thus when women experience either success or failure it is often assumed to be the result of sexism rather than of a decision which may actually be based upon universalistic criteria. The interpretative problem arises, of course, because there are in fact enough identifiable individual cases of actual discrimination on the basis of sex to make the explanation of sexism at least plausible, if not compelling. I am not referring here to ways that scientists psychologically adapt to failure. There is structurally induced ambiguity among both women and men about the basis on which women are being judged as scientists: Are they being judged on the quality of their role performance as researchers and teachers, or on the basis of the functionally irrelevant status, sex?[71] In a significant number of cases where decisions are based on relevant performance criteria, or even on some irrelevant characteristic such as religion, the verdicts are interpreted as decisions in which sex status was the determining factor. It must be stressed that this applies not only to failures but to achievements among women scientists.

When a particular status has become salient because of the *belief*, whether true or false, that it is used as a basis of evaluation, individuals who occupy that status can rarely be sure about the basis for any social judgment. Thus, the woman whose work is truly of the first rank and who is granted tenure at a distinguished science department finds it difficult to know if she is the "token" woman or if she has been promoted because of the quality of her work—or both. This ever-present ambiguity can "haunt" the lives of occupants of statistically rare status-set combinations.

71. Cynthia F. Epstein, "Positive Effects of the Multiple Negative: Explaining the Success of Black Professional Women," *American Journal of Sociology* 78, no. 912 (1973).

The consequence of this ambiguity is that interpretative confusion abounds, and the audiences to these decisions, both men and women, often get a mistaken impression of the frequency with which sex status is used as a criterion of judgment. In fact, once a status such as sex is believed to be the primary determinant of social actions, reactions, and judgments, it becomes a constant invisible party to all social interactions between the sexes and begins to haunt not only women but men as well.

One could argue that the perception of sex discrimination, whether empirically correct or not, has always haunted women scientists, with significant negative consequences in productivity. Believing that their work makes little difference in the rewards they may expect to receive, they may see little point to working diligently at their research. Once motivation subsides, the race is run—they in fact do produce few scientific papers and those are of little impact, and this performance provides a rational basis for negative evaluation. In short, a belief system without empirical foundation can create the first and essential condition for establishing a self-fulfilling prophecy.

The perception of greater discrimination than actually exists is intensified because members of the social group who hold modal status characteristics cannot claim, without a sense of absurdity, that judgments on their careers are based upon these same statuses. For example, if a white male scientist were denied tenure at a top department, he could not claim (until recently) that his racial or sex status affected the decision. More probably he would import other social and psychological attributes, such as his political views or abrasive personality, to explain his failure to be promoted or to get an appointment in a top department. Thus evaluative decisions about men and women scientists, even though based upon exactly the same universalistic criteria, may be variously interpreted. A survey asking men and women scientists about the bases on which they were evaluated would very likely show systematic differences between the sexes that would only be partly accurate.

CONCLUSIONS

The primary aim of this chapter was to examine a set of common beliefs about the treatment that women scientists receive in the academic-science community. Proving discrimination is no easy business. Apparent discrimination often turns out to be illusory—a consequence of the interaction of several factors with the variable that is perceived as the basis of discrimination. For example, I have suggested that studies of recognition or status attainment in science which do not adequately take into account variations in role performance can have little value in proving a pattern of systematic discrimination. I have been able to place some controls on both the initial talent and the research performance of the scientists in my sample.

Throughout this chapter, no extended attempt has been made to interpret the differences uncovered among different scientific fields. Can we conclude that there are systematic differences in the level of sex discrimination in different disciplines? There does exist some variability in the correlations between different fields. But these correlations are based upon widely different sample sizes. More important than the statistical sampling problems, those differences that do obtain rarely conform to any theoretically meaningful pattern. At this point it is premature to draw inferences about different patterns in different scientific disciplines.

In the light of the data presented here on the treatment of women scientists, is it necessary to alter previous conclusions about the extent to which the stratification system in science is fundamentally universalistic? I must reiterate that I am only considering the treatment of women scientists after the receipt of the Ph.D. First and foremost, sex status explains very little variance on recognition of all types. The zero-order correlations vary between zero and .35. Thus, even if sex discrimination were occurring, it would not explain much variance on rank in the stratification system. Of course, this raises the question: How strong must a correlation between sex status and forms of recognition be before we call it evidence for discrimination? Some might argue that any independent effect of sex, after controlling for performance variables, must be considered evidence for discrimination. Certainly we cannot expect to find sex status explaining most of the variance in our indicators of recognition. And in this study, when I examined the influence of sex status on promotion, I did find some sex-related particularism. Sex status, without controlling for other interpretative variables, explains some 10 percent of the variance in academic rank in 1965 and considerably less than that in 1970. Do we say *only* 10 percent, or *as much as* 10 percent?

The zero-order correlations between sex status and the forms of recognition discussed here seem to result from three basic factors. The first is self-selection. Women may to some extent be in less prestigious institutions and hold lower ranks than men because they have less geographic mobility. Many women scientists are geographically limited by the occupational contingencies of their husbands. The definition of a woman's job in the past as "less important" than a man's is certainly a type of discrimination, but it is not discrimination within science. If a married woman turns down a job at a university in a locale removed from her husband's place of employment, we cannot say this is evidence of discrimination in science. Because of reduced mobility, however, women may find themselves in poorer bargaining positions than men of equal talent at the same universities. Moreover, because women are less mobile, they may be less visible to other institutions. Consequently, their reduced visibility may result in fewer job offers. Even women who are visible may not receive job offers because it is assumed that they are "not moveable." When they are not offered jobs, their "market value" declines and bargaining position for promotion within their own department diminishes. Whether these speculations are, in fact, accurate remains for empirical examination.

The second reason for the small correlation between sex status and recognition is that women produce fewer and less frequently cited papers than men. All the traditional explanations for these productivity differentials remain as unsupported hypotheses. They do not seem to be a result of family status or of differences in institutional location. Why, then, do women produce less than men? Data on IQ's suggest that men and women have at least equal native ability. Although this must remain as speculation, productivity differences may ultimately be explained to some extent by differences in socially structured motivation. Due to socialization and the value system in the larger society, it is possible that in the past women have not been as committed to their careers and as driven to achieve as men.[72] Indeed, women may believe they must pay a high social price for career commitment, and aware of this price, be under heavy cross-pressure in pursuing career goals.

It is possible that receiving the doctorate has a qualitatively different meaning for men and women. For a man, the doctorate represents the "union card" necessary for entry into the profession, but it is in no way sufficient for achieving success in science. For a woman, obtaining the Ph.D., with the attendant struggles that the process entails, may place her in an elite position relative to her sex peers. These struggles are indicated by the higher proportion of female than male doctoral students who drop out of graduate programs.[73] Furthermore, women tend to take longer to finish their degree. To have received the degree may be viewed as, and indeed may in fact be for the woman, the end of an odyssey. To have earned the degree is in some measure a triumph. The motivation that is required to begin immediately producing new scientific research may not be the same for men and women doctorate recipients. And if women fail to be as productive in the years immediately following their degree, the social process of accumulative disadvantage may take over and contribute to their falling farther behind in the race to produce new scientific discoveries. In short, the scientific productivity of the female scholar may depend somewhat on her reference group and her associated motivation. If the comparative reference group of women scientists is other women scientists or women in the larger society rather than the entire scientific community, women may be less motivated and therefore less productive than men.

A third factor creating the small correlation between sex status and recognition is discrimination. In the past, there certainly was some discrimination against women in science, and indeed there still is some limited amounts of

---

72. A number of recent studies have discussed the impact of socialization on the achievement orientation of women. See Ralph Turner, "Some Aspects of Women's Ambition," *American Journal of Sociology* 72 (September 1966): 163–172; Rossi, "Women in Science: Why So Few?" *op. cit.*; P. A. Graham, "Women in Academe," *Science* 1969, (1970): 1284–1290; C. Bird and S. W. Briller, *Born Female: The High Cost of Keeping Women Down* (New York: David McKay, 1968).

73. A. Tucker, D. Gottlieb, and J. Pease, *Attrition of Graduate Students at the Ph.D. Level in the Traditional Arts and Sciences*, Publication no. 8 (East Lansing: Office of Research Development and the Graduate School, Michigan State University, 1964). In this study of twenty-four universities selected on the basis of size, region, and quality of their graduate program, it was found that "among female doctoral students, 54 percent dropped out, whereas among male doctoral students, the attrition rate was 36 percent" (p. 57).

discrimination. In my data, discrimination is most evident when considering the effects of sex on academic rank. In 1965 it took women slightly longer than men to be promoted from assistant to associate professor. In general, the data suggest that the measurable amount of sex-based discrimination against women scientists is small, at least when we consider the forms of recognition reviewed in this chapter. Certainly it is less than the "discrimination" faced by graduates—men and women—of low-prestige academic departments. Thus far, the data do not require that we modify significantly prior conclusions that the scientific stratification system is basically universalistic.

A CODA ON RECENT HIRING PRACTICES

Apparently, claims of widespread sexual discrimination in hiring are unfounded. If this is so for the 1958 cohort of Ph.D.'s, is it still true today? There persist, of course, claims of sex discrimination in hiring within the scientific community, and multiple lawsuits based on these claims. Changes in the social climate of opinion and the development of organized social protest movements in the 1960s might lead us to expect shifts in the sex distribution in college and university faculties. Pressure from women's caucuses of professional associations and from the Department of Health, Education, and Welfare's Office of Civil Rights, to cite just two sources, would suggest that sex discrimination in academic appointments currently exists, and that there exists a strong effort to eliminate it. Concomitantly, however, the national economic recession of the early 1970s and the fiscal retrenchment by universities and colleges has reduced opportunities in general, and thus has limited possibilities for significant social change.

These recent events, when coupled with our findings, permit a new question: If there was very little sex discrimination in appointments to academic departments in the 1960s when there was little organized pressure for change, and if there currently exists strong political and social pressure to increase the proportion of women hired, will this produce discrimination against men? To precisely determine the existence or extent of reverse discrimination would require elaborate data collection on educational training, on the social characteristics of male and female Ph.D.'s, and on their performance as scientists and scholars. Since these detailed data are not available, I have collected a limited amount of information on recent Ph.D.'s to find out whether men rather than women are currently discriminated against in hiring.

This analysis must be viewed as preliminary in the dual sense that it covers only two social science disciplines, sociology and anthropology, and that it is limited to simple multivariate relationships. In one respect the task is simplified. Focusing only on the first step in the careers of doctorate recipients—their first academic appointment—eliminates, for the most part, the need to control for quality and quantity of work when examining the relationship between sex status and job quality. Although I offer some data on citation rates of men and women at departments of varying quality, these data are of limited use, since most of these young Ph.D.'s have not published much.

The method used here to identify discrimination is a weak one, even though it is used by universities (and accepted by HEW) in framing affirmative action goals. At best it establishes a prima facie case. I compare the proportion of women who receive doctorates in each field with the proportion of women in their Ph.D. cohorts who are on the faculties of departments ranked in the American Council on Education's 1969 study of the quality of graduate departments (the Roose-Andersen study). Of course, such simple comparisons of marginals omit possible explanations based on demonstrated differences in talent and performance. But, as noted, this is less critical when considering first appointments.

The proportion of all doctorates earned by women in anthropology and sociology remained roughly constant from 1967 to 1972. National Research Council and Office of Education data indicate that roughly 18 percent of sociology doctorates in 1967 were earned by women, a figure which increased to about 22 percent by 1972. Comparable figures for anthropology were 26 and 28 percent.[74] In 1973, the figure in sociology climbed to 28 percent.[75]

To find the proportion of female doctorate recipients from 1970 to 1973 located at variously ranked sociology departments, I collected data from the 1974 American Sociological Association *Guide to Graduate Departments of Sociology*. With few exceptions, this publication lists the faculty members at each of the Ph.D.-granting departments of sociology and the year that they received the doctorate.[76] The American Anthropology Association *Guide* provides similar data, and was used to obtain data for anthropology departments.

Before attending to the academic location of male and female Ph.D.'s who received their degrees at roughly the same time, consider a prior question: What types of jobs do male and female Ph.D.'s obtain after receiving their degrees? Sociology Ph.D.'s of both sexes initially follow similar occupational paths. Among the men who received their doctorates in 1973 from one of the top 44 departments, only 6 percent received appointments in one of the top 12 universities; another 3 percent took posts in departments rated from 13th to 20th; and 6 percent were located at universities ranked 21st to 44th. The pattern among the women is similar: 7 percent in one of the top 12; 5 percent in those rated 13th to 20th; and another 8 percent in departments ranked 21st to 44th. Sixty percent of

74. Source: *Summary Report, Doctorate Recipients from U.S. Universities, 1967-70, 1972*; U.S. Office of Education [Earned Degrees Conferred, 1970-1971].

75. This figure is computed from the ASA *Guide*. The National Research Council discontinued its survey of Ph.D.'s in 1973. There is independent support for the 28 percent figure for sociology. See Maurice Jackson, "Affirmative Action—Affirmative Results?" ASA *Footnotes*, Dec. 1973, vol. 1, no. 9.

76. Four departments—Columbia, Pennsylvania, Chicago, and Ohio State—did not provide information on the job affiliations of their recent graduates. I tried to locate and trace the histories of these graduates through the ASA *Directory* for 1973-74 and through *American Men and Women of Science*. Thirty-two individuals could not be located. These cases are omitted from the analysis. Furthermore, some universities list graduates by the initials of their first names, making it difficult to establish sex. I again tried to trace these people, but 23 could not be identified and were also not used in the analysis.

the men, compared to 54 percent of the women, obtained jobs at lesser universities and colleges; 2 percent of the men and 8 percent of the women became research associates; and 7 percent of the men and 3 percent of the women took jobs outside the academic community.[77] Women are slightly more apt than men to take positions as research associates, but such small differences may easily be explained by self-selection processes. In sum, only minor differences obtain in the initial career locations of male and female sociology Ph.D.'s.

Turning exclusively to men and women in academic life, Table 3–6 presents the proportion of women receiving their degrees from 1970 to 1973 in each field who by 1973 were members of one of the 44 sociology departments or one of the 32 anthropology departments rated by Roose and Andersen. The results vary somewhat by field. Women sociologists were slightly underrepresented in 1970 and 1971 in rated university sociology departments. Although 18 percent of the 1970 Ph.D.'s were women, they represented only 14 percent of the faculty in the leading 44 departments. Two years later women continued to be slightly underrepresented in these departments. However, dramatic change occurred in 1973. While the proportion of females receiving sociology doctorates increased significantly, there was an even greater increase in the proportion of women obtaining their first jobs at one of the 44 rated departments. Thirty-three percent of the appointments were women, although they constituted about 28 percent of the Ph.D.'s.[78] Correlatively, male Ph.D.'s were slightly overrepresented in the lesser departments not rated by Roose-Andersen.[79] Data for anthropology indicate a similar if less sharply defined trend. A rise in the proportion of women located in rated departments appeared earlier, in 1972, when the ratio reached 43 percent. By 1973 the proportion had slipped back to the 1971 level. Nevertheless, despite this perturbation, from 1970 through 1973 the proportion of females at one of the 32 rated departments invariably exceeded the proportion of women Ph.D.'s produced by the field.

So much for larger patterns. Now let us consider how the quality of the Ph.D. department affects the placement of male and female Ph.D.'s. According to the American Sociological Association *Guide,* 365 sociology Ph.D.'s in 1973 took academic posts: 52 percent of the Ph.D.'s had been granted by the top 20 departments, 33 percent by other rated departments, and 16 percent by other departments listed in the *Guide.* Data were available for the 189 young sociologists who had received their degrees from one of the top 20 departments. Of these, 73 percent found first jobs in departments that were not even rated by Roose and Andersen—in short, less than distinguished departments. Most Ph.D.'s, regardless of gender or any other attribute, simply do not obtain jobs at

---

77. Job information was unavailable for 16 percent of the men and a comparable proportion of the women.

78. In fact, 46 university sociology departments were rated, but data were available for only 44 of these. Data were not obtained for MIT or The New School for Social Research.

79. The two criteria used for inclusion in Roose and Andersen's ACE evaluation study were that each department had to offer a doctorate and it had to have conferred at least one doctorate in each of the past ten years. These criteria served to exclude a number of outstanding but recently formed graduate departments. such as the one at Stony Brook (State University of New York).

TABLE 3-6. The Proportion of Recent Ph.D.'s in Sociology and Anthropology Located in Departments of Varying Quality[1]

	SOCIOLOGY			ANTHROPOLOGY		
Year of Ph.D.	% Female Ph.D.'s[6]	% Women in One of Top 44 Departments[2]	% Women in Sample of Unrated Departments[3]	% Female Ph.D.'s	% Women in One of Top 32 Departments[4]	% Women in Sample of Unrated Departments[5]
1970	18.3 (506)	14.5 (87)	15.2 (38)	28.0 (217)	17.0 (48)	18.2 (26)
1971	20.7 (574)	12.0 (93)	16.2 (43)		29.6 (35)	27.3 (42)
1972	21.6 (638)	18.9 (63)	22.2 (33)	28.0 (260)	43.3 (43)	38.5 (36)
1973	28.6	32.9 (105)	23.3 (37)		30.8 (34)	21.7 (28)

NOTES:

1. The quality ratings were obtained from the American Council on Education survey of graduate departments of Arts and Sciences, 1969, under the direction of K. Roose and C. Andersen.

2. A total of 46 sociology departments were rated. Our data cover only 44 of these, since data on MIT and The New School were not available.

3. A random sample of 30 universities not included in the Roose-Andersen study was selected from the American Sociological Association's Guide to Graduate Departments in Sociology for 1974. Every fifth university was selected.

4. These 32 departments represent those rated in the Roose-Andersen study.

5. A random sample of 37 universities not included in the Roose-Andersen study was selected from the American Anthropological Association's Guide to Graduate Study for September 1973. Every fifth university was selected.

6. Sources: 1970, 1972   National Research Council, *Summary Report: Doctorate Recipients from U.S. Universities*
          1970   U.S. Office of Education, Earned Degrees Conferred
          1973   ASA Guide to Graduate Departments

schools that are as distinguished as the ones at which they were trained. The data on the production of Ph.D.'s indicate that roughly 31 percent of those produced by the top 20 departments in 1973 were women. Of all those who received their degrees from one of the top 20 and gained appointments in similarly ranked departments, 33 percent were women. In other words, women graduating from high-ranking departments were placed in high-ranking departments in roughly the same proportion in which they graduated from them. The pattern is much the same for graduates of departments ranked 21st to 44th, of whom 40 percent were women. The data are presented in Table 3-7. Without even considering the research performance of these 1973 graduates, there is little support for claims that sex discrimination obtains in the hiring of new assistant professors of sociology. Correlatively, there is little evidence that reverse sexual discrimination is currently a problem of significant magnitude.

These comparisons would have little force, of course, if we found systematic differences in the talent or the performance of men and women doctoral candidates. One cannot emphasize too strongly the idea that if one group of sociologists is more able than another it should be overrepresented in high-ranking departments. For the cohort receiving their degrees in the late 1950s, we have already seen that women actually tend to publish fewer papers than men, and that their work is less frequently referenced in the work of other scientists. It is difficult to control for the quality of work produced by recent Ph.D.'s, since they have produced so little of it, and are often junior collaborators with older, more distinguished, colleagues. But I made a first approximation by collecting data on the quality of work produced by the 1970–73 cohort. Citation counts for

TABLE 3–7.   Placement of 1973 Female Ph.D.'s in Sociology by School of Origin and School of Destination* (Percent)

		HIRING INSTITUTION (RATED QUALITY)**		
		Top 20 Departments	*Other Rated Departments*	*All Other Universities and Colleges*
	Top 20 Departments	33 (36)	40 (15)	30 (138)
DEGREE-GRANTING INSTITUTION (RATED QUALTIY)	*Other Rated Departments*	40 (5)	23 (13)	19 (101)
	All Other Universities and Colleges	—	—	30 (57)

*The data presented in this table include only those men and women Ph.D.'s in 1973 who took academic jobs and for whom I had job information.

**The ACE study of quality of arts and sciences departments was used as the indicator of the quality of the sociology departments.

the first two-thirds of 1974 were collected, using the *Social Science Citation Index*. Of course, most of these men and women were never cited, but male Ph.D.'s were more likely than females to be cited, and they received, on average, more citations. Sixteen percent of the women located in the top 20 departments and 32 percent of the men received at least one citation to their work; 27 percent of the men and 12 percent of the women at departments rated 21st to 44th had been cited; 20 percent of the men against 12 percent of the women at unrated departments received citations to their work. On average, men in the top 20 departments received 2.4 citations; women, 2.2 citations. Similar patterns obtain for men and women at lesser departments. In sum, there is no evidence, however scanty, that the quality of work produced by young female sociologists is perceived by others in the discipline as superior to the work produced by young men.

Do these data alter our previous conclusion about the academic appointment process? Keeping in mind the limits to the data, the evidence runs counter to many beliefs, and further corroborates the finding that there is no relationship of any significance between the sex status of Ph.D.'s and appointments to departments of quality. Of course, in the matter of promotion to high rank, evidence for sex discrimination may continue to obtain. But the data suggest, if anything, that there is a slight tendency today—perhaps as a result of pressure placed on universities by HEW affirmative-action policy or internal pressure by women's caucuses—to favor women over men in hiring new assistant professors.

The belief, still widespread within the scientific community, that there is patterned and systematic bias and discrimination in hiring female Ph.D.'s is simply unsupported by data. Unless members of the academic-science community realize that the pattern of discrimination within science is not evenly distributed at all points in the career history of men and women, we are apt to find an increase rather than a decline in some forms of sex discrimination. But the ultimate product could be reverse discrimination in hiring.

Chapter 4

THE REPUTATIONS OF MEN
AND WOMEN SCIENTISTS

INQUIRIES INTO THE PLACE of women in the academic community have focused almost uniformly on positional forms of recognition, and have conspicuously neglected discussing the reputational standing of women scientists among
their peers. By now, there is a rapidly expanding number of papers and monographs reporting women's place in terms of academic recruitment, training, and
employment in general, and in particular their rate of admission to professional
schools, their affiliations and academic ranks, their productivity, their salaries,
their mobility from one university or college to another, their chances for promotion, and their opportunities for obtaining scarce scientific resources and
facilities.[1] Thus far I, too, have concentrated on these forms of recognition. But
there have been no studies of the reputations of women scientists that focus on
interest in, knowledge about, acquaintance with, and praise for their scientific
work.[2] The paucity of material on the subject does not accurately reflect, of
course, its importance to working scientists.

Indeed, a widespread and esteemed reputation is among the most coveted,

---

1. A review of the extant literature on these problems may be found in Harriet A. Zuckerman and Jonathan R. Cole,
"Women in American Science," *Minerva* 13, no. 1 (spring 1975): 82–102; Alice Rossi and Ann Calderwood,
eds., *Academic Women on the Move* (New York: Russell Sage Foundation, 1973); Cynthia Lloyd, ed., *Sex,
Discrimination, and the Division of Labor* (New York: Columbia University Press, 1975); John A. Centra, *Women,
Men and the Doctorate* (Princeton, N.J.: Educational Testing Service, 1974).

2. On forms of experience leading to various levels and kinds of knowledge, see William James, *The Meaning of
Truth*, 1885 (reprinted, New York: Longmans, Green, 1932), pp. 11–13; Robert K. Merton, "The Perspectives of
Insiders and Outsiders," in *The Sociology of Science: Theoretical and Empirical Investigations*, 1972 (reprinted,
Chicago: University of Chicago Press, 1973), pp. 99–136.

cherished, and well guarded of all scientific rewards. Testimony corroborating this can be drawn from a plethora of sources throughout the history of science. Suffice it to note only three personal reflections on the centrality of peer recognition in the lives of scientists of widely disparate rank. Charles Darwin said: "My love for natural science . . . has been much aided by the ambition to be esteemed by my fellow scientists."[3] Descartes expressed in a letter to his friend Mersenne his belief that Hobbes was intent upon pilfering his ideas: "I also beg you to tell him [Hobbes] as little as possible about what you know of my unpublished opinions, for if I'm not greatly mistaken, he is a man who is seeking to acquire a reputation at my expense and through shady practices."[4] And a relatively unknown female physiologist, interviewed by Helen Astin, stated: "If I am proud of any achievement, it is probably that despite my limited full-length publications my research has come to the attention of people in a number of foreign countries as well as in this one."[5]

Evidence that a reputation also structures others' perceptions of a scientist could be drawn from innumerable sources, but consider as only one example the autobiographical revelation of the young and frank James D. Watson, speaking of his first meeting with Max Delbrück:

> His visit excited me, for the prominent role of his ideas in "What is Life" made him a legendary figure in my mind. . . . Almost from Delbrück's first sentence, I knew I was not going to be disappointed. He did not beat around the bush and the intent of his words was always clear. But even more important to me was his youthful appearance and spirit. This surprised me, for without thinking I assumed that a German with his reputation must already be balding and overweight.[6]

Why is reputation an important part of the social reality of the scientific community? Reputation has multiple consequences for scientists; it interacts with scientific performance and with other forms of scientific recognition to influence significantly the direction of scientific careers. Although reputation is influenced by the quality of scientific research, it clearly in turn influences the quality of research. Since scientists with outstanding reputations have greater access to research facilities and resources than do those who are relatively unknown, it increases the possibilities that those who already have made discoveries will continue to do so.[7]

---

3. As quoted in Merton, "Priorities in Scientific Discovery," in *The Sociology of Science* (*op. cit.*), p. 293.

4. Descartes, *Oeuvres,* vol. 3, p. 320: as quoted in Merton, *op. cit.,* p. 313.

5. Helen S. Astin, *The Women Doctorate in America* (New York: Russell Sage Foundation, 1969), p. 133.

6. James D. Watson, "Growing Up in the Phage Group," in *Phage and the Origins of Molecular Biology,* ed. John Cairns, Gunther S. Stent, and James D. Watson (New York: Cold Spring Harbor Laboratory of Quantitative Biology, 1966), p. 240.

7. On the growing literature discussing accumulation of advantage and disadvantage, see, among others: Merton, *op. cit.,* pp. 273, 416, 439–459; Harriet A. Zuckerman, "Stratification in American Science," *Sociological Inquiry* 40, no. 1 (spring 1970): 235–257, esp. p. 245; Harriet A. Zuckerman and Robert K. Merton, "Age, Aging and Age Structure in Science," in Merton, *op. cit.,* pp. 497–449, esp. p. 532; Jonathan R. Cole and Stephen Cole, *Social Stratification in Science* (Chicago: University of Chicago Press, 1973), pp. 237–247; Paul Allison and John Stewart, "Productivity Differences Among Scientists: Evidence for Accumulative Advantage," *American Sociological Review* 39, no. 4 (Aug. 1974): 596–606.

Robert K. Merton, discussing this interaction between reputational standing and scientific resources, suggests how reputations play a central role in the opportunity structure of science:

> [Esteemed reputation among peers for scientific achievement] can be converted into an instrumental asset as enlarged facilities are made available to the honored scientist for further work. Without deliberate intent on the part of any group, the reward system thus influences the "class structure" of science by providing a stratified distribution of chances, among scientists, for enlarging their role as investigators. The process provides differential access to the means of scientific production. This becomes all the more important in the current historical shift from little science to big science, with its expensive and often centralized equipment needed for research. There is a continuing interplay between the status system, based upon honor and esteem, and the class system, based upon differential life-chances, which locates scientists in differing positions within the opportunity structure of science.[8]

Thus, the "Matthew effect," which "consists of the accruing of greater increments of recognition for particular scientific contributions to scientists of considerable repute and the withholding of such recognition from scientists who have not yet made their mark,'"[9] operates to accumulate advantages to those who are already esteemed. Merton's concept has been used to examine the parceling out of recognition among scientists whose reputations differ widely but who have been party to independent multiple discoveries. He discusses how location in the stratification system influences the disproportionate allocation of credit to the more esteemed party to the multiple, and how this process ironically serves some functions for the advance of scientific knowledge.[10] He is not concerned with the effects of ascribed statuses on the allocation of credit. If we recast Merton's formulation slightly and examine reputation as a dependent rather than an independent variable, we can ask whether ascribed statuses, such as sex, influence the reputations of men and women scientists who produce work of roughly equal quality.[11]

In Merton's passage quoted above, this allusion to Marxian "class structure" is undoubtedly chosen carefully. A lofty reputation not only gives a scientist access to the means of scientific production by legitimizing his request for facilities and resources; it places a scientist close to those means of production by structuring his associations. Reputation influences the types of others a scientist exchanges ideas and interacts with, on either a formal or an informal basis. Indeed, it influences the seriousness with which his ideas are considered. Beyond the conversion of these patterns of interaction into ideas, resources, and recognition, the admission into select groups of scientists makes scientific life that much more exciting and enjoyable.

8. Merton, *op. cit.*, pp. 442–443.

9. Robert K. Merton, "The Matthew Effect in Science," *Science* 199 (5 Jan. 1968): 55–63.

10. *Ibid.*

11. For an empirical test of Merton's idea on the Matthew effect, see Cole and Cole, *op. cit.*, pp. 191–215.

Recent research findings suggest that the reputation of academicians are influenced by their affiliations; it should also be clear that individual reputations significantly influence the larger reputations of academic departments and universities.[12] Studies of the reputational standing of arts and sciences departments and of medical schools indicate a strong correlation between the reputations of "stars" within a school or department and its general reputation.[13] In short, individual reputations have consequences for the reputations of colleagues, departments, and all the organizations with which a scientist is connected.[14] Whether an individual is a "borrower" or a "lender" of reputation affects his or her power and influence within a department, university, and discipline.

Reputation thus affects a scientist's bargaining power in the academic marketplace in two distinct ways. First, it partly determines his power in the marketplace of ideas, affecting his standing within the complex structure of authority in science.[15] Whether a scientist is consequential in shaping the cognitive content of a field, whether he serves as a gate-keeper of ideas and positions, whether he is instrumental in developing an organizational infrastructure and a professional identity for a research area or specialty—all are shaped in part by his reputational standing among his peers.[16] Second, in the nitty-gritty of everyday academic life, reputation affects the probability of a scientist obtaining other positional rewards, including offers from other universities, a higher salary in accord with his value in the marketplace, and large research grants from federal funding agencies.[17]

Reputation, then, not only is an intrinsic reward; it also is a commodity exchangeable into future opportunities for scientific productivity and positional recognition. Although reputation and quality of performance cannot be equated, the reward system operates to reinforce the positive correlation between them.[18] Because reputation is a critical part of the reward system of science, reputation with respect to women scientists becomes an important topic for inquiry.

---

12. On the interaction between individual and institutional reputations, see Peter M. Blau, "Structural Constraints of Status Complements," in *The Idea of Social Structure*, ed., Lewis A. Coser (New York: Harcourt Brace Jovanovich, 1975), pp. 117–138.

13. On reputations of arts and sciences departments, see Allan M. Cartter, *An Assessment of Quality in Graduate Education* (Washington, D.C.: American Council on Education, 1966); K. D. Roose and Charles J. Andersen, *A Rating of Graduate Programs* (Washington, D.C.: American Council on Education, 1970); on medical schools, see Jonathan R. Cole and James A. Lipton, "The Reputations of American Medical Schools," *Social Forces* 55, Nos. 3 (March 1977): 662–684.

14. Blau, *op. cit.*

15. On authority structure in science, see, among others: Michael Polanyi, *Personal Knowledge* (New York: Harper & Row, Torchbooks, 1964); John Ziman, *Public Knowledge* (Cambridge: Cambridge University Press, 1968); Merton, *The Sociology of Science* (*op. cit.*); Cole and Cole, *op. cit.*

16. A discussion of the relationship between reputation and these features of cognitive and social structures of scientific disciplines may be found in Jonathan R. Cole and Harriet A. Zuckerman, "The Emergence of a Scientific Specialty: The Self-Exemplifying Case of the Sociology of Science," and in Stephen Cole, "The Growth of Scientific Knowledge: Theories of Deviance as a Case Study"—both in Coser, *The Idea of Social Structure* (*op. cit.*).

17. Stephen Cole and I are currently examining the role that reputation plays in obtaining research support from government granting-agencies. This project is supported by the National Academy of Sciences, Committee on Science and Public Policy.

18. On processes that reinforce the association between quality of research and reputation, see Cole and Cole, *op. cit.*; Bernard B. Barber et al., *Research on Human Subjects* (New York: Russell Sage Foundation, 1973); Harriet A. Zuckerman, *Scientific Elite* (New York: Free Press, 1977).

In order to examine the reputational standing of women scientists, it is necessary to compare their standing with that of men. Therefore, the data reported in this chapter center on two questions: First, are there sex-related differences in the reputational standing of men and women scientists, and what social characteristics can explain any differences that do obtain? Second, do the social processes differ by which men and women establish reputations?

We actually have a fairly good idea by now of how reputations are built among male scientists. In a recent study of American physicists by Stephen Cole and myself, 61 percent of the total variance on visibility (one dimension of reputation, to be defined below) could be explained by three variables: the quality of a scientist's published research, the prestige of his highest honorific award, and the rank of his academic department.[19] Although the sheer quantity of research output did influence visibility independently of quality, the variables that most strongly influenced it were those associated with universalistic criteria for evaluating performance. In fact, reputation depended far more on universalistic criteria than did recognition through academic appointments. In general, we found that producing a large number of trivial papers had little value in gaining a reputation in physics. Quality of work, however, played an important instrumental role.[20] The principle of accumulative advantage operated in virtually every case in which the value of universalism was violated.

If the script for building reputations for male scientists is reasonably well established, it is far from articulated for females. To elaborate on the two central questions, we want to know, first, whether there is a "cost" to being a woman in terms of building a scientific reputation. Does sex status directly influence reputation; does it indirectly influence renown through its effect on other determinants of reputation; and does it interact with these other determinants to produce significant differentials in the reputations of men and women? Second, what is the overall distribution with respect to reputations of men and women scientists? Are the reputations of women as much as those of men influenced by their research performance? Will a highly productive female scientist have a greater, lesser, or equal reputation compared to a male who has produced less work that is of poorer quality? Do "halos" shine over the heads of women scientists located at highly prestigious departments, and if so, is the halo effect as great for women as for men? Will a woman who has produced high-quality work but is located at a less distinguished department have a poorer reputation than another woman, located at Stanford or Harvard, who has produced work of roughly the same quality? Indeed, for roughly equal research performance, will a female scientist at a quality university have the same chances of becoming prominent or esteemed as will a male scientist in a similar position? Are there social characteristics of

---

19. For an extended discussion of the visibility of male physicists, see Cole and Cole, *op. cit.*, pp. 99–110.

20. *Ibid.*; see also Stephen Cole, "Scientific Reward Systems: A Comparative Analysis," paper presented at the Annual Meeting of the American Sociological Association, 1973; Stephen Cole and Jonathan R. Cole, "The Reward Systems of the Social Sciences," in Charles S. Frankel, *Social Science Controversies and Public Policy Decisions* (New York: Russell Sage Foundation, 1976).

male and female scientists, such as their age or educational origins, that interact with gender to produce differential reputational standing?

These are the questions to be answered, and I will use status-set analysis as a conceptual frame for each of them. Some twenty-five years ago Robert K. Merton began to examine how the structure of status-sets affected the behavior of social groups and influenced the social responses to individuals who occupied statistically rare combinations of statuses. He defined status-sets as "the complex of distinct positions assigned to individuals both within and among social systems. ... It should be noted ... that, just as groups and societies differ in the number and complexity of social statuses comprising part of their structure, so individual people differ in the number and complexity of statuses comprosing their status-sets."[21] Since Merton's early explorations, little has been done to formulate empirical tests of the effects on behavior of status-set structures. The data reported in this chapter are examined in terms of the effects that variations in the status-sets of scientists have on the responses to their work. Status-set analysis is, of course, also multivariate analysis. The language of status-set analysis is particularly appropriate for this discussion because I focus on how social characteristics and structural positions of scientists, rather than their attitudes, beliefs, or psychological predispositions, determine their reputations. Personality surely plays its part in the construction of reputations, but specifying its effect remains for future inquiry.

Clearly, the 592 scientists whose reputations have been assessed, and who form the sample of scientists that will be considered in this chapter, occupy multiple statuses, and each has his or her distinctive status-set. As Rose Coser has perceptively pointed out, part of a person's distinctiveness and individuality derives from variations in his or her status-set.[22] This is particularly so for female scientists, who simply by virtue of gender have statistically rare status-set combinations. The range of statuses and social attributes for which I have collected data includes: scientific discipline, academic affiliation, age, honorific awards, academic rank, marital and family status, scientific output, educational background, and sex status.[23] Of course, each scientist holds, in varying degree, other positions about which I know nothing, and which, therefore, can not enter the analysis.

Simply put, I am trying to examine the effects on reputational standing of simple variations in the status-sets of scientists. In effect, I am simulating a controlled experiment in which the experimental and control groups differ only in

---

21. Robert K. Merton, *Social Theory and Social Structure*, 1949 (reprinted, New York: Free Press, 1968), pp. 434–435.

22. Rose L. Coser, "The Complexity of Roles as a Seedbed of Individual Autonomy," in Coser, *The Idea of Social Structure* (*op. cit.*), pp. 237–264. Of course, this idea was developed significantly in the earlier work of Simmel.

23. If we strictly define social status as a set of normative rights and obligations associated with a social position, many characteristics of scientists that we identify as statuses should in fact be thought of as social attributes. For example, the quality or quantity of research performance is not, strictly speaking, a social status, since it is not attached to a social position. Nonetheless, some social attributes such as research performance or educational achievement have much the same character as a social status. In this analysis of status-sets of scientists, I include the quality and quantity of research performance among the statuses held by men and women scientists.

gender. Although total randomization is never wholly achieved, the effect of gender on reputation is taken as its influence on social evaluations. How does variation in this one status control the social response to the work produced by otherwise similar scientists? If we find significant differences in reputations of males and females, after statistically controlling for all other statuses in the status-sets, an estimate is possible of the weight of sex status in the overall status-sets of scientists.

By way of illustration, contrast the status-sets of two scientists, for whom only four attributes are known: Each occupies a position at a distinguished department, holds high rank, has produced work of substantial impact, and is the same professional age. In short, their status-sets are identical; there is no contrast here. On the basis of such limited information, we would predict that the scientists had roughly equal reputations. Suppose, however, our knowledge is increased about these two scientists: One is male, the other female. Although virtual identity in their status-sets persists, one basic difference now exists that by hypothesis might affect reputational standing in the scientific community. But how much actual difference does this additional item of information make in appraisals of the two scientists' reputations?

In due course, this analysis will go beyond elementary experimental design by making the status-set configurations increasingly heterogeneous. Thus, if the quality of research produced by the scientists is also allowed to vary, the two status-sets, which had differed only in gender, will now vary in two ways. Then we may ask: What is the relative influence of gender and of quality of research in predicting reputations; is gender or quality the dominant element in the status-sets?[24]

METHOD

A short questionnaire was sent to a random sample of scientists working in sociology, psychology, and four biological-science specialties who were affiliated with American universities that grant doctorates. Only schools included in the Roose-Andersen study—that is, schools which had granted at least one doctorate in each of the past ten years—were included in the population from which individual faculty were sampled.[25] The questionnaire was designed to obtain appraisals of the scientific contributions made by a sample of men and women scientists working in the three disciplines.

The questionnaire had three parts. Part 1 asked scientists to evaluate the

---

24. Since I am considering academic scientists, I can avoid discussing the effects of "status-set typing," noted by Cynthia Epstein in *Women's Place* (Berkeley: University of California Press, 1971). Since the men and women scientists occupy many statuses in common, the social forces leading them to have fundamentally different status-set configurations are not relevant here.

25. Roose and Andersen, *op. cit.* The number of universities included in the study differs, of course, by academic field. Programs surveyed numbered 73 in sociology; 110 in psychology; 105 in biochemistry; 97 in botany; 115 in molecular biology; 88 in developmental biology; 102 in physiology.

importance of the work produced by a stratified random sample of scientists in their own field on one dimension of reputation: the perceived quality of the scientist's work. Scientists were asked: "By circling the appropriate number, please indicate the relative importance of the work of the following [sociologists, psychologists, biologists]: Has made very important contributions (6); Has made above-average contributions (5); Has made average contributions (4); Work has been relatively unimportant (3); Unfamiliar with work but have heard of this scientist (2); Have never heard of this scientist (1)."

Part 2 of the questionnaire asked the respondents to list the most important contributors in their discipline and in their specialty over the last decade. Part 3 requested information about the social background of respondents, such as their age, educational history, scientific specialty, current affiliation, and academic rank.

Each scientist was asked to evaluate the contributions of sixty members of his or her field; asking for more than this clearly was not practical.[26] The respondents were not informed of a particular interest in comparing the reputational standings of men and women scientists; they were simply asked to evaluate the work of a sample of individuals working in their field. Disclosure of the analytic intent would have increased the probability of normative responding. To avoid such bias, the proportion of the sixty names that were female reflected the sex compo-sition of Ph.D.'s in the field, roughly 25 percent. In fact, these three fields—rather than, say, physics or chemistry—were selected because of the relatively high percentage of women Ph.D.'s and other women working in them.[27]

Reproducing on the questionnaire the sex composition of the field created the problem of generating a large enough sample of women to be evaluated. To solve this, multiple forms of the questionnaire were designed.[28] Three forms were used

---

26. A number of difficulties were encountered in selecting the sixty scientists to be appraised in the biological sciences. First, it is often difficult to determine the specialty of biological scientists. For example, although an individual may list his or her specialty as physiology, there are a great number of subspecialties in physiology, many of which have little to do with one another. Consequently, in the biological sciences there tend to be lower visibility scores than in the social sciences, since scientists in one area of a discipline may have little awareness of work going on in other branches of the same discipline. Second, in combining molecular and developmental biologists for the purpose of appraisals, I further reduced the probability of high visibility scores. Although biologists' colleagues advised me that there was greater similarity between these two specialties than the others I was focusing on, the combination of the two was not ideal. To handle the problem of varying visibility scores, the analysis is based on standardized visibility scores. Each standardization transformation was handled within a field; all fields were then combined into a single variable.

27. A complete record of doctorate recipients within the United States (that is, 99 percent complete) had been maintained by the National Research Council until 1973, at which point the survey was discontinued. However, estimates of the proportion of women Ph.D.'s in various fields have been made for a group of forty-six universities that account for 75 percent of all doctorates to date in the United States. See Joseph L. McCarthy and Dael Wolfle, "Doctorates Granted to Women and Minority Group Members," *Science* 189 (12 Sept. 1975): 856–859, esp. Table 1. According to this survey, roughly 31 percent of sociology doctorates, 35 percent of psychology doctorates, and 22 percent of biochemistry doctorates were earned by women in the years 1972–1975. Only 4 percent of physics doctorates and 13 percent of chemistry doctorates went to women. Although these figures probably overestimate slightly the proportion of doctorates going to women, since the more prestigious universities have consistently produced higher proportions of female Ph.D.'s, the estimates are close to the actual figures.

28. In the past, I have made extended use of the multiple-form technique for enlarging sample size. The evidence consistently shows high inter-form correlations for names that are repeated on several forms. For other examples of utilization of this technique, see Cole and Cole, *op. cit.*; Cole and Lipton, *op. cit.*

for both sociology and psychology, and one for each of four biological specialties: botany, biochemistry, molecular and developmental biology, and physiology. Scientists' names were randomly positioned on the questionnaire. The
sociology and psychology questionnaires included four names common to each
form, to test for comparability of evaluation by scientists responding to the
various forms; for the biological sciences, of course, such a procedure made no
sense. Evaluations on the several forms were highly consistent. For example, a
female sociologist, one of the four names appearing on all three forms for her
field, had reputational scores (to be defined below) of 3.75, 3.79, and 4.00.
These differences were less than twice the standard error of the estimate, .28. In
short, the sample of respondents to each form of the questionnaire answered in
roughly the same way.[29] Each form of the questionnaire contained one fictitious
name, allowing for an estimate of guessing by responding scientists. There was
little evidence of guessing. On the different forms of the questionnaire, the
fictitious names were evaluated by only 1 to 6 percent of the responding scientists, the rest claiming never to have heard of them.[30]

For sociology and psychology, the work of a total of 173 scientists in each
field was evaluated. The sex breakdown was 130 men and 43 women. The
sample of rated scientists was drawn from a population of university scientists
stratified by prestige rank of their current department and by professional age.
The sample was not stratified by academic rank, since sex status is correlated
with rank, and I hypothesized that rank and reputation were positively associated. Older scientists in more distinguished departments were slightly over-
sampled. For example, 28 percent of all university sociologists at the time but
about 40 percent of the rated sociologists were affiliated with distinguished or
strong departments. Good and adequate departments, representing 33 percent of
the population, made up 30 percent of the sample; similarly, 38 percent of the
population and 30 percent of the sample were affiliated with lesser departments.

The departure from the population in terms of professional age was the slight
oversampling of scientists active in the field for more than twenty years (35
percent of the sample; 27 percent of the population), and correlatively, the
undersampling of recent Ph.D.'s (25 percent of the sample; 34 percent of the
population).[31] The sampling frames for psychology and sociology were identical.

Sampling differed for the biological specialties. The names of sixty scien-

---

29. It is now clear that independent samples of respondents to questionnaires such as those used here respond in
much the same way. Although it is difficult to determine the criteria that are used in making appraisals, independent
samples of scientists produce almost identical ratings of the same scientist's contribution to his or her field. The
similarity in ratings extends as well to ratings of science departments and medical schools. This is not surprising, of
course, since each sample of respondents has been selected randomly. There are no cases in this study where the
differences in form produce ratings that exceed twice the standard error of the estimate.

30. The level of guessing found here is similar to that found in previous studies of assessments of reputations. In the
study of physicists there was a slightly higher guessing rate when the fictitious name of the physicist was clearly
Oriental. In the present study there was no difference between fictitious male and female names: gender did not
affect the level of guessing.

31. The sample excluded all scientists who had received their degrees within the past five years, since I assumed
that younger scientists had not had enough time to establish reputations on the basis of their research performance.

tists, one of whom was fictitious, appeared on each form, but none of the names was repeated on other forms. Since each form of the questionnaire covered a different specialty, the sample of raters was restricted to scientists who worked in that specialty according to either *American Men and Women of Science* or the *Guide to Graduate Departments in Biological Sciences*. In all, 240 biologists were rated, of whom 48 were women.

Of the grand total of 582 rated scientists in the three disciplines, 135, or 23 percent, were women. The findings reported throughout this chapter are based only on the data obtained for these scientists.

In all, a random sample of 2,162 scientists were sent questionnaires; after two mailings 1,107, or 51 percent, had responded. The patterns of overall response from the three disciplines were similar: 53, 51, and 50 percent among sociologists, psychologists, and biologists, respectively. This overall response rate was consistent with rates obtained in other mail surveys of the academic population.[32] There was some variability in the response rate within biological specialties, ranging from 59 percent among botanists to 43 percent for molecular biologists. No social characteristics of responding scientists were significantly related to their evaluations. As examples. academic rank and prestige of affiliation were uncorrelated with appraisals. In sum, no systematic response biases were uncovered.

Two basic dimensions of reputational standing are examined in this chapter: the perceived quality and the visibility of scientific research, as evaluated by peers working in the same discipline.[33] An individual's visibility score is taken to be the percentage of all respondents who felt they had enough information to rate the person. It was obtained by dividing the total number of assessments of the individual, regardless of the quality ascribed, by the total number of respondents who returned the questionnaire. Perceived quality of work was measured by considering only the evaluations of respondents who actually assessed the individual's work; the perceived-quality score is the mean rating of these respondents. It has been noted that these scores can vary from a high of 6 ("Has made very important contributions") to a low of 3 ("Work has been relatively unimportant"). The number of responses on which perceived-quality scores are based varies among individuals of differing degrees of visibility.[34]

32. For example, two national surveys of the academic community by the American Council on Education (1972, 1973) received a usable response rate of 49 percent; a mail survey of the American Sociological Association membership by Sprehe (1967) had a response rate of 51 percent with two mailings; a survey of the American Political Science Association membership by Somit and Tanenhaus (1964) obtained a 52 percent response. See Albert Somit and Joseph Tanenhaus, *American Political Science: A Profile of a Discipline* (New York: Atherton Press, 1964); John Timothy Sprehe, "The Climate of Opinion in Sociology: A Study of the Professional Values and Belief Systems of Sociologists" (Ph.D. diss., Washington University, St. Louis, 1967).

33. Stephen Cole and I have examined in a number of papers the determinants of these two dimensions of reputation. See, among others: Cole and Cole, "The Reward Systems of the Social Sciences" (*op. cit.*); Cole and Cole, *Social Stratification in Science* (*op. cit.*); Stephen Cole, "Scientific Reward Systems" (*op. cit.*).

34. For example, if 100 scientists had enough knowledge of a scientist's work to evaluate it, the rated scientist's perceived-quality score would be based upon 100 ratings; however, if only 50 scientists were able to evaluate the contribution, the perceived-quality score would be based only on those 50 ratings.

These indicators of reputational standing have several limitations.[35] First, I am not examining *quality* of contributions per se, but only *perceived quality*.[36] This is a significant distinction. The perceptions of men and women scientists may have been based upon inadequate or false information; these perceptions may have been distorted or dated. The perceived quality of work may have reflected a "halo effect" of the larger university department at which the individual was located, or indeed, it may have been influenced by the sex status of the scientist.

Second, these visibility and perceived-quality scores are gross measurements. I asked for assessments of contributions in general, but did not specify dimensions to these contributions. Furthermore, the measurement taps only the research role of the scientist and equates that role with reputation. I do not examine here the multiple other bases of reputation, such as teaching, organization building, or association with other individuals. Third, the data are the product of a single measuring instrument administered just once. Multiple measurements might have altered the evaluations. Slightly varied wording of the questions might also have produced different results. Fourth, I do not deal with the problem of intersubjectivity. Scientists may have used differing subjective criteria and applied different levels of actual knowledge to reach the same evaluation that an individual "has made very important contributions." There is no way of knowing the relative weights assigned to evaluation criteria by respondents reporting their judgments.[37] These indicators should be seen, then, as first estimates of reputational standing. If it turns out that gender is related to the evaluations, we should be able to determine what accounts for the variations.

THE REPUTATIONS OF AMERICAN SCIENTISTS

In American society, the occupational "stars" have almost always been men. In every sphere of professional activity, men have dominated positions of esteem, trust, and responsibility.[38] Science is no exception. As noted in Chapter 3, men predominate among winners of honorific awards, among holders of high office, and among those few scientists who shape the cognitive structure of their discipline.

To reiterate: As late as 1974 there were only thirteen women members, or just over 1 percent, in the National Academy of Sciences;[39] and these few women reaped their reputational rewards later in life than male members.[40] I have

---

35. On limits to measures of reputational standing such as those used here, see Cole and Lipton, *op. cit.*

36. On the varied correlates of quality (as measured by citation counts) and of perceived quality, see Cole and Cole, *Social Stratification in Science* (*op. cit.*), Chapter 2.

37. The problem of standardized criteria for evaluation plainly requires more research. In fact, consensus on standardized evaluation criteria is a problem faced in many settings, but is particularly significant in the process of refereeing journal articles and in peer review decisions.

38. Two exceptions to this general pattern may be found: One is in the predominantly female profession of nursing and the other is in the entertainment business.

39. Zuckerman and Cole, "Women in American Science" (*op. cit.*).

additional evidence, drawn from the survey, that women scientists in general are not thought of as leading contributors to their field.

Influential Scientists

I asked the respondent scientists, you will recall, to list others in their field and specialty who had made the most important contributions within the past ten years.[41] Respondents were not offered a list of possible candidates; the choices were left entirely to them. Sociologists listed over 4,700 names, psychologists, close to 3,500; biologists, roughly 3,900. Of these, some scientists were named frequently; most were not.[42]

The results are extraordinary and consistent for the three fields. Women's names appeared rarely: Only 3 percent of the sociologists and psychologists and 2 percent of the biologists selected were women. When we remember that roughly a fifth to a quarter of those entering these fields are women scientists, these figures suggest the infrequency with which women are held in the highest esteem by their peers.

These data can also be examined in terms of the proportion of male and female respondents in each field who mentioned a woman scientist as a leading contributor. Since the raters were randomly selected, the absolute number of females responding to the questionnaire was low.[43] But the pattern of naming women does differ among men and women respondents in the three disciplines, even though the results, which are based upon small numbers, must be cautiously viewed. Seventeen percent of the male sociologists named at least one woman; but 46 percent of the women named one of their own sex. Virtually the identical pattern obtains among responding psychologists: 17 percent of the males and 50 percent of the females named a woman as an important contributor. Among biologists, however, men and women responded similarly: 14 percent of the males and 13 percent of the females mentioned another woman scientist.

Types of Scientific Reputations

Turning from the lists of most influential contributors, we now examine four ideal types of scientific reputations that may be described in terms of a scientist's

40. Zuckerman, *Scientific Elite* (*op. cit.*).

41. The exact wordings of the questions were: "Which five scientists have, in your opinion, made the most important contributions to your field as a whole in the last 10 years?" "Which three scientists have, in your opinion, contributed the most to the development of your specialty in the last 10 years?" The discussion in the text refers only to answers to the first of these questions.

42. We are studying the structure of these mentions in terms of the formation and maintenance of cognitive consensus in different scientific disciplines and specialties. The first report based on such data appear in Stephen Cole, Jonathan R. Cole, and Lorraine Dietrich, "Measuring the Cognitive State of Scientific Disciplines," in *The Metric of Science*, ed. Yehuda Elkana, Joshua Lederberg, Robert K. Merton, Arnold Thackray, and Harriet Zuckerman (New York: Wiley Interscience, 1977).

43. Although it was not possible to obtain a precise estimate of the number of female respondents, since some scientists chose not to place their names on the returned questionnaires, of those whose names were available approximately 12 percent of the respondents were female.

visibility and perceived quality of work.[44] (See the schematic presentation on p. 105.) The perspective shifts from concern with the sex distribution of elites to the distribution among reputational types of a larger sample of men and women scientists located throughout the scientific community. Type 1 is the *prominent* scientists, in the dual sense of having his or her work widely known and highly regarded within the discipline. Scientists in this category stand out above their peers and achieve distinguished reputations; they are more likely than others to be heard and heeded when they publish scientific papers or speak out on science policy.[45] At the other extreme, Type 4 is the *invisible* scientist, whose work is neither widely known nor highly regarded. Although scientists in this group empirically make up the majority of those active in any field, they have little reputational standing beyond their local environments.

The remaining two types point to the imperfect association between how well a scientist's work is known and how well it is regarded ($r = .68$). Type 2, the *notorious* scientist, has become widely known but is not highly regarded by his peers for his work. This is the scientist who is conspicuous, who is evident; he is well known by the general public on account of his work, but is not generally approved of or admired by other scientists. We are not here using notoriety in the sense of "notoriously evil, wicked, or vile; [or] held in infamy or public disgrace."[46] This type is essentially the mass-producing scientist who by sheer bulk of output and a keen eye for a public has succeeded in becoming well known.[47] Within this group may also fall, of course, some scientists who have become visible for activities other than their research, who have made valuable organizational or institution-building contributions, but whose research has failed to spark excitement. Finally, there is Type 3, the *esteemed* scientist, whose work is regarded as important among those who know it, but who either has chosen not to seek widespread renown or has yet to gain it. Scientists in this category are held in high regard by some sector of their scientific community. Among the Type 3 scientists we are likely to find a significant number of younger able scientists whose reputations are outstanding to members of their own special research area, but who have yet to permeate specialty boundaries. These are individuals who in time may become prominent. Indeed, Type 3 scientists remind us that movement from one type to another is continually occurring within the scientific community. In fact, the selection of these specific identifying markers for the four types is intended to signal that in the extreme the prominent scientist may become

---

44. Elsewhere Stephen Cole and I have developed a typology of research performance, which describes the relationship between the quantity and quality of scientific output. The four types generated there were the prolific; the perfectionist; the mass producer; and the silent scientist. As is noted below, there is a link between the two typologies. For an extended discussion of the four types of producers, see Cole and Cole, *Social Stratification in Science* (*op. cit.*), pp. 92 ff.

45. On the etymology and current meanings of "prominence," see the *Oxford English Dictionary*.

46. On the various uses of the term "notorious," see the *Oxford English Dictionary*.

47. In fact, the mass-producing scientist does achieve a level of reputational recognition in science beyond that which is merited by the quality of his or her work. In previous articles, Stephen Cole and I show that quantity of output as measured by citations has some independent influence on scientific visibility.

notorious, that the lines drawn between these types are frequently blurred in reality, and that these dimensions of reputational standing should be viewed as a continuum along which there is constant flux.

Four Types of Scientific Reputations Based on Perceived Quality and Visibility of Research Contributions

		Perceived Quality	
		High	Low
	High	Type 1	Type 2
		Prominent	Notorious
Visibility			
	Low	Type 3	Type 4
		Esteemed	Invisible

I have identified scientists in the three disciplines of the present sample in terms of these four reputational types. The division into high and low categories is accomplished here by dichotomizing the two dimensions of reputation at the mean of the standardized variables.[48] Men and women scientists are differentially distributed among the four types. At one extreme, 38 percent of the men but only 19 percent of the women are, by our statistical definition, "prominent." If truly exacting criteria were used for high visibility and perceived quality, the proportion of prominent scientists would be significantly lower. At the other extreme, 41 percent of the entire sample is "invisible," but the distribution among men and women differs. Among men, 37 percent are Type 4 scientists; among women, 56 percent. The remaining women are equally divided among Type 2 and Type 3 (13 percent in each); the remaining men are twice as apt to be among the esteemed as among the notorious scientists (17 compared to 8 percent).[49] Reference to these four ideal types will be made throughout this chapter and I will return to an explicit discussion of them below, in order to suggest how men and women with these reputations are distributed in the social system of science.

Whatever the explanation for these patterns, they suggest that women are not often perceived among their peers to have made truly important contributions, or

---

48. A few words are required about the use of standardized variables. Throughout this book, I have employed Z-score transformations to standardize variables to a mean of zero and a standard deviation of one. This is required when we deal with different scientific fields which have significantly different means and variances, and we wish to combine the fields into a single group. In order to avoid distortions resulting from these different means, we transform the variables into a common metric with similar parameters. Since the mean visibility and perceived-quality scores of sociologists, psychologists, and biologists differ, I have used standardized variables in the analysis, except where noted.

49. These four types are only roughly approximated by the data used in this study. If I had had a very large sample of men and women scientists, I would have been able to set more appropriate restrictions on the definitions of high perceived quality and visibility. Rather than selecting the mean value as the dividing point, which is, say, about 35 percent of the community for visibility, I would have used, say, 60 or 70 percent as the demarcation line between high and low. Indeed, a larger sample would have made it possible to identify a wider variety of types of reputational standing, which could not be done here for lack of a larger sample of female scientists.

to be among the most prominent and esteemed scientists in their field. This is the case for the apex of the stratification system of science. So often we do not go beyond analyzing the elite. But reputations, of course, are held by everyone; they lie along a continuum, with most scientists' reputations clustering toward the end that marks the "invisible" scientist. My aim is to probe beyond the scientific elite to see whether women throughout the social system of science are less visible and less esteemed than men in comparable locations.

THE VISIBILITY AND PERCEIVED QUALITY OF SCIENTIFIC WORK

I begin by considering the distribution of men and women scientists on the two dimensions of reputation. Respondents could judge the work of each scientist in terms of its perceived importance; the four available categories of evaluation ranged from one indicating a scientist had made a very important contribution (scored 6) to one indicating he or she had made a relatively unimportant contribution (scored 3). While the pattern of scores among fields does not vary greatly, there are significant differences between the sexes (see Table A-1).[50] As might be expected, the work by male scientists is consistently perceived to be of higher quality than is the work by females. In sociology, for example, the work of women scientists was rated, on average, 4.05 with a standard deviation of .47; men had average scores of 4.30, with a somewhat larger standard deviation. These are statistically significant differences. This pattern is repeated in the other two fields: Men have better reputations, according to this indicator, and there is greater variation among the reputations of men than of women.[51] While in several cases the perceived-quality scores of men approach the scale's upper bound, this is never the case among the women. The maximum score for women is 5.32 in psychology, 5.12 in sociology, and 5.11 in the biological specialties. These distribution data are further supported by a significant zero-order correlation of $-.18$ between sex status and perceived-quality scores for the entire sample of 582 scientists.[52] When the sample is subdivided by field, however, some variations emerge: $r = -.21$ in sociology; $-.22$ in psychology; $-.11$ in the biological specialties.

The work of male scientists, then, is in general held in higher esteem than that of women, and, it turns out, is also more widely known throughout the community. Treating the sample as a single group, the mean visibility of men is approximately 38, with a standard deviation of 26, compared to a mean of 28 and

50. Table A-1 refers to the first table in the Appendix. All tables in the Appendix will be identified as A-1, A-2, A-3, etc.

51. The standard errors of these estimates are quite low, .05 for the men and .07 for the women. Thus differences of more than twice these estimates may be accepted by statistical convention as representing statistically significant differences.

52. Sex status was coded: 1 = male; 2 = female.

a standard deviation of 24 for women (see Table A–2 for distribution of visibility scores). The pattern for the male scientists closely parallels that found in my earlier study of physicists.[53] The most significant differences in the visibility of men and women appear at the extremes of the distribution: 29 percent of the men but fully half of the women were known to less than one-fifth of the scientists who judged their work. Correlatively, 20 percent of the men had produced research known to at least 60 percent of the evaluators, whereas only 10 percent of the women had produced work that was equally visible. However, the variability of visibility scores is higher for women than for men.[54]

These distributions are summarized by the zero-order correlation between sex status and visibility for the entire sample, $r = -.19$. The association between scope of reputation and gender is stronger in the social sciences ($r = -.25$ and $-.22$ for sociology and psychology, respectively) than in the biological sciences ($r = -.08$) (see Table A–3). The data suggest that women scientists are less widely known and their work is less favorably evaluated by the community of scientists in the three fields investigated here. But whether these differences are ultimately attributable to gender cannot be determined until we begin to match the status-sets of the scientists. What appear at first to be sex-based differences in reputation may actually reflect other systematic differences in the status-sets of men and women, which are correlated with gender.

In order to examine the relative contributions of different social statuses to reputations, simple paired comparisons of several scientific statuses can be made. The primary focus is on comparisons between sex status and six other statuses: rank of current department, prestige rank of doctoral department, eminence (number of honorific awards), academic rank, chronological age, and quality and quantity of scientific research performance. These paired comparisons are followed by an estimate of the relative influence of each of these statuses in the larger status-sets of scientists.

---

53. For example, 29 percent of this sample as compared to 35 percent of the 120 physicists had visibility scores ranging from 0 to 19 percent. The only significant difference in the distributions is found at the upper level of visibility, where the sample of physicists tended to be better known. This was a result of a sampling decision to include a high proportion of eminent physicists. In short, the distribution of visibility is quite similar in the two studies.

54. The coefficient of variation is higher for female than for male scientists (see Table A–2). The other meaningful difference in visibility scores can be found among the various fields. Sociology and psychology have significantly higher average visibility scores than do the biological sciences. There are several possible interpretations of this result. First, the boundaries between the specialties in the biological sciences may be thicker than in the social sciences. It remains possible for a sociologist primarily interested in studies of juvenile delinquency to read and comprehend papers in the sociology of science. Indeed, the major journals of the field are eclectic in their specialty representation. Although it may favor quantitative over qualitative work, any issue of the *American Sociological Review* is likely to contain a wide range of subject matter representing work in many specialties. Consequently, reputations, at least insofar as they are captured by simple visibility measures, are apt to permeate the boundaries of specialties and of special research areas within specialties. Second, the two social sciences that I am studying operate on a much smaller scale than do the biological sciences. This size differential is reflected in organizational structure. Sociologists and psychologists tend to be located in single departments; the annual meetings include representatives from all of the specialties. Not so in the biological sciences. Perhaps more than any of the other sciences, the life sciences have the most peculiar organizational history. It is not uncommon to find some departments of biological science housed in the medical school of a university, while other departments are located in the arts and sciences. The structural barriers to interaction, even among individuals working on related matters, may well make for reduced interaction and visibility of work produced by other biological scientists.

Rank of Current Department: A social system dominated by achievement principles will reward excellence wherever it is found. Good scientific work, whether produced at Stanford or at the University of South Carolina, should be recognized and rewarded in equal measure.[55] Of course, this ideal is at best roughly approximated; good work produced at lesser universities and colleges is, in fact, less apt to become widely known than work of similar quality produced at a distinguished university or laboratory.[56] In the study of 120 academic physicists, structural location influenced significantly not only the diffusion of scientific ideas, but also the visibility of physicists among their colleagues. Thus, even after controlling for the quality of their work, the reputations of two physicists—one working at Stanford, for example, and the other at a less prestigious university—differed significantly. In fact, for physicists structural location had an impact on visibility roughly equal to the quality of scientific work.[57] That is, the study indicated that scientists tend to evaluate the "same" work—or work that is roughly equal in quality—more highly if it is produced by a member of a distinguished department.

The foregoing analysis can be extended by determining the relative effects on reputations of sex status and rank of current department, and thus whether the advantages and disadvantages of location accrue equally to men and women.

For the 582 scientists, rank of department makes a significant difference in reputational standing. Men and women who are located at elite schools are more likely to be known ($r = .30$) and highly regarded ($r = .29$) than scientists at less distinguished places. Although the association reported here is somewhat lower than that found among the 120 physicists, it remains a significant predictor, independently of other social statuses occupied by the scientists. Consider its influence relative to sex status. The regression of perceived quality on gender and rank of department estimates their relative predictive influence. The partial regression coefficients in standard form for both statuses are significant ($b*_{pqr.s} = .28$; $b*_{pqs.r} = -.16$), although rank of department is the stronger predictor of perceived quality.[58] However, the regression suggests that even at similarly ranked science departments, women are less apt than men to be prominent or esteemed. This pattern is equally apparent when the focus shifts to visibility or scope of scientific reputations (see Table A–4, paired comparison 1).[59] These

55. For further discussion of the influence of current affiliation on achievement, see Chapter 5. On the basic scientific normative structure, see Merton, *The Sociology of Science* (*op. cit.*), pp. 228–280; Ian Mitroff, *The Subjective Side of Science* (New York: American Elsevier Publishing Co., 1975).

56. See Cole and Cole, *Social Stratification in Science* (*op. cit.*), pp. 96–105. In this chapter I use the Roose-Andersen ratings of arts and sciences departments as the indicator of quality. The ratings have been standardized, and the results presented are based upon the computed standardized variable.

57. Cole and Cole, *Social Stratification in Science* (*op. cit.*), pp. 96–105.

58. The following subscript notation is used throughout this chapter: pq = perceived quality; v = visibility; r = rank of current department; s = sex status; a = eminence as measured by the number of honorific awards listed after a scientist's name in *AMWS* (including postdoctoral fellowships); d = rank of doctoral-granting department; g = chronological age; q = quality of scientific research as measured by citation counts; p = rate of production of scientific research papers per year since the doctorate; o = index of scientific research performance; b = academic rank.

59. The effects of rank of department on reputational standing are strongest in sociology ($r_{pq.r} = .44$, $r_{vr} = .38$) and weakest in the biological sciences ($r = .17$ and .26, respectively).

data corroborate the earlier findings, but specify them by pointing to the different effects for men and women. In short, benefits of a superior structural location accrue to both sexes, but at every level in the social structure men benefit significantly more than do women. Of course, we have yet to examine the influence of performance variables on these relationships.

Referring, as I have been, to a standardized metric makes it somewhat difficult to clearly assess the real "costs" and benefits of being a woman at, say, Harvard rather than at a less distinguished university. If we examine the unstandardized metric of the regression equation, however, the intercept gives the estimated visibility if values on the independent variables are zero, and the B scores are the simple estimate of the effect on visibility that results from a change in one unit of an independent variable while holding constant the other variables in the equation. For our problem, every 1-point change in visibility, up or down, represents a loss or gain in awareness among the scientific community of 1 percent. Thus, in concrete terms, what reputational price do scientists pay for being female, after controlling for structural location?

First, if a scientist is at a distinguished department rather than a lesser one, it significantly increases visibility, regardless of gender. Prominent scientists, whether male or female, are more likely to be located at strong departments, and correlatively, part of their prominence may be attributable to their location. Second, and for immediate purposes more important, even when men and women are located in exactly the same setting, being female means that a scientist will be known, on average, to 8 percent fewer members of her discipline. In essence, that is the cost of sex status to visibility, if only a scientist's sex and structural location are known.[60] In each of the paired comparisons, the B weights represent the net loss or gain in visibility that results from being female. Again, final inferences about these costs must await my analysis of the effect of quality of role performance.

Rank of Doctoral Department: Sociologists of science have repeatedly demonstrated the influence of sponsorship and educational training on the mobility patterns of scientists.[61] A scientist's educational background has an important impact on the quality of his or her first job. It represents the initial phase in the process of accumulating advantages that aids students who have attended the best graduate departments and have studied under the established scientific elite. What is the relative influence of sex status and rank of doctoral department in determining reputations? The men and women in this sample received their training in comparably ranked institutions; no correlation exists between rank of doctoral department and sex status.[62] Therefore, the associations between the two

60. Untransformed variables are used for the analysis of unstandardized regression coefficients so that changes in percentages may be related to the mean visibility scores presented above.

61. See Diana Crane, "Scientists at Major and Minor Universities," *American Sociological Review* 30 (Oct. 1965): 699–714.

62. For this analysis, the rank of doctoral department, as measured by the Roose-Andersen study, was transformed into a standardized variable.

TABLE 4-1.   The Relationship between Sex Status and the Quality and Quantity of Scientific Output*

FIELD	SCIENTIFIC OUTPUT		INDEX OF SCIENTIFIC PERFORMANCE
	Quality	*Quantity*	
Entire Sample	−.14	−.24	−.23
Sociology	−.13	−.27	−.23
Psychology	−.16	−.30	−.29
Biological Sciences	−.14	−.18	−.18

*Correlations for the entire sample are based on standardized variables; those of individual fields are based on unstandardized variables.

Indicators of Output:

    Quality—Total number of citations received by scientist in *SCI* for 1961, 1965, 1967, 1970, 1972. For sociologists the years 1972–74 were used. For the entire sample the data are standardized by discipline.

    Quantity—The average number of papers per year published by a scientist since receiving his or her Ph.D. Data for the entire sample are standardized by discipline.

    Index of Scientific Performance—After standardizing the quantity and quality indicators, the index of scientific research performance is the simple addition of the two standardized variables. The index gives equal weight to both quality and quantity of output.

statuses and indicators of reputation tell the entire story: Sex status is a slightly better predictor of perceived quality of work ($b^*_{pqs.d} = -.19$; $b^*_{pqd.s} = .13$) and is of equal importance in determining the visibility of work ($b^*_{vs.d} = -.19$; $b^*_{vd.s} = .21$). These findings parallel those obtained for rank of current department. All scientists, regardless of gender, benefit by attending superior graduate schools, but beyond the influence of source of training, sex differences make for additional differences in reputations.

Again, think in terms of the cost of being a woman scientist. If a man and a woman are trained at identically ranked institutions, one can predict that the woman will be known to 9 percent fewer of the colleagues in her discipline than will the man. Although it is more difficult to intuitively comprehend, the same cost is apparent when predicting quality assessments (see Table 4-1). A pattern apparently is beginning to emerge, then, of substantial reputational costs of being a female scientist. But once more, each of these relationships must ultimately be examined in the light of differentials in performance.

Eminence (Number of Honorific Awards):   Honorific awards are given in science for high-quality performance.[63] Scientists who have been honored are apt

63. For the correlations between number of honorific awards and quality of research performance among physicists, see Cole and Cole, *Social Stratification in Science* (*op. cit.*), pp. 93–99; for correlations for five other scientific fields, see Stephen Cole, "Scientific Reward Systems" (*op. cit.*). In recent work Stephen Cole and I have found the correlation between the number of awards received by a scientist (as listed in *AMWS*) and the quality of research performance to be weaker than it was for the 120 physicists. This may be the result of different types of samples.

to have better reputations than those who have not, and this indeed is the case $(r_{pqa} = .27 ; r_{va} = .28)$.[64] Comparatively, formal recognition is a stronger determinant of reputation than sex status. The regression of perceived quality on sex status and eminence produced significant partial regression coefficients for eminence $(b^*_{pqa.s} = .26; b^*_{va.s} = .27)$.[65] (See Table A-4.3.) However, since sex status is virtually unrelated to eminence as measured by number of awards $(r_{as} = -.07)$, gender continues to influence reputational standing independently of the extent of formal recognition.[66]

The cost of being a woman continues to be notable: If the number of honorific awards remains constant, women are known to 8.4 percent fewer members of their discipline, and there is a similarly significant cost in terms of prominence and esteem. These data apparently further illustrate the impediments to high reputational standing facing women simply because of gender. They reinforce the emerging double-edged pattern: On the one hand, as expected, both men and women benefit by their social location whether it be in terms of departmental affiliation or number of awards; on the other, when rank or other statuses are accounted for, men have significantly better reputational scores than do women.

Academic Rank: The data presented in Chapter 3 indicate that women are disproportionately found among the lower-ranked positions in science, even after considering differences in labor force participation and scientific research performance. In this sample, the rank differentials are only modest $(r_{bs} = -.14)$, in part because the men and women scientists were stratified by professional age before a sample was selected. Academic rank is, of course, related to both visibility $(r_{vb} = .35)$ and perceived quality of work $(r_{pqb} = .30)$.[67] The regression of these two indicators of reputational standing on sex status and academic rank suggests again that both statuses contribute to prominence and esteem. High-ranked professors, regardless of their gender, are more likely than those of lower rank to be known and well thought of in their field, but among scientists in any particular rank men continue to have reputational advantages (see Table A-4.4). Specifically, the net effect of gender on perceived quality $(b^*_{pqs.b} =$

There was greater variation among physicists in the number of awards they received than among the samples of social scientists and biologists. In the study of physicists we also had data on the prestige of a set of ninety-seven awards given to physicists. For that example the correlation between prestige of highest award and number of awards was strongly associated, $r = .78$. We concluded that a rough but adequate indicator of eminence could be obtained by simply counting the number of awards a scientist had received. That conclusion should be examined in further detail for the social and biological sciences. Although I continue to use the total number of honorific awards and postdoctoral fellowships as an indicator of eminence, additional research is required to find out how adequate such counts are in dealing with the social and biological sciences.

64. These correlations do not vary greatly among the different fields examined here.

65. Again, there are only slight variations among the social and biological sciences.

66. Note the similarity in the correlations between gender and number of awards for this sample and for the sample of 1957–58 doctorates discussed in Chapter 3, $r = -.05$. However, note also that Reskin found women more apt to receive postdoctoral fellowships than men. Of course, here I include in the measure of eminence all awards received by the scientists. Barbara F. Reskin, "Sex Differences in Status Attainment in Science," *American Sociological Review* 41, no. 4 (Aug. 1976): 597–612.

67. Four categories of academic rank were used: instructor, assistant professor, associate professor, full professor. The ranks were obtained from *AMWS* and professional-association directories.

$-.14$) and on visibility ($b^*_{vs.b} = -.15$) is roughly equal to the first-order associations. Comparatively, however, academic rank remains the stronger influence on reputation ($b^*_{pqb.s} = .28$; $b^*_{vb.s} = .33$).

The cost pattern is further exemplified in this paired comparison. The predicted visibility of females is roughly 6 percent lower than that of males, and their perceived quality is significantly lower after taking into account the prediction of reputations resulting from rank differentials.[68] The paired status data suggest that women are treated differentially from men in terms of peer appraisals, even after accounting for other social statuses.

Chronological Age: Although older scientists should be more prominent, notorious, or esteemed by simply having been around longer, it turns out that there is no correlation between age and perceived quality ($r_{pqg} = .01$), and indeed only a modest one between it and visibility ($r_{vg} = .17$). Apparently older scientists are somewhat more widely known but not better thought of than younger ones. These correlations mask a curvilinear relationship, however, and therefore are misleading. To examine the nonlinearity, I divided the sample into five age groups. The results, reported in Table 4–2, indicate that the perceived quality of scientific work is lowest among scientists less than 35 years old, then rises steadily and peaks between 45 and 54, only to begin a decline after 55. In fact, the perceived quality of work among scientists over 65 is roughly the same as that for scientists in their late thirties and early forties. The visibility of scientists follows the same curvilinear pattern. Note, however, how differences in perceived quality and visibility scores peak at varying ages. There is a temporal lag between the peakings of perceived quality and of visibility: the peak score for perceived quality precedes the highest average visibility score by roughly ten years. If we treat these data as a synthetic cohort and assume that these patterns are not tapping actual differences in the quality of work produced by the several age cohorts, the data suggest that at least for the scientific community there is no "ratchet effect" in building reputations.[69] Although the dictum "once a Nobel laureate, always a Nobel laureate" may hold at the apex of the stratification system, below it there is an apparent loss of reputational standing after the middle years.[70]

---

68. Since the zero-order associations in the various fields are similar, the regressions also turn out to have similar effects on the association between sex status and reputational standing. Consider two examples. Within sociology the regression of perceived quality on the two statuses produced partial regression coefficients of $-.13$ for sex status and $.29$ for academic rank; within the biological sciences the comparable coefficients were $-.07$ and $.30$. The same pattern obtains when visibility is the dependent variable. In short, within all fields the initial relationships are not affected appreciably.

69. The ratchet effect suggests that when a person has reached a particular level of eminence, he or she can never fall below that level, but may rise above it. In fact, this remains an open empirical question in respect to scientists: Do older Nobelists who have not gone on to produce significant science following their prize receive the same deference and command the same esteem as those who have continued to "justify" their award? Or are they actually held in less regard than before their awards were won? For a more extended discussion of the ratchet effect in science, see Merton, "The Matthew Effect in Science" (*op. cit.*), pp. 442 ff.; Zuckerman, *Scientific Elite* (*op. cit.*).

70. In the study of 120 physicists, Stephen Cole and I found a curvilinear relationship between age and visibility

TABLE 4–2.   Mean Perceived Quality and Visibility Scores for Men and Women Scientists

1. Entire Sample of 582 Scientists:

AGE	PERCEIVED QUALITY	VISIBILITY	N
65 and over	4.22	32.6	26
55–64	4.39	42.4	113
45–54	4.44	39.4	211
35–44	4.25	28.9	191
Less than 35	4.09	27.2	38

2. Male and Female Scientists Separately:

AGE	PERCEIVED QUALITY		VISIBILITY	
	Men	*Women*	*Men*	*Women*
65 and over	4.29 (13)	4.15 (13)	46.3	18.9
55–64	4.49 (93)	3.95 (20)	45.7	27.2
45–54	4.51 (164)	4.19 (46)	40.7	34.8
35–44	4.27 (144)	4.19 (47)	30.5	23.9
Less than 35	4.04 (32)	4.31 (6)	25.6	35.7
Total	4.39 (446)	4.16 (132)	37.5	28.2

NOTE: The total number of cases in this table is 578; age information was not available for the four other scientists.

So much for age and reputation standing for all 582 scientists. Examining the same relationship for men and women separately reveals a number of significant differences (see Table 4–2.1). First, among men the curvilinear pattern between age and perceived quality is well defined, but among women there is no discernible pattern whatsoever. The perceived-quality and visibility scores for successive age groups fluctuate seemingly randomly. Although we must unfortunately rely on a limited number of cases, there is an interesting aspect to the group of scientists under 35 years old. The men have the lowest perceived quality relative to other ages; the women have the highest. Moreover, among these youngest of scientists, the general pattern of appraisal of men and women is reversed; the work produced by women is regarded more highly than that of men. When the sample size is increased to include all scientists under 45, there is no difference in the reputational scores of men and women.[71]

How might this relationship be plausibly interpreted? Assume that only part of the pattern is explained by the aging process itself. The data suggest a cohort

(*Social Stratification in Science* [*op. cit.*], pp. 107–109). Physicists over 65 had somewhat lower mean visibility scores than those who were younger. However, when we subdivided the sample by the quality of research produced by the scientist, there was some specification of the result: For those scientists who produced high-quality work there was little difference in their visibility at different ages; for those who produced less distinguished work there was a sharp decline in visibility after their late fifties.

71. The mean perceived-quality score for men in this group was 4.23; for women it was 4.20.

effect, perhaps the result of social pressure from women's organizations and from the application of affirmative-action policies. The increased proportions of females entering the professions in general, and science in particular, may mean that the newest cohort of young women scientists has better opportunities to demonstrate its talent.[72] Today women scientists who publish have a better chance of obtaining jobs at quality universities and positions of high rank—in short, structural positions that increase the probability of peer recognition. Furthermore, young women may be comparatively more competent than older women in their fields. It is possible, although it must remain conjecture, that reputations in a discipline are in some measure sex-specific; that is, the work of a woman is judged relative to that of other women, but not to men. Another, and no less significant, feature of this cohort effect is the changing response by the scientific community, largely male, to work produced by young women. The criteria used by men to judge work produced by women are probably changing, and the benefits of this change are being felt mostly among the younger women who are currently publishing and trying to make their mark.

Consider three additional observations about the data presented in Table 4–2. First, with the important exception of the youngest age group, the difference between the visibility of men and that of women increases as the scientists get older, until the average visibility of males over 65 is more than twice that of females in the same age cohort.

Second, the general pattern of visibility of women fluctuates curiously: It is highest among the youngest cohort, declines for the cohort from 35 to 45, peaks again in middle age, and sharply declines again after 55. The vicissitudes of visibility parallel in a remarkable way, I suggest, the typical career pattern of female academics. In the early part of their scientific careers, as we have seen in Chapter 3, they are almost as productive as their male peers. But within the first decade after receiving their degrees, a significant proportion of women Ph.D.'s, according to data reported by Helen Astin, interrupt their careers for an average of roughly one year, typically for bearing and raising children.[73] This interruption typically comes in the middle or late thirties. It affects these women's chances for promotion, cuts down their productivity markedly, creates some measure of obsolescence of skills and knowledge, reduces their salaries upon reentry into the labor force,[74] and apparently may negatively affect their reputational standing. The drop in average visibility scores, from 35.7 among women under 35, to 23.9 for those between 35 and 44, suggests a possible further negative consequence of interrupted careers.[75] After they return to the academic commu-

---

72. On age, period, and cohort effects, see the definitive work by Matilda White Riley, Marilyn Johnson, and Anne Foner, *Aging and Society,* 3 vols. (New York: Russell Sage Foundation, 1968, 1969, 1972).

73. On career interruptions, see the discussion in Chapter 2 and in Astin, *op. cit.,* p. 65.

74. An interesting discussion of the economic consequences of female career interruptions appears in Solomon W. Polachek, "Discontinuous Labor Force Participation and Its Effect on Women's Market Earnings," in Lloyd, *op. cit.,* pp. 90–124.

75. Of course, to test this hypothesis we would need data on the career interruptions of women scientists enabling comparisons of the visibility of those who do and those do not interrupt their careers.

nity, women's reputational scores improve markedly. If these later years are not the prolific ones, they are frequently, especially among social scientists, the age at which scientists produce their most mature work. Astin also points out that "women are much more likely than men to have retired by the age of 60–65."[76] The pattern of retirement would not affect this sample, all of whom are currently members of science departments, but if earlier average retirement is preceded by a decline in scientific activity in the late fifties, then the data on visibility of women over 65 might be a reflection of this pattern.

Third, the earlier observation about the ratchet effect must now be modified in the light of comparisons between men and women. The visibility of men consistently rises throughout their career, and there is, in fact, no drop in visibility to accompany the fall in perceived quality. The decline observed for the sample treated as a single unit is the result of the sharp decline in the visibility of women in the older cohort.[77]

To reiterate the general conclusions from the paired comparisons made thus far: First, sex status has a significant independent influence on the prominence, esteem, and notoriety of scientists; second, although gender makes a difference, the other statuses in the status-set are either roughly equal or stronger predictors of reputational standing.

If analysis were to stop here we might conclude that sex status is one among a set of significant determinants of scientific reputations. But one important paired comparison remains to be examined. Although quality and quantity of research performance is more a social attribute than a social status, in science the type of research a person produces takes on the character of a social status with its attendant rights and obligations, and so we shall treat it as one.[78]

As I have said repeatedly, when science is operating in harmony with its universalistic ideal, the quality and the quantity of scientific research are the most important determinants of all forms of recognition—including, of course, reputational standing. In accordance with this expectation, I hypothesized that the scope and quality of reputations is predominantly a function of one basic scientific status: research performance. By comparing the pair of statuses, gender and research performance, we may make a first test of this hypothesis.

Quality and Quantity of Research Performance: Data reported in Chapter 3 indicate that men publish more scientific papers and that their output is more

---

76. Astin, *op. cit.*, p. 65.

77. As noted in footnote 70, the pattern obtained here is somewhat at variance with that obtained for the physicists. Here among the males there is no reduction in visibility, and the curvilinear pattern originally observed is wholly a result of the significant downward turn in the visibility of older female scientists. Actually the results are not significantly different between fields. When I control for quality of research, there is a slight downturn in the mean visibility score of male scientists who have not produced outstanding research. This is offset by higher visibility scores among the older men who have produced exceptional research. These patterns are consistent in the several fields.

78. The distinction between social status and social attribute remains ambiguous in the literature. For a discussion of some related problems of definitions, see Merton, *Social Theory and Social Structure* (*op. cit.*).

frequently cited.[79] If the 582 men and women in my sample display a similar unequal pattern of productivity, the differences in their reputations may simply be a function of their research output. To test this, data on the production of papers were collected for each scientist, either from the time each one obtained the doctorate or, in the case of older scientists, as far back as 1950. For psychologists and biologists, these data came from *Abstracts*; for sociologists, from curricula vitae.[80] Thus, I had publication information for all the scientists except for 17 percent of the sociologists.[81] I constructed measures of annual and total productivity as well as productivity rates.

The quality of scientific output was measured by a count of citations to the work of each scientist.[82] For the biological sciences, which presented no special difficulties, I collected citation data for five separate years: 1961, 1965, 1968, 1970, and 1972; data covering the same years were collected for psychologists.[83] Several problems were encountered in collecting data for the psychologists and sociologists. Until 1970 the *Science Citation Index* (SCI) did not abstract many social science journals; there were some psychology journals in its file, but few sociology journals. The publications in 1973 of the *Social Science Citation Index* (SSCI) dramatically changed this situation, and today we have available an extensive file of abstracted social science journals.[84] Nonetheless, it remains difficult to obtain adequate measures of quality for the 1960s. Citation data were so scanty in the 1960s for sociology that I collected them only for the years 1972–74.[85] When the 582 scientists are discussed as a single group, the total number of citations to an individual's work in the *SCI* or *SSCI*, standardized within each field, is used as the indicator of quality of research output.

---

79. See Table 3-3, page 64.

80. All of the publication data for the psychologists were obtained from *Psychological Abstracts;* for the biologists, from *Biological Abstracts*.

81. Publication data for sociologists were obtained from vitae sent to me by the sampled sociologists. Each sociologist was sent a letter outlining the general purposes of the study (although no mention was made of the gender comparisons) and requesting a full, up-to-date vita. A postcard follow-up was used to obtain vita from nonrespondents; telephone calls were made to those not responding to the second inquiry. The final response rate was 83 percent.

82. There is now a substantial literature dealing with citations as indicators of quality of research output. For a summary of the validity of this method and problems associated with its use, see Cole and Cole, *Social Stratification in Science (op. cit.)*, Chapter 2. It should be noted, however, that while for large samples of scientists citations may be a good rough indicator of the relative quality of research, as assessed by peers, there has been a tendency to reify the measure. This is wholly inappropriate, given the current validation of the counts as fine measures of research quality. For example, to the extent that citation counts are used to compare two individuals who are applying for a single job or are up for one position of higher rank, it is being misused. The claims that may be made for the use of citations as rough indicators of quality can only be made for their use in fairly large samples, and not for comparisons between individuals. In this chapter, as elsewhere in the book, I have used a straight count of citations that omits all self-citations.

83. These data were obtained from the *Science Citation Index* published by the Institute for Scientific Information, Philadelphia.

84. In 1974 the *SSCI* abstracted a total of 1,278 social science–related journals. The average number of citations per cited author was 3.40; the mean number per cited item was 1.35. The index covers all of the social science disciplines. Since these data were collected the Institute for Scientific Information has begun to produce back editions of the *SSCI*. Volumes now go back to 1969.

85. The citation counts have, of course, been standardized within each field and aggregated for the entire sample. This eliminates the patterned differences in citation practices between the several fields.

This sample of scientists conforms to the general scientific productivity pattern described in Chapter 3: Sex status is correlated with the production and reception of scientific papers. For the 582 scientists the zero-order correlation between sex status and the quantity of output is $-.24$, insignificantly lower than the $-.30$ reported in the previous chapter for the sample of 1957–58 Ph.D.'s. The correlation is somewhat lower in the biological sciences ($r = -.18$) than in either sociology ($r = -.27$) or psychology ($r = -.30$). The association between gender and quality of work is $-.14$ for the entire sample, with little variability among fields.

Let us turn to the third critical set of zero-order associations: between indicators of scientific output and of reputational standing. A moderate correlation does obtain between the quantity of scientific output and perceived quality. Yearly average production of scientific papers and perceived quality are correlated .45 when all fields are treated as a single sample.[86] There is virtually no difference in the relationship between scientific productivity and perceived quality for psychologists and biologists ($r = .38$ and .39, respectively); for sociologists the association is stronger ($r = .71$), in part because of my method of computing sociologists' productivity scores.[87] The correlation between productivity rates and assessments of work is stable over time. Among biologists, for example, the association for 1955–59 is .36; it is .33 for 1970–73. Similar consistency obtains for the two other disciplines. The correlation for the total number of years that the scientist has been active in the field and his or her perceived quality is somewhat higher than the associations for any particular five-year period ($r = .56$).[88] These first-order associations are statistically highly significant.

The association between productivity rates and visibility is .40, ranging from .56 for sociologists to .37 for biologists.[89]

The correlation between quality of research as measured by citations and

---

86. The total sample here is 544 scientists; it covers the years 1950–73. Sociologists who did not supply their vitae are not included.

87. Since productivity data for the sociologists were based on vitae, books were included in the counts and were given weights of 15; articles published in the four leading sociological journals were given weights of 5; other scientific articles were weighted as 4; edited books were weighted as 4; nonscientific papers were weighted as 1. There is, of course, a conflation here between sheer productivity and quality. This affects the correlation between quality and quantity in sociology, but should not significantly alter the relationship between productivity and the two indicators of reputation. Furthermore, since the productivity measures have been standardized within fields, the correlations between reputation and productivity are unaffected by the different method of computing research productivity counts in sociology.

88. There is a significant amount of variability in the correlations for the biological specialties, ranging from a high of .58 for botanists to .30 for molecular and developmental biologists. Part of this difference results undoubtedly from the problems of correctly classifying the specialties of biologists. Molecular and developmental biology are not really the same specialty, and there are a significant number of specialized research areas within both larger specialties. The lower correlation between productivity counts and perceived-quality scores for the molecular biologists may be tapping a boundary problem, discussed in note 54. It is probably the case that the more specialized a field is, the lower the average visibility of scientists working in it.

89. The association is somewhat higher if I use the absolute number of published papers: $r = .65$ for sociologists, .61 for psychologists, .40 for biologists.

perceived quality is .48 for the sample of 582 scientists: .63 for sociology, .43 for psychology, and .45 for the biological sciences.[90]

Finally, there is a moderate correlation between quality of work and visibility: $r = .41$ for the entire sample, with no significant variations in the three disciplines.

In sum, scientific research performance clearly is a significant predictor of reputational standing. Scientists who achieve prominence and esteem within their fields are those who produce work that is used widely by their peers. Correlatively, the invisible scientist is one who has published little of note. But how will these relationships between research performance and reputation affect the apparent influence that sex status has in predicting prominence and esteem? With this third set of relationships established, the original effect of gender on reputation will be modified when scientific research performance is controlled. The paired comparison will estimate this modification.

Regression models that consider scientific recognition and that include quality and quantity of research among a set of independent variables produce coefficients that are difficult to interpret—particularly when it comes to estimating the relative effects of the two performance measures. Since those scientists who produce good work are also apt to produce a lot of it, and correlatively, those who do not produce good work tend not to produce much of it, there is often substantial multicolinearity in regressions including both variables. In fact, in every equation to be discussed, quality of research was the more important determinant of reputational standing, although the sheer amount of published work did positively affect reputations independently of quality. To simplify the analysis, and to avoid the multicolinearity problem, I constructed and used a simple index of scientific research performance, by standardizing both the quality and quantity measures and combining them by adding the standardized scores together. Since I am not here particularly interested in making fine distinctions between the relative effects of quality and quantity of research on reputations, beyond what I have just noted, I will employ the index throughout the remainder of the chapter when referring to research performance.[91]

The idea that research performance is a critical status intervening between antecedent statuses and reputational standing is being tested here. The zero-order correlations suggest that this is so empirically; the issue before us now is the force that performance has in reducing the effect of sex status. The regression analysis leads to two strong conclusions. First, research performance is the primary determinant of reputation for both sexes. The "perfectionist" or "pro-

---

90. It makes little difference whether I use total citations for five years or for 1972; the correlations are virtually the same, since the association between the five-year and single-year indicators is .95. Within the biological sciences, there are some significant differences: For example, $r = .67$ for biochemists; $r = .45$ for physiologists.

91. Although the data on scientific output were standardized within fields, this will not eliminate a skewed distribution of research output. Since the original distribution of productivity and citations is not normal, the resulting standardized distribution is skewed. When the sample is divided by sex status, there is a skewed distribution of output: Few female scientists are more than a standard deviation above the mean.

lific'' research scientist, whether male or female, is also the one who is esteemed or prominent.[92] Correlatively, if a scientist is ''silent,'' he or she will remain relatively invisible. Second, and here is the significant difference from all the other paired comparisons, **sex status has only a negligible effect on reputation once the scientist's research performance is known.**

Statistically controlling for performance reduces the effects of gender on perceived quality ($r_{pqs} = -.18$) to one-third of its initial strength ($b^*_{pqs.o} = -.07$); correlatively, there is almost no change in the strength of the effect of research performance ($r = .56$; $b^*_{pqo.s} = .54$). Research performance alone accounts for almost all of the 32 percent of the total variance that the two variables can explain on perceived quality. The same pattern of effects appears when visibility of scientific work is the dependent variable. The zero-order correlation between sex status ($r = -.19$) and visibility is reduced to $b^*_{vs.o} = -.07$; the zero-order association between research performance and visibility remains virtually unchanged ($r = .56$; $b^*_{vo.s} = .54$). The pattern is the same regardless of discipline. In sociology, a correlation of $-.21$ between gender and perceived quality effectively disappears ($b^*_{pqs.o} = -.02$); it is cut in half in psychology and almost totally ''washed out'' in the biological sciences. The same results obtain within the specialties when visibility is the dependent variable. Sex status is no longer a reliable predictor of reputational standing.

What do these strong findings mean for the pattern of ''costs'' identified earlier? In brief, they require a marked alteration of the interpretation. It is not so much gender that has an independent and significant influence on prominence or esteem, but the scientist's research performance. Scientific performance seems to be the crucial status, and the important substantive finding is that there is only a negligible reputational cost of being female once a scientist's research output is known. The apparent effect of gender can be almost entirely understood as a function of research differentials; its influence is indirect, as a predictor of research performance.

Let us carry the analysis a step further. I have examined the expected reputations of scientists whose research performances lie along a continuum: The average man or woman gains 10 points in visibility for every standard deviation from the mean productivity level, but a woman loses 3 points simply because of her sex. But do there remain costs of being female at the extremes of research performance?

Since research performance appears to be a critical attribute in determining reputational standing, I formulated two hypotheses: (1) as the productivity of scientists approaches zero, the differences between the reputations of men and women will attenuate; (2) as scores on research performance approach the upper bound—that is, are significantly above the mean—the differences between the visibility of men and women will increase. The regression equation in fact

---

92. For an extended discussion of these types, see Cole and Cole, *Social Stratification in Science (op. cit.)*, Chapter 4.

predicts a 7 percent difference in male and female visibility scores among those who are in the top 2.5 percent of productive scientists.

There is no reason to remain speculative here. The actual visibility differences among scientists whose productivity varies can be obtained with the data in hand. I used the entire sample of 582 scientists for this analysis, standardized their average yearly productivity and citations to make the rates in different fields comparable, and subdivided them in terms of gender and standardized productivity and quality score.[93] (See Table 4–3). For the invisible, or silent, scientists, who have produced almost no papers and who are at least half a standard deviation below the mean productivity score, there are only minor sex differences in visibility. Indeed, women are at least as well regarded for their work as men. This seems to confirm the first hypothesis. The visibility of men and women scientists is actually quite similar until the most prolific scientists are compared. Among the top quarter of productive scientists—that is, the 30 percent of men and 16 percent of women who are half a standard deviation or more above the mean—there are sex differentials in visibility: Men are known to roughly 8 percent more of the scientists in their discipline.[94]

Comparing the visibility of male and female scientists who are cited at the same rate yields similar results. Increased differences in visibility between men and women parallel increased rates of citation: At half a standard deviation above the mean, men are known to 12 percent more of the scientific community.[95]

While it is more difficult to comprehend the meaning of changes in perceived-quality scores—since the metric, unlike that for visibility, is not self-evident—the analysis of citation patterns in Table 4–3 produces the same story. At lower levels of citation no meaningful differences are observed; at the highest levels of citation there is significant difference between the scores of men and women scientists. How can we interpret such patterns?

Why are women scientists of the first rank—the Maria Goeppert Mayers and the C. S. Wus—on average even slightly less well known than men, holding other statuses constant?[96] I think in the end this is another product of accumulative disadvantage. First, women produce less science that makes a significant impact in their fields. Second, women are assumed not to be geographically mobile; consequently, they meet fewer people who work in their discipline. Because they have produced less in the past, they are less apt to be invited in the present to special conferences or be asked to give special lectures and papers.

---

93. Since the variables presented in Table 4–3 have been standardized to have a mean of zero and a standard deviation of one, the table categories are presented in terms of the standardized metric. These standardized units do not refer to either number of papers or citations.

94. Again, the reason for the different proportions of men and women located at least a standard deviation above the mean is that men and women scientists have markedly different productivity patterns.

95. The same pattern obtains when we examine the perceived quality of scientists. Those comparisons are not specified because the metric is difficult to interpret.

96. These two women are selected as examples of prominent female scientists without knowledge of either their perceived-quality or their visibility scores. They were not in any of my samples.

TABLE 4–3.   The Reputational Standing of Men and Women Scientists by Productivity and Quality of Scientific Output: Mean Perceived Quality and Visibility Scores

QUALITY OF WORK* (STANDARDIZED)	PERCEIVED QUALITY		VISIBILITY	
	Men	*Women*	*Men*	*Women*
Less than −1.0 to −.5	4.05 (53)	4.09 (22)	18	22
−0.51 to 0.0	4.20 (259)	4.05 (92)	33	26
0.01 to 0.50	4.70 (68)	4.57 (14)	49	44
0.51 or higher	5.01 (67)	4.77 (7)	60	48
QUANTITY OF WORK* (STANDARDIZED)				
Less than −1.0	3.76 (24)	3.92 (18)	20	16
−1.0 to −.50	4.12 (92)	4.04 (52)	24	24
−0.51 to 0.0	4.30 (110)	4.14 (30)	33	29
0.01 to 0.50	4.53 (87)	4.45 (13)	41	37
0.51 or higher	4.65 (134)	4.43 (22)	51	43

*The indicator of quality is total citations for each scientist, standardized within disciplines, with a mean of zero and a standard deviation of one. Quantity is the average number of published papers per year standardized by discipline. It is clear that the distribution of both quality and quantity is somewhat skewed.

The accumulating benefits of productivity and informal activity aid men in obtaining rewards and resources for research, which can be translated into scientific papers. Women receive less attention within their disciplines as well as less public coverage for the work they do produce, and are infrequent members of national advisory panels, peer review panels, and so on. This limits in turn their opportunities to become editors of leading scientific journals, to hold elective office in local and national societies, and ultimately, to receive honorific recognition by election to the National Academy of Sciences and other honorary societies. Correlatively, as time spent in national activities is minimized, time given over to local departmental and nonscientific activities increases.[97] Although this is not a zero-sum game, it all adds up; it is all self-reinforcing. What is perhaps most remarkable is that the disadvantages confronting women do not have even greater effects on their actual standing.

Consider a second possibility: Female scientists are not as successful, for as-yet-unknown reasons, in converting their productivity and discoveries at a given stage in their careers to scientific returns at a future stage. Perhaps highly productive women do not use their research performance as effectively as men. They may be less familiar with the mechanisms by which scientists convert one form of achievement into other forms of reward; they may choose not to attempt

---

97. This is speculation and requires more research to establish its basis in fact. It may turn out that this is not a zero-sum phenomenon and that activities of the two types are independent of each other.

to convert scientific performance into reputational standing; or they may be structurally constrained from translating success of one type into that of another. The coinage that men and women deal in is the same, but the rate of exchange may be different for them. Women, like men, may obtain accumulating advantages from their structural location and their research performance, but the accumulation may have a significantly different slope than that found for males in similar situations.[98]

The analysis relative to this point suggests that research performance is more important than sex status in predicting reputational standing. But is performance a dominant status in the status-set of scientists? By dominance I mean: Is one status continually acceded to when two or more statuses might predict conflicting behavior toward an individual?[99] Specifically, what status in the status-set dominates evaluations of scientists' work when the statuses produce cross-pressure on an assessor to rate a scientist in different ways? For instance, if being a woman tends to produce negative evaluations by scientists but high-quality work tends to produce lofty appraisals, which characteristic is more important when scientists are appraising a female who has produced outstanding work? Will the assessment be based solely on performance criteria, solely on gender, or on some measure of both? In the paired comparison of sex status and performance, dominance was clear; in paired comparisons of gender with all the other scientific statuses, both significantly determined reputation. But will research performance reduce to insignificance the effects on reputational standing of all the other scientific statuses? For example, is it actually rank of department or academic rank that influences reputation, or is it that scientists at better departments and in higher academic ranks produce more and better work?

Indeed, is it possible to identify a set of conditions under which female scientists will actually have reputational advantages over males? If we examine information on all of their statuses simultaneously, are there situations in which the original effect of gender not only is washed out but in fact is reversed?[100]

The extent to which research performance dominates a scientist's status-set is estimated by regressing the indicators of reputation on the characteristics paired with gender (see Table A-5 for correlation matrix). Table 4-4 presents the standardized partial regression coefficients for each status. The results clearly

---

98. Although these is increasing interest in studying the phenomenon of accumulative advantage, little work has been done to date on structural constraints to the accumulation of advantages. There may be particular characteristics of scientists—ascribed characteristics most probably—that do not totally impede the accumulation of advantage but do slow down the rate of accumulation. Identification of characteristics of scientists that facilitate or hinder accumulation is worth study.

99. The concept of a hierarchy of statuses has been developed in the lectures of Robert K. Merton at Columbia University. He discusses dominant statuses in detail. The substance of these lectures has yet to reach print.

100. It is appropriate to raise the more general issues here of causal direction and reciprocal causation. In this analysis the forms of scientific recognition are interrelated. It is impossible to establish a clear causal direction because the variables interact with and feed back on each other. When temporal data are available, as in Chapter 5, the reciprocal influence of variables can be examined, but even there it is difficult to accurately estimate the strength of the reciprocal causation. Here I am dealing only with a prediction equation. When the analysis shifts to a causal model, assumptions about the temporal sequencing of effects are necessary.

TABLE 4–4. Predicting Reputational Standing from Six Statuses of Men and Women Scientists: OLS Regression Analysis

SCIENTISTS' STATUS-SET	VISIBILITY*				PERCEIVED QUALITY OF WORK*			
	r	B	Beta Coefficient	Std. Error of B	r	B	Beta Coefficient	Std. Error of B
Sex Status	-.19	-1.81	-.05	2.23	-.18	-.05	-.04	.05
Rank of Current Department	.30		.18		.29		.18	
Rank of Ph.D. Department	.21		.12		.13		.04	
No. of Honorific Awards	.28		.11		.27		.10	
Academic Rank	.35		.22		.30		.17	
Scientific Research Performance	.56		.41		.56		.44	
Coefficient of Determination		.65				.62		
R^2		.43				.39		

*Both dimensions of reputational standing were standardized within the three fields and combined for computations for the entire sample of men and women scientists.

indicate the relative but far from total dominance of performance in predicting reputational standing. Research performance is at least twice as strong a predictor of perceived quality and of visibility as any other status in a scientist's status-set. The interpretative effect that performance has on the zero-order relationships between other statuses and reputation is noticeable. Sex status had little impact in reducing the correlation between these other statuses and reputation, nor was it significantly reduced when paired with each of them. Research performance, however, does significantly alter the strength of the relationships between these other statuses and reputational standing. For one thing, take the influence of rank of current department, which is correlated .29 with perceived quality; it is reduced to $b^* = .18$ after controlling for the other scientific statuses. Of these other statuses, it is research performance that is primarily responsible for the reduction in the zero-order correlation.

Furthermore, examining the entire status-set indicates that the independent influence of gender is reduced to where it simply is no longer an aid in estimating the reputations of these scientists. Other than research performance and structural location, academic rank and eminence continue to have modest but significant predictive effects on both dimensions of reputation. In other words, even after we know that two scientists have published work of roughly equal quality, the one located at Harvard is more apt to be prominent, esteemed, or notorious than the one at a less distinguished university or college; the one who is a full professor will have higher standing than the one holding lower rank; and the one who has received forms of honorific recognition will have a better reputation among peers than the one who has not. But holding all of these other statuses constant, there is no significant difference in the reputations of men and women. This set of findings is exactly the same for sociologists, psychologists, and biologists.[101] In fact, if there is any notable difference among the fields, it is the reversal among sociologists of the effect of gender on reputational standing, after controlling for these other social statuses; women sociologists are slightly more likely than men to have high visibility when these other statuses are equalized.[102]

Thus far I have been examining the relative influence of several statuses in predicting two dimensions of scientific reputational standing. I have avoided causal modeling, since it is in fact difficult to determine the causal sequence of research performance variables and several forms of institutional recognition.

---

101. In each field the sex status coefficient is reduced to significantly less than twice its standard error. In substantive terms, sex status becomes an ineffective predictor of reputational standing in all fields under study.

102. Consider the results for sociology more closely. The zero-order association between sex status and perceived quality is $-.21$. After controlling for research output and the five other scientific statuses, the zero-order association is eliminated; the partial regression coefficient is in fact now positive: $b^* = .07$. After controlling for these other statuses, women sociologists actually have somewhat higher perceived-quality scores than men. The story is much the same when examining the prediction of visibility scores. An initial association between sex status and visibility of $-.25$ is reduced to a beta of .00, when I control for the statuses in the status-set. In psychology the pattern follows closely that found for the entire sample. An initial correlation between sex status and perceived quality of $-.22$ is reduced to $-.07$; the same pattern is found when visibility is the dependent variable. For biologists the $-.12$ association between gender and perceived quality is reduced to a beta of $-.03$; the $-.12$ association between sex status and visibility is reduced to a beta of $-.02$ after controlling for the other scientific statuses.

The processes that determine reputational standing are undoubtedly mutually interactive and self-reinforcing. Those factors that make for prominence or esteem are in turn influenced by it. As I have repeatedly emphasized, there are highly articulated feedback mechanisms in science which lead to the sharp differentiation of rewards. When these feedback mechanisms operate to reinforce talent, regardless of functionally irrelevant traits of scientists, the social system of science efficiently supports the people who are most able and most likely to produce important discoveries.[103] When it makes incorrect assessments of talent, or bases the distribution of resources on irrelevant characteristics, it departs from maximum efficiency and produces inequities.

Consider how the sharp differentiations in facilities, resources, and "track records" tend to maintain existing reputational distinctions.[104] Past outstanding research performance by a scientist increases the likelihood of support for current and future projects. But a high level of funding also makes for high productivity and encourages the production of original ideas, especially in the age of Big Science. The combination of resources and past performance largely determines a scientist's current reputational standing and increases the probability that he or she will be offered positions at centers of scientific activity. Once an academician is located securely among the scientific elite, with ample resources in hand, and with a reputation of some scope, the chances for obtaining continuing support and for producing additional research are somewhat increased. Thus there is continual feedback between past recognition and track record and future performance and recognition.

To construct an adequate causal model of the formation and maintenance of reputational standing within the scientific community would require data for a single cohort at several points during the scientific career. While career data are available to explore the temporal sequences which lead to positional recognition (see Chapter 5), data with several independent measurements of reputational standing are not currently available. If we acknowledge that a feedback process produces stability of the current arrangements for each new cohort of scientists, and if we assume no cohort differences, then for analytic purposes a tentative causal model of the determinants of reputational standing at one point in time can be created. Table 4–5 presents a model of the determinants of perceived quality

---

103. On this feedback process, see Merton, "The Matthew Effect in Science," *op. cit.*; Zuckerman, *Scientific Elite* (*op. cit.*).

104. In this chapter I have not discussed in detail the way that reputations may alter a scientist's chances for obtaining scientific resources, particularly research grants from government agencies such as the National Science Foundation and the National Institutes of Health. The place of resources in the picture of accumulating advantage has recently been examined by Stephen Cole, Leonard Rubin, and Jonathan R. Cole, *Peer Review at the National Science Foundation* (Washington, D.C.: National Academy of Sciences, 1978). We found that a scientist's track record in terms of publications produced from prior grants only affected slightly his or her chances of obtaining further support. In fact, most status characteristics of scientists had only a minimal effect on the probability of being funded. Peer review ratings largely determined whether or not an applicant received funds. Status characteristics and track record had only small affects on peer review ratings. In sum, at least at the NSF, there was little evidence in support of the hypothesis of accumulation of advantage. More recent analysis of the peer review data resulted in moderate to strong correlations between measures of reputational standing and peer review ratings.

TABLE 4-5. A Causal Model of Influences on Reputations of Men and Women Scientists

INDEPENDENT VARIABLES	DEPENDENT VARIABLES			
	Perceived Quality of Work (beta coefficient)	Quality of Current Department (beta coefficient)	Academic Rank (beta coefficient)	Visibility (beta coefficient)
Research Performance	.55	−.04	.15	.23
Sex Status	−.05	.01	−.07	−.03
Perceived Quality of Work		.26	.24	.47
Quality of Current Department			.12	.12
Academic Rank				.14
Residual	.83	.96	.94	.69

NOTE: The OLS model is estimated for 582 men and women scientists in the biological sciences, sociology, and psychology.

and of visibility in American science. The model includes only a limited number of scientific statuses; it is not exhaustive and does not estimate all influences on either perceived quality or visibility. Rather, it presents a truncated portion of the complex set of relationships and social processes that determine reputational standing.[105] Indeed, I have limited the model to the set of scientific statuses that have been considered throughout this chapter.

Before we consider the details of the model, note the significant conceptual change made explicit in the relationships shown in Table 4–5. Rather than treat both dimensions of reputational standing as determined by the same set of scientific statuses, I have indicated that perceived quality is a cause of visibility and rankings on other scientific statuses. Elsewhere Stephen Cole and I have discussed the relationship between actual quality of research performance and perceived quality:

> The effect of quality on rewards is mediated by the perceived quality of work. The independent effect of quality [and quantity] on positional and reputational success is slight. This allows us to make a subtle but important modification in analyzing the way in which universalism works in science. There can be little doubt that the quality of work *as it is perceived by other scientists* is the most important variable in determining the allocation of rewards. There is also little doubt that to have one's work highly evaluated you must actually produce work that other scientists find useful, that is, work which is highly cited. The production of work deemed useful is a necessary condition for earning a reputation as a good scientist.[106]

Perceived quality of work is hypothesized to be a critical intervening variable in the model, mediating the effects of actual performance on three forms of recognition. It also suggests the reciprocal nature of some of these relationships, a set of interactions that may be tentatively estimated by altering the causal assumptions implied by this model.

In general the model allows research performance, defined by the index of scientific output, and sex status to be intercorrelated and to be determinants of the perceived quality of work. The model omits, however, both the source of graduate training and chronological age as causal statuses. The first of these statuses will be discussed in some detail below, and age is omitted because of its complex curvilinear relationship with reputational standing. Additional models of the scientific reward system might include estimates of these two statuses as well as other characteristics of scientists, such as IQ.[107] Perceived quality has a

---

105. It is important to be aware of the limitations to the path analytic technique. Beyond the various assumptions of linearity in the models, beyond the variables that are omitted from the models, beyond the problems of obtaining estimates for overidentified and underidentified models, there is the added problem of understanding the meanings of the paths themselves. There are important but unidentified social processes that lie behind each path. The production of a path model with its attendant regression coefficients must be viewed as a first step in understanding the actual detailed social mechanisms which produce correlations among a set of variables. Nonetheless, as a first step the models are effective ways of getting a handle on a set of indirect links between the variables. But they are not theories and they do not represent the end point in analysis.

106. Cole and Cole, *Social Stratification in Science (op. cit.).* p. 119.

107. For a model that examines the influence of research output, source of doctoral training, perceived quality, and rank of current department on visibility, see Cole and Cole, "The Reward Systems of the Social Sciences" (*op.*

direct effect on visibility and an indirect effect through its influence in determining the scientist's structural location and his or her academic rank. Quality of department is determined by the antecedent statuses and influences visibility directly and indirectly through academic rank. Professional rank is determined by all of the statuses other than visibility, while visibility is directly influenced by each of the other scientific statuses.

Turning to the results, note the total absence of any meaningful influence of gender on the formation of reputation. Once we know a scientist's research performance and his or her structural location, it makes no difference whatsoever whether the scientist is male or female in explaining reputational standing. The only influence of gender in the model is an indirect one on reputations that results from its association with actual research performance.[108]

It is equally apparent that perceived quality is the critical factor in the model.[109] The path coefficient from perceived quality to each form of scientific recognition is at least twice the strength of any of the others. Research performance remains a significant status; it continues to directly influence a scientist's academic rank and visibility, and to indirectly influence rank of department and visibility because it effectively controls the perceived quality of a scientist's work.

Several additional features of the model are notable. First, I assumed that perceived quality precedes and partially determines visibility. The causal chain positing that actual research performance determines the perception of work, which in turn affects the scientific community's knowledge of it, seems on its face more plausible than assuming that visibility of a scientist would determine perceived quality. Interestingly, even if we accept the less plausible causal structure—visibility affects structural location, rank, and perception of work— the results are almost identical.[110] A reversal in the causal position of perceived quality and visibility does not alter significantly any of the coefficients, and it suggests what the outcome of a panel analysis might look like: a stable and self-reinforcing system within cohorts that builds and maintains reputations. Such an outcome suggests a feedback mechanism that increases inequality of reputational outcomes in science while maintaining a state of equilibrium.

Second, more of the total variance on the two indicators of reputational

---

cit.); for a more extended discussion of this model for biochemistry, chemistry, physics, psychology, and sociology, see Stephen Cole, "Scientific Reward Systems" (op. cit.).

108. The path coefficient from sex status to academic rank is likely to differ somewhat from that found for an alternate sample. In this sample the zero-order correlation between sex status and academic rank is lower than expected, because I sampled on chronological age, which, of course, is strongly correlated with academic rank.

109. The analysis of the five fields by Stephen Cole (see note 107) produced results almost identical to those obtained in this independent study. Perceived quality was the critical intervening variable in that study.

110. In order to illustrate the influence of reversing the causal order of perceived quality and visibility, I have estimated the coefficients for a model that reverses the causal sequence. I have simulated a temporal sequence in which research performance, perceived quality, and two forms of positional recognition determine visibility at T_2; then visibility and research performance influence perceived quality at T_3. The regression coefficients from perceived quality ($b^* = .45$) and from research performance ($b^* = .26$) to visibility are virtually the same as the coefficients from research performance ($b^* = .22$) and from visibility ($b^* = .45$) to perceived quality. In short, the model suggests a high degree of stability in the reward system for each cohort of scientists after the first ten years of activity in the field.

standing can be explained than on either form of positional recognition. This corroborates earlier findings that scientific rewards that are less dependent on personal choice, such as reputations, have a stronger universalistic basis than those that are more affected by individual tastes and personality variables, such as appointments and promotions in university departments.[111]

Third, although the results reported in Table 4–5 are for the entire sample, the results are no different when computed for each of the three fields. There are, of course, a number of small deviations in the coefficients, but the relative effect of each of the variables within the model is identical regardless of discipline.

Finally, the findings are probably generalizable to other scientific disciplines. A model similar to this one, which includes the source of doctoral training but not sex status and academic rank, has been discussed elsewhere by Stephen Cole.[112] The results obtained there parallel those reported here. For example, the coefficient for perceived quality with visibility as the dependent variable was .48 in S. Cole's study, compared to .47 in Table 4–5; the independent effect of rank of department on visibility was .14 in the other study and .12 here; the net effect of perceived quality on current department was .33 in the study of comparative reward systems and .26 in this one.[113] The comparability of results obtained in two studies that used identical indicators of reputational standing and of scientific statuses suggests that inferences drawn from the data on these three fields may be applied to the larger society of science.

What can be concluded, then, about the reputational standing of men and women scientists? Within the status-sets of scientists, sex status is significantly associated with both perceived quality of scientific work and its visibility. When paired comparisons of the relative impact of the several statuses on reputations are made, in each pairing, with one strong exception, gender and the other statuses have largely independent influences on reputational standing. But once the scientific research performance of scientists is known, earlier differences in the reputations of men and women are fully accounted for. When we compare men and women who have produced roughly equal amounts of research which has had a similar impact in their field, there is almost no cost of being female. The greatest negative effect of gender is found among men and women who have been prolific in their research productivity: Here women are significantly less well known than men. Scientific output is the significant force in building scientific reputations. It far outweighs the influence of all other statuses. In the causal model discussed above, actual research performance and the way in which that

---

111. The study of the physics community, in which Stephen Cole and I examined various types of recognition, probably did not emphasize sufficiently that different degrees of universalism obtain for the several forms of recognition. For example, for hiring and promotion decisions in general, and specifically for tenure decisions, there seems to be a higher level of particularism in the reward system of science than when we consider peer appraisals and honorific awards. The correlations between quality of research and the latter forms of recognition are consistently higher than for the positional forms. Further developments of this difference can be found in the doctoral dissertation of Leonard Rubin, Department of Sociology, Stony Brook (State University of New York), 1975.

112. Stephen Cole, "Scientific Reward Systems" (*op. cit.*).

113. *Ibid.*, Figure 1.

performance is perceived turn out to largely determine the scope of scientific reputations.

It should be emphasized and underscored, however, that I am not suggesting that gender is irrelevant with respect to reputational standing. That research performance effectively "washes out" the first-order association with both perceived quality and visibility only means that we have interpreted the relationship between gender and reputational standing. To the extent that gender remains a significant factor in producing differentials in research performance, it remains part of the causal chain leading to reputational standing. The estimates of the indirect influence of gender on reputations may be found by subtracting from the zero-order correlation between gender and the indicators of reputation the product of the indirect paths from sex status to perceived quality and visibility.

All of this leads me to a remark bordering on a prescription. To understand sex differences in science careers, more work must be focused on research productivity. The existing arguments and explanations for these acknowledged differences simply are not adequate or are empirically unsound. For example, marital and family statuses as explanatory factors do not seem to help much, nor does knowledge of the structural location of female scientists. Perhaps explanations which rest more on models of socialization and the psychological traits of females in science, such as those discussed by Matina Horner or Eleanor Maccoby, will help explain some of the publication differential.[114] But we should not assume that there are no structural roots of differences in performance. In fact, I think there are such sources, which are not easily captured by the types of data presented here.

For example, a detailed examination of the processes of sponsorship in science might uncover significant differences in the actual training of women. We should not simply try to establish whether women are not treated formally in the same way as men—whether they are discouraged, for instance, from selecting dissertation topics they are interested in. The search for gross patterns of bias is not likely to produce fruitful results. Rather, it is time to describe in detail and to analyze the informal structure of activities of young men and women scientists that set in motion and sustain an accumulation of advantages and disadvantages. Are women less apt than men to spend extended amounts of time in informal scientific discourses with their teachers—and if so, why? Are they less likely to assimilate the style of scientific work which is so important to later productivity? Are they, for lack of informal discussions and observation, denied the opportunity to acquire "scientific taste" for an interesting problem, if not for a solution? Do science professors attempt to place women protégées in laboratories and departments where there is a significant chance for them to work on ongoing projects or programs, to develop their own ideas, and to have the facilities

114. On a large number of areas of potential psychological differences between the sexes, see the extraordinarily useful and interesting book by Eleanor Emmons Maccoby and Carol Nagy Jacklin, *The Psychology of Sex Differences* (Stanford: Stanford University Press, 1975).

necessary to translate ideas into publishable papers? Perhaps even more important for research performance, do senior professors engage in scientific collaboration with junior female colleagues; do they ask them to join their workshops; do they get them invited to conferences at which they describe their work to groups of scientists who have already gained some prominence and are influential in determining the next generation of esteemed and prominent scientists; do they aid in obtaining for their female students and junior colleagues invitations to publish papers in journals, in conference reports, and in proceedings from symposia that are not subject to the refereeing process? Of course, the simple answer is that some do and some do not; but a more detailed inspection of the problem should aim to establish whether for women the probabilities are lower than for men of having these informal experiences that eventually are reflected in productivity rates and higher reputational standing. Furthermore, do differential probabilities exist among the most talented young male and female scientists, for whom these opportunities might actually be significant?

The role of sponsorship in science is becoming increasingly misunderstood. In recent work, sponsorship increasingly has been contrasted with meritocratic mobility, and has come to symbolize the predominance of particularistic over universalistic criteria of judgment. Used this way, sponsorship suggests that scientists don't get jobs on the basis of their talents and merit but on the basis of whom they know and who is singing their praises. Of course, this is a thoroughly simplistic way of addressing a far more complex and, I daresay, more interesting problem. Such treatment of the sponsorship idea subverts the useful distinction between the contest and sponsored types of mobility developed originally by Ralph Turner and elaborated upon by Hargens and Hagstrom.[115] To put an obvious observation to one side immediately, it must be clear that if the most able individuals are also the most sponsored, the social system of science is not abridging its value of universalistic criteria of rewards. To the extent that sponsorship facilitates the careers of the most talented junior scientists, it only furthers the rational allocation of scarce resources and the effective pursuit of new knowledge.

Actually, sponsorship has a central position in the development of scientific knowledge, and discussion of it should not be confined to its role in the marketplace. The history of sponsorship goes as far back as science itself. This is no place to trace that history, but I will briefly relate it to our main concern: how sponsorship may influence differential productivity among male and female scientists. There are few scientists of note who did not have an identifiable sponsor. Indeed, autobiographical accounts almost invariably contain homage paid to a sponsor relationship, even if it was not always devoid of ambivalence. Whether we examine Auguste Comte's relationship with Saint-Simon; Fermi's with Cor-

---

115. Ralph Turner, "Sponsored and Contest Mobility and the School System," *American Sociological Review* 25 (1960): 163–72; Lowell Hargens and Warren O. Hagstrom, "Sponsored and Contest Mobility of American Academic Scientists," *Sociology of Education* 78 (May 1973): 1381–1402.

bino; Segré's with Fermi; Otto Hahn's with Rutherford; Lisa Meitner's with Otto Hahn; Schwinger's with I. I. Rabi; J. D. Watson's with Luria; Mary Whiton Calkins' with William James; or the thousands of other master-apprentice relationships, we are dealing with one important mechanism of transmitting a scientific tradition from one generation to another. Even a cursory scan of the autobiographical histories of scientists reveals poignantly how sponsorships involve jealousies, ambivalence, conflict, admiration, and love—and how they are always essential in producing in young scientists a sense for a good question or a key problem, a style of doing research or theorizing, a critical stance, and a way of teaching their own future intellectual progeny.

In fact, there is no need to rely solely on impressionistic materials and autobiography to corroborate these points. Harriet Zuckerman's recent work on Nobel Prize recipients demonstrates the genealogical tree linking the great figures of modern science.[116] There is hardly a physicist who has received a Nobel who has not been linked in one way or another, frequently as a student or apprentice, to one or two great teachers of physics. For example, the number of outstanding figures in contemporary physics who are linked to Rabi is extraordinary.[117] This has led some observers to sarcastically claim that the best way to win a prize is to study with a former winner. And indeed, if you penetrate the cynical facade of this comment, it might be asked: Why is this so, putting aside the possibility of out-and-out conspiracy? Physics does not stand alone; inquiries into all of the sciences, including the social sciences, will undoubtedly turn up similar patterns. Unfortunately, to date there is a striking paucity of systematic study of the impact of sponsorship on the developing cognitive structures of scientific disciplines.

Of course, sponsorship has its practical importance as well, and not simply in terms of job location, promotions, and salaries, although these consequences should not be minimized. It places men and women with great potential in the intellectual arena with others who either have already made great contributions or are about to; it puts one set of junior scientists in touch with other sets who are intellectually exciting and who are doing important work that will be recognized in short order.

It is the informal quality of the sponsorship system that is clearly the most important and indeed the hardest to capture in a quantitative analysis. Who could or would want to try to measure the profound influence of walks along the beaches that Fermi describes taking with his students and his teachers, or the effects of his talks with Corbino:

> We had almost daily conversations and discussions which not only clarified many
> of my confused ideas, but aroused in me the deeply felt reverence of the pupil for

116. For an extensive discussion of the relationship between great masters and great apprentices in science, see Zuckerman, *Scientific Elite* (*op. cit.*).

117. This section has benefited from discussions with my colleagues Gerald Holton and David Tyack at the Center for Advanced Study in the Behavioral Sciences.

the master. This reverence steadily increased during the years I was privileged to work in his laboratory.[118]

Or the impact of sitting by a fire with William James that Mary Whiton Calkins describes in her autobiographical writing:

> I began the serious study of psychology with William James. Most unhappily for them and most fortunately for me the other members of his seminar in psychology dropped away in the early weeks of the fall of 1890; and James and I were left not, as in Garfield's vision of Mark Hopkins and himself, at either end of a log but quite literally at either side of a library fire. The *Principles of Psychology* was warm from the press; and my absorbed study of those brilliant, erudite, and provocative volumes, as interpreted by their writer, was my introduction to psychology. What I gained from the written page, and even more from tête-à-tête discussion, was, it seems to me as I look back upon it, beyond all else, a vivid sense of the concreteness of psychology and of the immediate reality of "finite individual minds; with their thoughts and feelings."[119]

Or the contextual effect, to use the contemporary parlance, of keeping the company of Rutherford that impressed Otto Hahn so greatly as a young scientist:

> Rutherford's enthusiasm and his restless energy rubbed off on us all. The resumption of work in the laboratory after supper was the rule rather than the exception.... Quite often the evening was spent in Rutherford's house where the conversation consisted almost exclusively of shop talk, not always according to the wishes of our hostess, who preferred to play the piano but could rarely get her husband to listen to her.[120]

Or, to cite only one more among scores of alternatives, consider J. D. Watson's observations about the close working relationship he had with S. E. Luria in the early days of the Phage Group:

> I looked forward greatly to the forthcoming summer (1948) when Dulbecco [another graduate student of Luria's] and I would go with the Lurias to Cold Spring Harbor. Delbrück and his wife, Manny, were coming for the second half while, before they arrived, there was to be the phage course given by Mark Adams.... As the summer passed on I liked Cold Spring Harbor more and more, both for its intrinsic beauty and for the honest way in which good and bad science got sorted out.[121]

How can such things be measured? These are the indelible marks that are imprinted upon apprentices by masters, by scientists who sponsor their juniors because they see in them potential—the next generation to carry on the scientific tradition. Or more simply, because they feel such sponsorship is normatively prescribed in science.

Today we wait in our laboratories or offices for the arrival of potential

---

118. As quoted in Emilio Segré, *Enrico Fermi: Physicist* (Chicago: University of Chicago Press, 1970), p. 26.

119. Quoted from the autobiography by Mary Whiton Calkins as published in Carl Murchison, ed., *History of Psychology in Autobiography*, vol. 1 (Worcester, Mass.: Clark University Press, 1930), p. 31.

120. Quoted from *Otto Hahn: A Scientific Autobiography* (New York: Charles Scribner's Sons, 1966), p. 32.

121. Watson, "Growing Up in the Phage Group," *op. cit.*, p. 241.

students; the structuring of large graduate programs has made this possible, if not wholly desirable, and perhaps has oversimplified the process of identifying exceptional beginners. But what this surely represents is a recent historical shift. No more than forty years ago in Europe, great scientists would go out in search of young boys with exceptional talent who might become the next great generation of physicists or biologists ("boys" is used advisedly, as I shall make clear shortly). Surely it is self-evident, and does not require formal spelling out, that since relatively few scientists are going to produce any science of note, the informal process of sponsorship can be enormously consequential for the rate and quality of scientific productivity.

Which brings me finally back to the original problem: How do women fit into this sponsorship process? Throughout the history of science, how often do we find the kinds of sponsorship of young women that we do of young men, with all of the attendant advantages that accrue to those elected? How often did the great European scientists comb the gymnasia for young women with potential to do superb physics or chemistry? How often did it even cross their minds to do so? One can, of course, think of the notable exceptions, such as Franz Boas, who deliberately recruited and trained exceptional young female anthropologists. Indeed, he had quite a record of success, being able to count among his former apprentices the likes of Erna Gunther, Elsie C. Parsons, Ruth F. Benedict, Margaret Mead, Fredrica de Laguna, Ruth M. Underhill, Rhoda Métraux, and May M. Edel. Such men of science come easily to mind because of their exceptional record in tutoring able women.

Even today, how are young women science students perceived and treated in these important informal domains which may make all the difference between being good and being of the first rank in terms of what is actually produced? Do women today have to go it alone? In short, how much of the marked productivity differentials, especially at the extreme, can be explained by patterns of sponsorship?

Even when established scientists are inclined to seek out and sponsor young women, there are cultural and structural constraints to acting on the inclination. First, putting aside presumptions made by scientists about the capability and devotion of female students, there is the problem of the ease of interactions between male professors and female students. What start out as impersonal relationships in which sex status has no particular significance can and often do move into relationships that are sexually involved. The complexities of such entanglements may give male professors pause in initiating relationships in the first place. Furthermore, the assumption by scientists' role partners that a close sponsorship relationship with a female student must also involve a sexual relationship may deter many scientists from subjecting themselves to rumor, insinuation, and innuendo.

The higher proportion of females in graduate programs today increases the chances of their sponsorship as well as the attendant role-set problems. Significantly, the entanglements of male professors with young female students have in recent years made their way into the academic novel, which heretofore had limited love affairs on the campus to professors or to a professor and a colleague's wife. But with the publication and generally warm reception of Richard Stern's *Other Men's Daughters* and Alison Lurie's *The War Between the Tates,* among perhaps lesser works, students have been added to the cast of characters. Indeed, not only does Stern's professor of biological sciences find happiness in his student-lover, but the affair seems to alter his angle of scientific vision. At one point Professor Merriweather remarks: "I feel about her the way Galileo did about the telescope. My feelings for her enlarge my feelings for other things."[122] Of course, all is not so blissful for the professor-lover in these novels. But Stern and Lurie concentrate on the obvious conflict between the professors and their wives, and largely neglect the multiple consequences that such affairs may have for departmental relations and for the research, if not the sexual, performance of faculty members.

An examination of sponsorships of young women by eminent male professors should not overlook the perceived and actual threat of altering an instrumental relationship into an expressive one. Identifying this and all other constraints and conditions, particularly those involving sponsorships, that make for lower rates of scientific research performance among female scientists remains one of the key problematics in a research program trying to understand the place of women in the scientific community.

Then there are these additional questions: Do women tend not to enter "hot" or prestigious scientific research areas which are the focus of attention within the broader discipline? Either as a result of self- or social selection, do they turn away from "hot" research problems, and if so, why? Finally, are they trained more often to be experimentalists and empirical researchers than to be theorists? In all of the sciences for which evidence exists, the rolls of the most esteemed and prominent members of the elite, with relatively few exceptions, are dominated by theoreticians. If women do not begin to do theoretical work in a field, they are already at a reputational disadvantage. Presently there are no solid answers to any of these questions, yet they call for detailed investigation.

REPUTATION BUILDING

Let us now pass to the second major division of the discussion and examine the question of whether the process of acquiring a reputation is the same for both sexes. We shall now consider the separate determinants of reputation for men and women. In particular, does research performance have a roughly equal impact in building reputations for women relative to other women as it does for men

122. As quoted in Aristides [pseud], "Sex and the Professors," *The American Scholar* 44, no. 3 (summer 1975): 358.

relative to other men? Do structural location and educational origins play a similar role in producing reputations for both sexes?

Before answering these questions, let us return briefly to the four reputational types discussed earlier in the chapter: What are the status-set profiles of the men and women treated as separate groups? Specifically, how do the prominent, the notorious, the esteemed, and the invisible scientists differ by sex? These profile summaries appear in Table 4–6.

The prominent female scientist is by far the most visible of all the women, but there is virtually no difference in the perceived quality of work of prominent and esteemed women. Of course, this is in part simply an artifact of the definitions of these two types, but the closeness of their perceived-quality scores and the remoteness of their differences in visibility is not predetermined. Concretely, as shown in Table 4–6, prominent women have perceived-quality scores of 4.70 compared to 4.69 for esteemed women, but the former are known to 62 percent of their respective communities, while the latter are visible to only 18 percent. The prominent women scientists are the most productive of the four types; they are most likely to be found among the more distinguished departments and to have received the most honors. Correlatively, the invisible scientists are in almost every way the least distinguished. These results are expected. The more interesting patterns are found in comparing Type 2 (notorious) and Type 3 (esteemed), the two types that point to the imperfect association between perceived quality and visibility.

First, the esteemed woman scientist with lower visibility is also on average two years younger than the notorious scientist. Furthermore, the scientific research performance of esteemed women is significantly better than that of notorious scientists. Nonetheless, they have received fewer honorific awards and are less likely to be found in the highest-ranked departments. Indeed they are less apt to have been trained at the best departments. There is no need to detail the story for males, since the scenerio for the four types is virtually the same as for females.

To examine the sex-specific processes of building a reputation, I again analyze the relationships among a set of scientific statuses. Using a set of zero-order correlations between selected statuses, the pattern of associations with visibility suggests some significant variations between the sexes (see Table A–6). There is one similarity: Current academic affiliation is associated with visibility in equal strength for both genders; otherwise, the correlations differ in several ways. First, the association between the two indicators of reputational standing differs; it is somewhat stronger for the men. Second, research performance, the key ingredient in building reputations, is more strongly associated with visibility for men than for women. In fact, all of the associations between scientific statuses and visibility are stronger among men than among women, with the exception of the source of doctoral training.

These associations provide the first clue that the process of establishing visibility may be quite different for men and women. Turning to perceived

TABLE 4-6. Profiles of Four Ideal Types of Reputations among Men and Women Scientists

MALE SCIENTISTS

Reputational Type	X̄ Perceived Quality	X̄ Visibility Score	Research Performance	Age	Current Affiliation	Academic Rank	Rank of Ph.D. Department	No. of Honorific Awards
I Prominent	4.88	61	1.39	51	.371	3.9	0.22	1.48
II Notorious	4.12	46	−0.21	50	.199	3.7	0.21	0.81
III Esteemed	4.62	21	−0.13	46	−.069	3.4	−0.17	0.77
IV Invisible	3.84	20	−0.68	46	−.301	3.3	−0.20	0.54
F =	177	189	44	9	16	26	7	16
df =	(3,443)	(3,443)	(3,416)	(3,442)	(3,440)	(3,443)	(3,441)	(3,417)

FEMALE SCIENTISTS

Reputational Type	X̄ Perceived Quality	X̄ Visibility Score	Research Performance	Age	Current Affiliation	Academic Rank	Rank of Ph.D. Department	No. of Honorific Awards
I Prominent	4.70	62	−0.09	48	0.187	3.5	0.44	1.12
II Notorious	4.14	46	−0.74	49	0.005	3.6	0.30	0.76
III Esteemed	4.69	18	−0.30	47	−0.060	3.4	0.13	0.59
IV Invisible	3.83	15	−1.00	52	−0.267	3.2	−0.25	0.69
F =	82	71	9	2	1.17	1.4	3.5	1.3
df =	(3,131)	(3,131)	(3,120)	(3,128)	(3,129)	(3,130)	(3,130)	(3,122)

137

quality of work, we note somewhat greater similarity of association for both sexes. Although formal recognition is more strongly correlated with perceived quality among men than among women, the association between research performance and perceived quality is almost identical for both genders. Rank of current academic affiliation is similarly associated with perceived quality for men and women, but there are notable differences between the sexes in the associations with age. In fact, there is a moderate and positive relationship between age and both forms of reputational standing for men and a negative one for women.

So much for simple correlations. To estimate the relative impact of these variables on the formation of opinions of quality and on renown, I regressed the two measures of reputation on them. The regression coefficients, presented in standard form, appear in Table 4-7. The results indicate again the predominant role that research performance plays in the estimates of quality, and this is so for both sexes. The partial regression coefficients indicate that research performance has a slightly stronger independent effect in the building of female reputations. However, current affiliation and receipt of honorific recognition is more instrumental in producing good opinions of men's work than of women's. The regression for men may actually be summarized as follows: The quality and quantity of research output is by far the most significant predictor of appraisals, but a male scientist's institutional location, his academic rank, and his honorific recognition also play a significant role in predicting his reputational standing. In terms of perceived quality of work, being at an eminent as opposed to an undistinguished university makes a considerable difference for men, after accounting for the influence of research performance. Not so for women. Performance is still the main predictor; honorific recognition has no predictive value, while age, academic rank, structural location, and source of doctorate all have some, but limited, influence in predicting reputational assessments independently of research output.

TABLE 4-7. Regression of Reputational Standing on Scientific Statuses of Men and Women (Regression Coefficients in Standard Form)

SCIENTIFIC STATUSES	PERCEIVED QUALITY*		VISIBILITY*	
	Men	Women	Men	Women
Research Performance*	.427	.455	.430	.303
Current Department*	.197	.111	.190	.159
Rank of Ph.D. Department*	.017	.121	.085	.240
Academic Rank	.167	.114	.170	.155
No. of Awards	.111	.019	.119	.043
Age	.047	−.111	.133	−.051
R	.627	.567	.680	.509
R^2	.393	.321	.462	.259

*Each of these variables was standardized before the regression.

These data further support the idea that women do not transform their institutional achievements into reputational standing as easily as men. This impediment is even more recognizable when we shift focus to predictors of visibility. As anticipated in discussing the zero-order correlations, research performance does not influence the visibility of women as much as it does that of men. While current affiliation has much the same effect for both sexes, educational origin makes more of a difference for women than it does for men. Rank of doctoral department is the second strongest predictor of visibility and perceived quality among women, and not much less of an influence on visibility than is research performance. Recall that the source of the doctorate had little independent influence on visibility when we considered the sample of men and women as a single group. Furthermore, there was no correlation whatsoever between sex status and rank of doctoral department. Yet here we see significant interaction between gender, educational origin, and reputational standing: Source of the doctorate does have a significant effect for women but not for men. What does this finding signal?

In building reputations men probably can start off with greater institutional disadvantages than women and still make out fairly well. If they publish a lot of work and it is fairly well received, if they find themselves after a time in a department with a strong reputation, they will transform these achievements into esteem or prominence; or, correlatively, the social system of science converts these status ranks into reputation. The same apparently does not hold for females. Once they begin the race at a disadvantage, the race is pretty much run. The data suggest that a functional prerequisite for eventual prominence or esteem for the female scientist is to start out at a first-ranked institution. Surely this is not sufficient to gain a lofty reputation, but without proper initial location, renown or esteem within a discipline is an unlikely social outcome. These data also lend support to the argument about the significance of sponsorship in the careers of women. Since initial placement within the community is critical in their scientific careers, sponsorship makes for greater differences in women's ultimate reputational standings, given their lower scientific output.

Another notable interaction is found in the relationship between gender, chronological age, and reputational standing. Even after controlling for the influence of rank differentials, age is negatively associated with perceived quality of work among women. Quite the opposite is so for men. The data suggest that the age and talent structures may be related among women. Younger female scientists are simply more productive and have produced more significant work than have older female scientists: The correlation between chronological age and scientific research performance among women is $r = -.19$. The opposite is true among men ($r = .10$). Reiterating an earlier point, whether these correlations signal a differential in the average ability of females now entering science or increased opportunities to publish (or, indeed, both) remains a moot question.

Finally, the statuses in these truncated status-sets can predict significantly more of the total variance in the reputational standing of male than of female scientists. While this difference is not striking for perceived quality ($R^2 = .39$ for men; .32 for women), it is for visibility ($R^2 = .46$ to .26). What additional characteristics of scientists and of their environments would improve our ability to predict reputational standing must await further inquiry.

Turning again from prediction to causal models of the influences of scientific statuses on reputations, I have removed sex status from the model, and added educational origin and age as determinants of research performance and the several forms of recognition.

The results, summarized by the regression coefficients in Table 4–8, indicate several differences in the way men and women achieve scientific prominence and esteem. Let us see how research performance influences reputational standing. The object of the analysis is to better understand the correlations between statuses. We are trying to see how much of the zero-order association remains as an independent, or direct, influence, and how much of the original association is, in fact, a product of the association between several statuses in the model. For the statuses shown in this model, research performance is the only effective determinant of perceived quality. Temporal data, if available, would undoubtedly show that prior positional recognition had an additional influence on perceptions of quality. The effect of research performance on quality assessments is the same for male and for female scientists; those who produce a lot of good work are prominent and esteemed. The zero-order associations tell much of this story. The story becomes more interesting, however, when we try to understand the very different correlations for men and for women between research performance and visibility. The regression estimates suggest that the difference in association is a result of the stronger effect for men of research performance on visibility independently of the mediating factor of perceived quality. Together the direct effect of performance on visibility and its mediated effect through perceived quality account for 85 percent of the original correlation for men, but 92 percent for women. In short, the remaining 15 percent for men and 8 percent for women results from the interconnections between research performance, rank of current department, and honorific recognition.

Although female scientists do manage to convert opinions of their work into visibility, they are much less apt to translate high regard into positional success. The net effect of perceived quality on rank of current department and on academic rank is stronger for men than for women. The model further supports the idea that intellectual origin has almost no independent influence on either form of reputational standing among men, but continues to influence the degree of prominence of women after holding constant the other scientific statuses.

There is one other set of findings that suggests different processes by which male and female scientists achieve reputations. I wanted to consider the extent to which reputational standing as judged in 1973 could be predicted by scientific performance at varying stages of the scientific career. To do this, I divided the sample by sex status and looked only at scientists who were at least forty-five years of age, in order to consider only those scientists who had a significant

TABLE 4–8. Model of Reputational Standing of Men and Women Scientists (Regression Coefficients in Standard Form)

| | | DEPENDENT VARIABLES | | | | | | | | |
| | $X_1 =$ Visibility | | $X_2 =$ Academic Rank | | $X_3 =$ Rank of Dept. | | $X_4 =$ Perceived Quality | | $X_5 =$ Performance | |
INDEPENDENT VARIABLES	Men	Women	Men	Women	Men	Women	Men	Women	Men	Women
$X_2 =$ Academic Rank	.10	.10								
$X_3 =$ Rank of Current Department	.10	.09	.05	−.15						
$X_4 =$ Perceived Quality	.45	.47	.17	.11	.28	.13				
$X_5 =$ Research Performance	.26	.12	.14	.27	.15	.02	.53	.52		
$X_6 =$ Age	.16	−.05	.53	.51	−.04	−.10	.13	−.06	.10	−.19
$X_7 =$ Rank of Ph.D. Department	.08	.19	.00	.01	.18	.14	.06	.14	.10	.02
R^2	.57	.40	.40	.32	.13	.09	.32	.30	.02	.04
Residual	.65	.77	.78	.82	.98	.95	.82	.84	.99	.98

history of publications. For each sex, I computed the zero-order correlation between 1973 reputational standing and earlier scientific productivity at five-year intervals (as measured by total number of papers published). The results go still further to suggest that females are not as able as males to convert their performance into reputation, and that there may well be institutional impediments to this exchange. The association between males' scientific productivity from 1950 to 1954 and their 1973 visibility was $r = .45$; for the three subsequent five-year intervals the correlations were .52, .51, and .55, respectively. By knowing the productivity of male scientists very early on in their careers, we can make a fair prediction about how well known they will be at least twenty years later. Not so for females. Consider the comparative correlations for women scientists: For the 1950–54 years, $r = .16$; for the subsequent three intervals the associations were .28, .44, and .40. The predictability of a female's reputation by her early scientific productivity is weak until the 1960–64 interval. Predictions for perceived quality follow much the same general pattern. For example, the association for men between research productivity in 1950–54 and later perceived quality was .34; it was $-.01$ for women. By 1960–64, the correlation for men was .38, but only .20 for women.

In summary, the preceding analysis reveals that research performance is the most significant determinant of reputational standing for men, far overshadowing the influence of positional recognition. While it helps to be located at a top-ranked department, to have received honors, to have graduated from an outstanding department, or simply to have been around for a while, what really matters is the quality and quantity of the science produced. It is indeed quite simple: Men who produce good science that is widely used in other research become prominent, and there are few structural barriers to achieving lofty reputations if the functional performance criteria are met. Correlatively, those who do not meet high standards of research performance are not widely known or esteemed, regardless of their other scientific statuses. The picture for women is no less clear but is somewhat different. Scientific output is also the critical factor in eliciting positive evaluations and, indirectly, renown for women, but a different set of other statuses, most notably rank of Ph.D. department, continues to have some lingering influence on reputational standing.

Chapter 5

IQ AND ACHIEVEMENT IN AMERICAN SCIENCE

MORE THAN A DECADE AGO, Michael Young wrote a provocative fable, *The Rise of the Meritocracy*.[1] In it he envisioned an increasing tendency in Western society to recognize excellence. Intelligence, especially measured intelligence, was increasingly being socially rewarded.

> Today all persons, however humble, know that they have had every chance. They are tested again and again. If on one occasion, they are off color, they have a second, a third and fourth opportunity to demonstrate their ability. But if they have been labelled "dunce" repeatedly they cannot any longer pretend.... Are they not bound to recognize that they have inferior status ... because they are inferior? ... For the first time in human history the inferior man has no ready buttress for his self-regard.[2]

A similar sense of fairness in social judgments based upon intelligence is argued by Harvard psychologist Richard Herrnstein.[3] He, too, speculates that IQ will increasingly be highly correlated with social standing and recognition. The roll of others who see a similar pattern evolving in Western industrial societies is becoming quite long.[4]

Most of these observations, however clever or well reasoned, remain largely

---

1. Michael Young, *The Rise of the Meritocracy* (Baltimore: Penguin Books, 1958).

2. *Ibid.*, pp. 107–108.

3. Richard Herrnstein, *IQ in the Meritocracy* (Boston: Little, Brown & Co., 1971).

4. See among many others, Daniel Bell, *Post-Industrial Society* (New York: Basic Books, 1973).

speculations or predictions. But in this chapter, I want to deal with the present. Let us put aside the large and monumental issues of heritability of IQ, the role of self-fulfilling prophecies and labeling processes in creating seemingly meritocratic social systems, and the problem of measuring the concept of intelligence.[5] I want to deal with what we currently know about the relationship between measured intelligence and occupational achievement in American science. Before moving to this main concern, I would like first to review briefly what we know about this relationship in two other contexts: within the larger American occupational structure and within specific occupational groups.

Allow me one brief digression, however, on the issue of testing, since we will be considering the relationship between a test score, conveniently called IQ, and social recognition. I happen to believe there is substantial evidence which indicates that IQ tests are good rough measures of certain basic verbal and manipulative skills. Yet this evidence does not speak to the issue of how well test scores predict performance and creativity. As a culture, we tend to make this inferential leap quite easily and perhaps mistakenly. Americans seem to put great stock in IQ, SAT, GRE, civil service, and many other tests of aptitude and skill:

> Probably 250 million standardized ability tests are administered in the American school system each year. . . . In modern American society intelligence seems increasingly to be singled out from the many characteristics of man and elevated to a position of high importance. We seem to be moving toward a society that is organized on the basis of standardized intelligence test scores, and in which the manpower conception of man prevails.[6]

Americans believe or have been persuaded to believe that these tests are good predictors of performance or capacity. The descriptive labels associated with performance on aptitude tests suggest a striking level of confidence in them. For example, if a student in college has gotten very high College Board scores in high school but is now making rather low grades, he or she is almost invariably labeled by self and others as an "underachiever." Correlatively, a student with low Board scores who does very well is an "overachiever." But why can't we invert the angle of vision? Would it not be just as valid to label the first an "overtester" and the second an "undertester"?[7] Scores on these standardized tests are often reified. This is an error. These scores are imperfect indicators of native abilities. The results reported in this chapter should be taken, therefore, as only suggestive of the possible role of intelligence in the processes of social stratification. With this initial cautionary remark in mind, let us consider the first

---

5. Each of these problems has an important indirect relationship to the problem addressed in this chapter. However, the answers to these large issues are not solidly established. Raising questions about heritability in this chapter can only open a Pandora's box of auxiliary problems. For discussion of the heritability question, see, among a plethora of other works, Arthur R. Jensen, "How Much Can We Boost I.Q. and Scholastic Achievement?" *Harvard Educational Review* 30 (winter 1969):1–123; Herrnstein, *op. cit.*; John C. Loehlin, Gardner Lindzey, and J. N. Spuhler, *Race Differences in Intelligence* (San Francisco: W. H. Freeman, 1975); N. J. Block and Gerald Dworkin, eds., *The IQ Controversy* (New York: Pantheon Books, 1976).

6. Orville G. Brim, Jr., David Glass, John Neulinger, and Ira J. Firestone, *American Beliefs and Attitudes about Intelligence* (New York: Russell Sage Foundation, 1969), p. 1.

7. I believe that Paul Lazarsfeld in a private conversation first pointed this labeling phenomenon out to me.

issue: what we know about the influence of IQ on achievement in American society.

IQ AND ACHIEVEMENT IN AMERICAN SOCIETY

I shall briefly review what we as sociologists have come to know about the association between measured intelligence and three forms of social recognition in American society: occupational status, income, and education. Intelligence is moderately correlated with individual occupational achievement in American society ($r = .45$).[8] Duncan, Featherman, and Duncan suggest that there is a tautological element in this correlation. They note that the content ,of IQ tests probably was influenced by cultural definitions of the kinds of ability required to be successful at highly technical jobs.[9] All projective or predictive tests, if designed well, will contain, of course, a strong tautological element. However, even if the tests were designed with verbal and symbolic manipulation in mind, this would not imply that the correlation between IQ and occupational prestige is perfect. In fact, all studies that have been done indicate a moderate to strong, but not perfect, correlation.

There are two standard and alternative ways of examining correlations between ability and occupational prestige. One is to compute the average IQ within each occupational classification and then correlate these means with the rated prestige of the occupation. This tends to yield extremely high associations in the area of .8 to .9.[10] Needless to say, there is considerable variation in intelligence within any given occupation. Obviously, for example, some lawyers have extremely high IQ's; others may not be far above average. A second way is to correlate individual IQ scores with individual occupational rankings. Lower correlations, from .40 to .45, are found when the correlation is based upon individuals rather than aggregates.[11] Native ability, as measured by IQ, explains about 20 percent of the variance in occupational status at the individual level, without considering the effects of other intervening variables, and as much as 70 percent or 80 percent of the variance when considering aggregated data.

Perhaps the meaning of these correlations can better be understood when translated into standard IQ scores (mean 100; standard deviation 15 to 20) for several occupational classifications. Johnson reports the following mean IQ's: 120 for professional men; 113 for semiprofessional and managerial workers; 108 for clerical, skilled, and retail workers; 96 for semiskilled workers; 95 for both farm and nonfarm laborers; and 94 for farmers.[12]

8. Otis Dudley Duncan, David Featherman, and Beverly Duncan, *Socioeconomic Background and Achievement* (New York: Seminar Press, 1973).

9. *Ibid.*, pp. 77–79.

10. Jensen, *op. cit.*, pp. 15–16; Duncan, Featherman, and Duncan, *op. cit.*

11. Duncan et al., *op. cit.*, p. 89.

12. D. M. Johnson, "Applications of the Standard Score I.Q. to Social Statistics," *Journal of Social Psychology* 27 (1948):217–227, as reported in John B. Minor, *Intelligence in the United States* (New York: Springer Publishing Co., 1957), p. 73.

A large literature in psychology uniformly shows that IQ is positively associated with occupational achievement. The studies range in methodological sophistication from descriptive case studies to statistical analyses of 10,000 careers. Terman's study of gifted children identified a sample of extremely bright youngsters and examined their occupations at about age thirty.[13] More than 70 percent of the males were in either professions or semiprofessions, or higher business occupations. This percentage is more than five times the proportion in these occupations at a similar age in the general population.[14]

Until recently, however, there have been almost no studies in either psychology or sociology which examine the influence of native ability on various forms of social achievement while simultaneously accounting for other status characteristics. Blau and Duncan, in their detailed examination of the American stratification system, presented a basic model of status attainment for American males. But they did not include intelligence as an independent variable.[15]

In the follow-up to the Blau and Duncan study, Duncan, Featherman and Duncan introduced intelligence as an additional independent variable, prior in time to both educational and occupational achievement.[16] In their model, intelligence (as measured at age twelve) is treated as a background characteristic that influences education, occupation, and income. Intelligence is moderately correlated with each form of achievement: .54 with education; .41 with prestige of 1964 occupation; .28 with 1964 earnings.[17] The authors then did a multivariate analysis of the effects of background characteristics, including intelligence, on these three forms of achievement. They used both regression and path analytic techniques to arrive at their estimates. It turns out that intelligence has only a limited direct effect on occupation ($b* = .08$) and on income ($b* = .10$).[18] Its influence on mobility is a result of the effect it has in determining the level of an individual's educational achievement. In fact, intelligence is a considerably stronger determinant of educational achievement than either father's education or father's occupation.[19] The authors summarized the effects of intelligence on occupational achievement: "The direct path from intelligence to occupation . . . is only .08, whereas the indirect path via education is (.40) (.52) = .21, or more than twice as large. The sum of the two, .08 plus .21 equals .29, is . . . the entire effect of intelligence upon occupation, apart from joint effects with the other three background variables."[20] The inclusion of intelligence as a background

13. L. M. Terman, *Genetic Studies of Genius I: Mental and Physical Traits of a Thousand Gifted Children*, 2nd ed. (Stanford, Calif.: Stanford University Press, 1926); L. M. Terman and M. H. Oden, *The Gifted Child Grows Up: Twenty-five Years' Follow-up of a Superior Group* (Stanford, Calif.: Stanford University Press, 1959).

14. Florence P. Goodenough, *Mental Testing: Its History, Principles, and Application*, 1949 (reprinted, New York: Holt, Rinehart & Winston, 1969).

15. Peter M. Blau and Otis Dudley Duncan, *The American Occupational Structure* (New York: John Wiley & Sons, 1967), p. 170.

16. Duncan et al., *op. cit.*, Chapter 5, pp. 69–105.

17. *Ibid.*

18. *Ibid.*, p. 102.

19. *Ibid.*; see Figure 5.11, p. 103.

20. *Ibid.*, p. 102.

variable improves our knowledge of the processes affecting educational achievement. It is a stronger determinant of education than is socioeconomic background, and appreciably increases the explained variance on educational achievement.

In "Ability and Achievement," Otis Dudley Duncan decomposes the explained variance in educational achievement in American society.[21] Three family statuses taken together account for roughly 15 percent of the variance and intelligence accounts for about 16 percent, without considering their joint effects. The joint influence of all the background factors accounts for an additional 11 percent of the variance. Duncan reaches the following conclusion:

> Of the 42% of variance in education accounted for by all four variables (three family statuses and intelligence) 16% is due to intelligence alone. Of the 28% of the variance in occupational status explained by the multiple regression, 9% is attributed solely to intelligence; and of the 11% of earnings variance due to multiple regression, 5% is accounted for by intelligence.... The American ideal of equal educational opportunity is realized in the wide population studied to the extent that progress through the grades of the school system is influenced at least as much by how bright you are as by "who you are"; that the latter, indexed by measures of family size and status, does make a substantial difference in educational outcome apart from the correlation with intelligence, is an indication that the ideal is far from being completely realized at this time. We have quite a way to go before the stratification system will have evolved to the pure meritocracy that Michael Young . . . prophesied.[22]

It turns out, then, that early intelligence has a more substantial independent effect on all three achieved statuses than do father's occupation, father's education, or family size. The data reported by Duncan and others can be interpreted as reflecting a social system in which achievement is significantly determined by native ability. It must also be noted, however, that the very same numerical findings can be variously interpreted. For example, Christopher Jencks, reinterpreting the Duncan data, comes to a very different set of conclusions.

> Economic success seems to depend on varieties of luck and on-the-job competence that are only moderately related to family background, schooling, or scores on standardized tests. . . . Economic success has a rather modest relationship to test scores, and . . . even this relationship derives largely from the fact that standardized tests measure skills that are useful in getting through school, not skills that pay off once school is over. This suggests several things. First, economic success, as measured by occupational status and income, depends on a variety of factors besides competence. Second, competence depends on many things besides basic cognitive skills. Third, standardized tests do not measure basic cognitive skills with complete accuracy.[23]

21. Otis Dudley Duncan, "Ability and Achievement," *Eugenics Quarterly* 15 (1968):1–11.

22. *Ibid.*, p. 10.

23. Christopher Jencks et al., *Inequality: A Reassessment of the Effect of Family and Schooling in America* (New York: Basic Books, 1972), pp. 8, 53. Jencks, in contrast to Duncan, chooses to focus on the variance which cannot be explained by social background characteristics, including native ability. However, by adopting this perspective, and by attributing in effect all of the residual vaiance to "luck," Jencks's argument is conceptually flawed. The data

Many scholars, taking a more circumscribed approach to the problem than Duncan and his colleagues, have examined the relationship between early signs of talent or ability and later achievement among subsets of the overall population. Their curiosity about the relationship between native ability and later achievement is often heightened by the number of magnificent anomalies that they continually encounter. We have all heard of the eminent physician, scientist, or author who was termed hopelessly dull as a child. Attempts have been made to examine the relationship between signs of early talent or native ability and later eminence. Hudson, for example, examined early academic records of a group of fellows of the Royal Society or Doctors of Science. He found that

> the classes of degrees gained by Fellows of the Royal Society or Doctors of Science did not seem to differ significantly—at Cambridge, at least—from those gained by contemporaries of the same age and sex, who took the same subject in the same year, and went into research at the same time, but did not become Fellows of the Royal Society or Doctors of Science. . . . Among research workers, in other words, there seemed no positive relation between degree class and scientific accomplishment.[24]

Donald MacKinnon comes to much the same conclusion in a detailed study of creativity among a small group of architects and scientists.

> As for the relation between intelligence and creativity, save for the mathematicians where there is a low positive correlation between intelligence and the level of creativeness, we have found within our creative samples essentially zero relationship between the two variables, and this is not due to a narrow restriction in the range of intelligence . . . scores on this measure of intelligence (Terman Concept Mastery Test) correlate $-.08$ with rated creativity . . . above a certain required minimum level of intelligence which varies from field to field and in some instances may be surprisingly low, being more intelligent does not guarantee a corresponding increase in creativeness. It just is not true that the more intelligent person is necessarily the more creative one.[25]

The data sets on which these recent studies have been based are all limited. For example, path estimates in the Duncan studies are based upon a subpopulation in the occupational structure—the white, male population between twenty-five and thirty-four years old as of 1964. Although there have been no detailed analyses of the relationship between intelligence and status attainment for other subpopulations, Hauser reports for blacks which indicate close parallels with the white population.

---

that Jencks analyzes are correlations based upon individuals. But Jencks infers that the relationship between variance in education and variance in income on the societal level is equivalent to that on the individual level. This is untenable, as James Coleman and others have pointed out. Coleman, for example, notes: "Their [Jencks et al.] own data, if analyzed at the societal level, suggest that the relation between educational inequality and income inequality is strong indeed, exactly counter to their argument" [James S. Coleman, *American Journal of Sociology* 78 (May 1973): 1525].

24. Liam Hudson, "Intellectual Maturity," in *The Ecology of Human Intelligence,* ed. P. Hudson (Middlesex, England: Penguin Books), p. 268.

25. Donald W. MacKinnon, "The Nature and Nurture of Creative Talent," *American Psychologist* 17 (1962): 487–488.

The correlation between Armed Forces Qualification Test scores and attainment are virtually the same for white and Negro men, and the regression of AFQT scores on attainment is four-fifths as steep for Negro as for white men. This suggests that blacks are not markedly less able than whites to transform years of schooling into competence.[26]

Furthermore, the total absence of samples of women in the studied population represents a significant limitation in our current knowledge. We do not know yet whether the processes of stratification found for men are the same for women, although much current sociological research focuses on this problem.[27] There are almost no published data on the relationship between ability and achievement for women. Finally, on a purely technical level, the correlation coefficients used to estimate path coefficients are drawn from a variety of studies. Their use in a single correlation matrix assumes, of course, that the sample data are drawn from the same population.

What can we in fact conclude about the association between native ability and achievement in Western societies, and in particular American society? If achievement is loosely defined as gaining high social rank through prestigious occupations, high income, and substantial education, then there does seem to be a demonstrably significant influence of native ability as measured by IQ. The additional knowledge gained by considering intelligence in the stratification system is not so much in terms of additional variance which it explains in occupation or income as in the understanding gained about processes that lead to social standing. And this additional knowledge comes primarily in the effects of intelligence on educational achievement independently of other social background characteristics.[28] Intelligence influences educational attainment more than do other social background characteristics, and indirectly affects occupation and income primarily through the payoff for years of schooling. The data suggest that individuals who have extraordinary ability can overcome the disadvantages of their social origins, and that the road to achievement passes primarily through the educational system.

The data do not allow us, however, to assert that meritocratic processes of status attainment predominate in the American stratification system. For one thing, it remains unclear what dimensions of ability are captured in IQ scores, and correlatively, what dimensions are not. Further, the data on which current analysis is based do not cover important subpopulations, females and minority groups being only two that remain unexamined. In addition, while intelligence explains as much variance on achievement as any other identifiable and measur-

---

26. Robert Hauser, "Educational Stratification in the United States," in *Social Stratification: Research and Theory for the 1970's,* ed. Edward O. Laumann (Indianapolis: Bobbs-Merrill Co., 1970), p. 112.

27. There are several major research projects in progress that will shortly be able to fill these gaps. Robert Hauser and David Featherman are doing a continuation and replication of the Blau-Duncan study. This will include, I am told, data for women. Also Donald J. Treiman is currently working on status attainment of women. See Donald J. Treiman and Kermit Terrell, "Sex and the Process of Status Attainment: A Comparison of Working Women and Men," *American Sociological Review* 40, no. 2 (April 1975): 174–200.

28. Duncan et al., *op. cit.,* p. 91.

able variable, none of these variables explains a great deal of variance in forms of recognition. Clearly, social characteristics such as race continue to have substantial independent effects on such rewards as income, even after controlling for education. Otis Dudley Duncan has recently estimated that

> at least one-third of the income gap [between white and Negro men] arises because Negro and white men in the same line of work, with the same amount of formal schooling, with equal ability, whose families are the same size and socioeconomic level, simply do not draw the same wages and salaries. This $1,200, analogous to what Paul Siegal has called the "cost of being a Negro," is in no meaningful sense a consequence of the inheritance of poverty.[29]

Surely the modest coefficients described by Duncan and others are somewhat lower than they would otherwise be because of measurement error. But correcting for measurement problems would not substantially alter the size of these correlations. Plainly, more research is called for on both the direct and indirect influence of measured intelligence on social rewards.

IQ AND ACHIEVEMENT WITHIN OCCUPATIONS

Thus far I have briefly reviewed what is known about the influence of intelligence on achievement in the overall American stratification system. I have said nothing about its influence within particular occupations. Actually, little is known about this relationship. The evidence that is available indicates there are substantial differences in the average IQ's between occupations, but that within occupations the correlation between "success" and intelligence is either near zero or quite low.

Tyler presents data confirming this point for clerks and mechanics who took intelligence and ability tests.[30] Thorndike and Hagen in their book *10,000 Careers* show that occupational groups differ in the patterns of native ability, but "that the degree of success a person will attain within an occupation cannot be predicted from his test scores."[31] The authors used the air force test scores for 10,000 men in 1943 as indicators of ability. Aside from identifying their occupations, the authors assessed individual achievement in terms of salary, advancement, and professional qualifications. They found that "the correlations between success ratings in various occupations and scores on the tests were almost all in the near-zero range."[32] Thorndike and Hagen conclude: "As far as we were able to determine from our data, there is no convincing evidence that aptitude tests or biographical information that was available to us can predict success within an

29. Otis Dudley Duncan, "Inheritance of Poverty or Inheritance of Race?" in *On Understanding Poverty*, ed. D. P. Moynihan (New York: Basic Books, 1968), p. 108.

30. Leona E. Tyler, *Tests and Measurements* (Englewood Cliffs, N. J.: Prentice-Hall, 1963).

31. R. L. Thorndike and Elizabeth P. Hagen, *10,000 Careers,* as quoted in Tyler, *op. cit.*

32. *Ibid.*, pp. 63–64.

occupation insofar as this is represented in the criterion measures that we were able to obtain.''[33]

Ghiselli and Brown, who were concerned with the use of intelligence tests to select personnel, collected data in 1948 from professional journals as well as unpublished sources on the relationship between intelligence tests and on-the-job performance.[34] One hundred eighty-five reports of the effectiveness of the tests were discovered. Performance criteria included proficiency ratings by supervisors, actual production figures, and other measures of role performance. The authors computed correlations between intelligence and job proficiency within different occupational categories. For clerical workers, supervisors, skilled workers, and salesmen intelligence tests had moderate predictive value; they had less value in predicting the performance of sales personnel and unskilled workers.[35] In a later work ''Ghiselli found that intelligence tests correlate on average in the range of .20 to .25 with ratings of actual proficiency on the job.''[36] In 1971 Siegal and Ghiselli reported the relationship between intelligence and salaries among 293 middle-level managers working in five large corporations.[37] Controlling for age, the authors found significant positive correlations between intelligence and annual salaries. Among managers under 35 a correlation of .45 was obtained; for managers between 35 and 39 there was a correlation of .47. For managers over 45 there was no real correlation.[38] This age gradient may be a result of ceiling effects on salaries of middle-managerial personnel within the corporations studied.

For the most part, impressionistic evidence is more abundant than hard data on this relationship. Even sports teams in recent years have given prospective players batteries of intelligence tests in hope of finding a correlation with athletic performance. Even if the data have been analyzed in some systematic way, they have not been made public. Yet coaches have spoken to the issue. For example, Chuck Fairbanks, former coach of the NFL's New England Patriots, has stated:

> If you get a kid from a small school in the South with an I.Q. of 80, you figure maybe with some exposure he can get it up to 100. Now I've seen players who were bright who could not relate their intelligence to football. They weren't very instinctive about it. And I've seen players who were not so bright perform brilliantly without mental mistakes. However, in pro football it just stands to reason that a guy with an I.Q. of 120 is going to do better than one with 80, especially if he comes from a different culture. Yes, they can grow. But if a player from let's

33. *Ibid.*, p. 50.

34. E. E. Ghiselli and C. W. Brown, ''The Effectiveness of Intelligence Tests in the Selection of Workers,'' *Journal of Applied Psychology* (1948):575–580.

35. *Ibid.*

36. Quotation from Jensen, *op. cit.*, p. 16. The original study is by E. E. Ghiselli, ''The Measurement of Occupational Aptitude,'' *University of California Publication in Psychology*, vol. 8, no. 2 (Berkeley, California: University of California Press, 1955).

37. Jacob P. Siegal and E. E. Ghiselli, ''Managerial Talent, Pay, and Age,'' *Journal of Vocational Behavior* 1 (1971):129–135.

38. *Ibid.*, p. 131.

say the University of Michigan tests out with an I.Q. of 80, you forget him because he's not going to get any smarter.[39]

Ann Roe, some twenty years ago, addressed this general problem in her study of sixty-four eminent physical, biological, and social scientists,[40] all of whom were members of the National Academy of Sciences or the American Philosophical Society. She administered to these scientists especially devised and unusually difficult aptitude tests that had verbal, mathematical, and spatial-relations sections. The median score on the verbal test was roughly equivalent to an IQ of 166; the scores ranged, however, from roughly 120 to 180. Scores on the mathematics and spatial-relations sections were also extremely high, although a definite range was obtained. Roe concludes: "It seems quite clear, that having chosen your general field, the particular kind of work you do in it is weighted in some degree to your particular capacities. How well you do in the field is partially a function of your capacity for that particular field, but even more a function of how hard you work at it."[41] Ann Roe may have been a bit hasty in reaching this sweeping conclusion. Aside from the methodological problems Roe faced in administering her tests and drawing inferences from her small sample to the general population, there are other neglected aspects of her work. First, she assumes for the purposes of her analysis that all of the scientists studied are equally eminent. This cannot be the case, since there clearly is a continuum of eminence. Unfortunately Roe does not correlate her test scores with additional independent measures of eminence, such as scales of honorific recognition, or quality of research as measured by peers. Furthermore, since she deals only with eminent scientists, it is impossible to tell whether there is a significant correlation between ability and achievement within the overall scientific community. Clearly, the median scores of the eminent scientists were considerably higher than average, and the existence of a few deviant cases of eminent scientists with relatively low scores hardly offers convincing proof that in the aggregate innate intelligence is not a key factor in scientific achievement.

Other than these few studies that I have mentioned, little evidence speaks to the correlation between ability and achievement within specific occupations. An increasing amount of data on the association between aptitude test scores and job performance is becoming available as a result of Title VII litigation. For cases involving the selection of individuals partly on the basis of standardized tests that measure, *inter alia,* cognitive and verbal skills, the defendants are required to demonstrate that there is a correlation between the test scores and on-the-job performance (*Albemarle Paper Co.* v. *Moody,* discussed in Chapter 2) or between scores and performance in training programs (*Washington* v. *Davis,* also discussed in Chapter 2). This requirement in cases where a prima facie case of

39. Quoted from *New York Times,* Sunday, 15 July 1973.
40. Anne Roe, *The Making of a Scientist* (New York: Dodd, Mead & Co., 1952).
41. *Ibid.,* p. 170.

discrimination has been established has led to the collection of a significant number of data sets that can be used to examine the relationship between predictive tests and actual performance.

I now turn to the problem of ability and achievement within the scientific community. If intelligence is to affect performance within any set of occupations, scientific occupations should be a strategic set upon which to focus. After all, the major goal of scientists is to contribute to the advance of knowledge through original discoveries. The kinds of intelligence scientific work requires are probably more similar to the types of verbal and abstract skills that IQ tests are designed to measure than are the skills required for most other occupations. In other words, the correlation between IQ and achievement may be near zero for clerks and mechanics but significantly strong for scientists. It is not unreasonable, therefore, to hypothesize that scores on tests of native ability would be significantly correlated with the quality of work and ultimate rewards received by men and women scientists. A secondary hypothesis holds that measured intelligence will have a significant independent influence on scientific recognition, above and beyond the effects of other social characteristics of scientists.

Before testing these hypotheses, let us first consider the type of population we are dealing with, and its relative position on the scale of intellectual ability. The average IQ score on almost all intelligence tests is, of course, standardized at a mean of 100. Thus, about 50 percent of the American population has scores in excess of 100. Since the standard deviation on these tests usually ranges from 15 to 20 points, about 67 percent of the population has IQ's ranging from 80 or 85 to 115 or 120.[42] Only about 2.5 percent of the entire American population has IQ's in excess of roughly 130. At least in a statistical sense, individuals with scores higher than this are part of the intellectually gifted elite of the nation. A second and perhaps more appropriate way of setting guidelines for the range of IQ scores is to examine the scores of individuals with varying levels of education. The average AGCT score (the army IQ test) for high school students is 105; for high school graduates, 110; for those entering college, 115; for college graduates, 121.[43] Thus, the average IQ of college graduates is about one standard deviation above the mean.

If college graduates are among an intellectually select group, then scientists are truly members of the intellectual elite. I confine my discussion to those scientists who have received the doctorate, although this degree surely is not a prerequisite to becoming a working scientist. Individual doctorate recipients, on average, have a score of 130 on the AGCT scale. As Wolfle points out, "Less than 7% of the population scores as high, and only slightly more than a quarter of college graduates scores so high."[44] Derek Price reports some variations by

42. Different IQ tests have different standard deviations. The Stanford-Binet has a standard deviation of 15 points; the army IQ test has a standard deviation of 20.

43. Dael Wolfle, *America's Resources of Specialized Talent: A Current Appraisal and a Look Ahead* (New York: Harper & Brothers, 1954), p. 146.

44. *Ibid.*, p. 182.

TABLE 5-1. Mean Intelligence Scores[1] for Several Academic Disciplines

FIELD	PRICE* AGCT	HARMON** N=11,834	COLE*** All Ph.D.'s x̄	s.d.	n	Academics x̄	s.d.	n
Physics	140.3	69.1	70.18	9.01	43	71.5	9.49	22
Mathematics	138.2	69.5						
Engineering	134.8							
Geology	133.3	65.1						
Arts and Humanities	132.1							
Social Sciences	132.0							
Sociology		64.2	65.62	7.71	24	65.21	8.33	19
Psychology		65.5	67.14	8.99	185	67.52	9.24	85
Economics		65.4						
Natural Sciences	131.7							
Chemistry	131.5	65.5	68.03	8.26	63	67.90	8.10	30
Biology	126.1	62.6	64.70	8.57	208	64.72	8.46	140
Biochemistry		64.2						
Education	123.3							

NOTE: 1. To obtain scores that are roughly equivalent to the AGCT, the Harmon and Cole IQ scores can be doubled.

Sources: *Derek J. de Solla Price, Little Science, Big Science (New York: Columbia University Press, 1964), p. 52.

**Lindsey R. Harmon, as reported in John K. Folger, Helen S. Astin, and Alan E. Bayer, Human Resources and Higher Education (New York: Russell Sage Foundation, 1970), p. 446.

***Data adapted from subset of Harmon's.

field, ranging from approximately 140 for physics, mathematics, and statistics to 123 for education (see Table 5-1). Another set of IQ estimates was compiled by Lindsey R. Harmon at the National Register.[45] Since scientists did not all take the same IQ test when they were in high school, the Harmon IQ scale was standardized with a population mean of 50 and a standard deviation of 10. Harmon's data were obtained directly from the high schools attended by the scientists. They were not self-reported estimates. Consequently, the amount of bias in these scores is minimal. Further, that these test scores were obtained at an early point in the educational process reduces the probability of their reflecting the effects of education.[46] Rough approximations of the AGCT are obtained by doubling the National Register figures. The results are much the same. The mean score for approximately 12,000 doctorate recipients was 65.5 with a standard deviation of 8.6. Scores ranged from a mean of 69.5 in mathematics and statistics to 62.6 in the biological sciences (see Table 5-1). In all fields, the averages are extremely high compared to the general population. By and large the differences in the means of the several fields are relatively minor.[47]

Finally, the IQ's of academic scientists are also high on average. Consider the IQ's of the 602 men and women scientists who received their doctorates in 1957 and 1958 and who were academically employed in 1965.[48] Actually, IQ data were available for only roughly half of the academicians, 296 of the 602. Throughout this chapter the IQ correlations reported for academic scientists are based upon this subsample. As we shall shortly see, the average scores of the 296 academicians and those of the larger sample of 523 scientists (both academics and nonacademics) in the five fields under discussion are almost identical. Physicists had a mean IQ of 71.5; chemists, 67.9; biologists, 64.7; psychologists, 67.5; and sociologists, 65.2.[49] Clearly the measured IQ's of academic scientists place them at the upper limit of the intelligence continuum.

But thus far we have only examined mean intelligence scores. If there is no variation to speak of in the IQ of scientists, then the question of the relationship

---

45. Lindsey R. Harmon, *High School Ability Patterns: A Backward Look from the Doctorate*, Scientific Manpower Report no. 6 (Washington, D. C.: National Academy of Sciences–National Research Council, Aug. 1965); Lindsey R. Harmon, "High School Backgrounds of Science Doctorates," *Science* 133 (10 March 1963): 679–688. These data are also reported in John K. Folger, Helen S. Astin, and Alan E. Bayer, *Human Resources and Higher Education* (New York: Russell Sage Foundation, 1970), p. 446.

46. Actually, IQ scores stabilize after about the age of seven. The correlation between IQ test scores for the same individual taken at approximately age seven and again at any point in the teens or later years is extremely high, in the area of .95.

47. Folger et al., *op. cit.*, pp. 444–446.

48. The IQ data that I report is a subset of Harmon's larger data set. The distinction lies in the specification of my sample to academics only.

49. The reliability of IQ scores at the extreme upper limit of the scale could pose a problem affecting the results obtained from this analysis. There is a question about both the validity and the reliability of such scores. Furthermore, some of the variability within the group of scientists may simply result from regression effects. However, the patterns of results obtained are so consistent for both the entire sample and various subsamples that it is unlikely that these measurement problems are seriously distorting the results. In fact, the standard errors for the intelligence scores are very low, for the most part lower than those found for many of my other variables. Moreover, the measurement errors would attenuate the true correlations, making the associations that are found all the more plausible.

between native ability as indicated by IQ and relative achievement in science becomes moot—ability could not be related to achievement. There is, however, significant variability in IQ scores of academic scientists. The standard deviation of IQ within the five fields that we shall consider averages 8.67 points, and is roughly the same in each field: physics, 9.49; chemistry, 8.10; biological sciences, 8.48; psychology, 9.24; sociology, 8.33.

Although I have chosen to concentrate on academic careers, the IQ data suggest that there are no significant differences between the scores of academic and nonacademic scientists (see Table 5-1).

Scientists, then, are clearly among the intellectually elite of the nation. This should surprise no one. The data suggest, however, that there is not a great deal of variability in IQ's of practitioners of the different physical, biological, and social science fields, and this is somewhat less expected. Nevertheless, there is a good deal of variation in IQ within each of the fields. I turn now to the most fundamental question: How is native ability related to scientific role performance and to scientific recognition?

IQ AND RECOGNITION IN SCIENCE

Recent work by Warren Hagstrom,[50] Lowell Hargens,[51] Jerry Gaston,[52] Harriet Zuckerman,[53] and Stephen Cole and myself,[54] among an expanding number of others, indicates that the social system of science adheres quite closely to its universalistic standards of reward. Quality of research is the primary determinant of various forms of honorific and positional recognition.

In our book *Social Stratification in Science* Stephen Cole and I presented a model of the scientific stratification system. In that model we suggested that quality of research, as measured by citations to scientific work, has a strong independent effect on the receipt of honorific awards, the visibility of a scientist's research, and the prestige rank of academic affiliation.[55] The model does not, however, include estimates of the influence of native ability on either research output or the three forms of recognition. The data relating ability to achievement simply were not available. The current data set that I will presently describe has measures for all of these variables except for visibility of research. It is possible, therefore, to expand some aspects of our earlier analysis.

50. Warren Hagstrom, *The Scientific Community* (New York: Basic Books, 1965); "Inputs, Outputs, and the Prestige of University Science Departments," *Sociology of Education* 44 (fall 1971): 375–397; "Competition in Science," *American Sociological Review* 39 (Feb. 1974):1–18.

51. Lowell Hargens and Warren Hagstrom, "Sponsored and Contest Mobility of American Academic Scientists," *Sociology of Education* 40 (winter 1967):24–38; "Patterns of Mobility of New Ph.D.'s Among American Academic Institutions," *Sociology of Education* 42 (winter 1969):18–37.

52. Jerry Gaston, *Originality and Competition in Science* (Chicago: University of Chicago Press, 1973).

53. Harriet Zuckerman, *Scientific Elite: Nobel Laureates in the United States* (New York: Free Press, 1977).

54. Jonathan R. Cole and Stephen Cole, *Social Stratification in Science* (Chicago: University of Chicago Press, 1973).

55. Cole and Cole, *op. cit.* See Figure 3, p. 20; Figure 4, p. 121; and the hypothetical model, p. 117.

Data

The data set examined in this chapter has already been described in Chapter 2. To refresh the memory, it consists of roughly 600 academically employed men and women scientists in the physical, biological, and social sciences, who received their doctorates from American universities in 1957 and 1958.[56] Males and females were matched in terms of year of doctorate, university where the Ph.D. was earned, field, and specialty.[57]

A portion of the career data for the women was obtained from Helen Astin; data on selected social characteristics and social mobility were collected for the entire sample from *American Men and Women of Science;* data on publications and citations came from abstracting journals and the *SCI*.[58] The measured intelligence of the subsample of these scientists was obtained from the Office of Scientific Personnel. These various data files were then merged.

Intelligence in the Process of Scientific Stratification

If the social system of science operates universalistically, it should channel the most able young scientists into the better graduate departments for their training and provide them with the best available resources. Is this in fact the case? As I noted in Chapter 3 and at the outset of this chapter, IQ scores are at best an imperfect indicator of scientific aptitude or ability. Nevertheless, IQ scores are probably as good a general indicator of the native aptitude required to do science as is available. Accepting this limitation, we can state that the social system of science does appear to place, on average, the more able students at the better graduate departments. The correlation between IQ and the quality of graduate schools as measured by Cartter is .26 for the total sample of scientists who are later employed within the academic community. The product-moment correlation for all students in the five fields examined here is .25. There are a number of minor variations in this association when subpopulations are examined. For example, the correlation in the physical and biological sciences is slightly lower ($r = .22$) than in the social sciences ($r = .33$). The association for each of the fields is presented in Table 5–2.[59]

---

56. A preliminary report on these data on women scientists appeared in Chapter 5 of *Ibid.*

57. The matching procedure involved a number of contingency operations. These are fully described in Chapter 3.

58. Citation data were collected for six years: 1961, 1964, 1965, 1967, 1969, 1970. In all counts, self-citations are excluded.

59. A word is in order about the field categories used in this chapter. For the most part, data which are subdivided by field are classed in one of five categories: mathematics and physics, chemistry, biology, psychology, and sociology. At times I employ a dichotomy between the natural and the social sciences. The distributions that are found for each field are a result of the proportion of female Ph.D.'s in these various fields. Physics and chemistry produce a far lower percentage of female Ph.D.'s than do biology or psychology. Since this sample was used for multiple purposes, the distributions between fields are skewed. There is clearly an arbitrary element in the field classifications. In chemistry and biology, for example, there are numerous specialties that have little to do with each other in either substance or method. However, the reward processes may be similar within all of the specialties and fields. There is some cumulating evidence that this is so. Thus, the classificatory scheme is a rough one. It is

TABLE 5–2. IQ and Prestige of Graduate Department Where Doctorate Was Obtained: Zero-Order Correlations

	R (TOTAL SAMPLE)	N	R (ACADEMICS ONLY)	N
All Fields	.253	523	.260	296
Physics and Mathematics	.225	43	.204	22
Chemistry	.208	63	.140	30
Biological Sciences	.279	208	.254	140
Physical and Natural Sciences	.255	314	.219	192
Social Sciences	.236	209	.333	104

NOTES:

1. Product-moment correlations are based upon pairwise comparisons. The numbers listed next to the correlations in this table are the numbers on which the correlation is based.

2. Prestige of graduate department was based upon ratings reported by Allan Cartter in 1965.

Recall that I have already shown in Chapter 3 that the measured intelligence of women Ph.D.'s is on the average slightly, although not significantly, higher than that of male Ph.D.'s; that at every level of doctoral department prestige, women Ph.D.'s have on the average slightly higher IQ's than their male colleagues; and that there is less variability in average IQ's between departments of varying ranks for women than for men (see Table 3-2 in Chapter 3, p. 62).[60] While these differences are not always statistically significant, they are strikingly uniform. In terms of correlation coefficients, there is virtually no difference in the correlation between IQ and quality of Ph.D. department for men ($r = .23$) and for women ($r = .28$). On the whole, then, the more able men and women scientists (who later receive their Ph.D.'s) attend the better scientific training centers. Since the selection of graduate schools by students involves a strong component of individual choice, and is not always based solely on the quality of the department that accepts the student, the moderate association between ability and graduate training site that does obtain suggests that high native ability is more likely than not to be funneled into the better sites for scientific training.

These results have implications for the process of accumulative advantage. The dilemma in dealing with accumulative advantage lies in ferreting out the independent effects of native ability and later achievement. How much of the achievement of scientific recognition is due to superior native ability of honored scientists; how much to superior training and access to capital and human resources; and how much to the interaction between types of social advantages and superior ability? First rough estimates of the independent influence of training and IQ, if not of native scientific ability, can be made with the data in hand. We shall see if quality of graduate department has a strong independent influence on achievement independently of native ability.

Individual intelligence is correlated with quality of doctoral departments, but is it related to status attainment during scientific careers? My analysis is limited to the academic-science community, since I have no data on the reward systems within government and industrial organizations. Measured intelligence is positively correlated with positional success at several points in the first thirteen years of the scientist's career. But it is not correlated with the first job experience. For physics, chemistry, the biological sciences, psychology, and sociology considered as a single group, IQ is uncorrelated with the prestige rank of the scientist's first job ($r = .04$). However, after eight years in the community, roughly in 1965, the correlation between native ability and achievement through appointment to prestigious departments increases dramatically to .24. By 1970, some thirteen years into the careers of the scientists studied, this correlation is reduced slightly ($r = .20$) from the 1965 level. Both of these correlations with rank of

---

employed primarily for practical reasons of dealing with sample size. If I subdivide the sample into molecular biology, for instance, there are simply too few cases to analyze. For the moment, then, I must be content with these rough approximations.

60. Harmon, *High School Ability Patterns* (*op. cit.*). The data are summarized in Folger et al., *op. cit.*, pp. 280-304.

department in 1965 and in 1970 are statistically significant at the .001 level. These data suggest that intelligence may be a significant determinant of achievement of prestigious positions. The same pattern of results is found when the sample is subdivided into men and women. For men the association between native ability and prestige of first job is .02; for women the correlation is .05. Prestige of 1965 academic affiliation is significantly correlated with intelligence for men ($r = .19$) and even more so for women ($r = .30$). Finally, in 1970 the association between IQ and rank in department is only .11 for men but a far more substantial .31 for women. The pattern of association appears to differ somewhat for men and women scientists. IQ is more strongly and continually associated with prestige of department for women than for men scientists. These zero-order correlations are summarized in Tables 5-3 and 5-4. We will want to examine these correlations in greater detail to determine the independent influence of native ability on positional recognition, and why the pattern of influence differs for men and women scientists.

The relationship between IQ and department prestige is slightly stronger within the natural sciences than within the social sciences. In the natural sciences the relationship is minimal at the point of first academic job ($r = .03$), increases to .28 by 1965, and is slightly modified by 1970 ($r = .25$). For the social sciences the relationship is less variable: $r = .13$ for the first job, .16 in 1965, and .15 in 1970.

The general pattern of correlations, then, for the entire sample of scientists—for subpopulations of men and women, and for scientists in the physical and biological sciences as opposed to the social sciences—indicates that native ability is positively associated with the academic positions achieved by scientists in the scientific community.

Honorific recognition may also be influenced by native ability through similar social processes. Actually, the zero-order correlation between IQ and honorific recognition in 1965 (measured simply by counting the total number of honorific awards, including postdoctoral fellowships, that a scientist had received) is modest ($r = .09$). Again some variation is found when scientists are subdivided by sex and type of field. The correlation is .13 for men and .08 for women. In the natural sciences the association is .17; for the social sciences it is virtually zero (.02). The positive association between IQ and honorific recognition for the five fields is, in fact, a result of the correlation between the two variables within only two of the five fields.

Does this mean that meritocratic processes operate in science; that the natively more intelligent reside in the top strata in science? Thus far we have a beginning idea of what the effect of IQ is, but we do not know why it has an effect. Before we can draw inferences about the level of meritocracy, we need to determine the social process through which native ability is translated into positional recognition.

Stage 1   I now want to begin to build a truncated model of occupational career lines in science, by examining the multivariable relationships that influence place

TABLE 5–3. Selected Background and Achievement Variables: Scientists Receiving Ph.D.'s in 1957–58 (Correlation Coefficients)

$\bar{X}$	S.D.	Variables		I	E	O_1	P_1	Q_1	O_2	P_2	Q_2	O_3	P_3	Q_3
									VARIABLES					
66.26	8.89	Intelligence	(I)	—	.261	.037	-.025	.089	.236	.041	.025	.201	-.008	.060
5.53	1.11	Education: Rank of Ph.D. Dept.	(E)		—	.325	.094	.166	.302	.155	.169	.232	.157	.192
3.23	2.70	Occupation at t_1 (1958)	(O_1)			—	.208	.184	.622	.222	.179	.447	.233	.114
*		Productivity at t_1 (1961)	(P_1)				—	.415	.187	.738	.452	.123	.502	.417
*		Quality of work at t_1 (1961)	(Q_1)					—	.070	.337	.791	.006	.261	.529
3.66	2.48	Occupation at t_2 (1965)	(O_2)						—	.232	.125	.764	.258	.154
*		Productivity at t_2 (1965)	(P_2)							—	.431	.213	.653	.433
*		Quality of work at t_2 (1965)	(Q_2)								—	.034	.388	.792
2.91	2.41	Occupation at t_3 (1970)	(O_3)									—	.167	.181
*		Productivity at t_3 (1970)	(P_3)										—	.462
*		Quality of work at t_3 (1970)	(Q_3)											—

*The scientific output measures have been standardized to have a mean of zero and a standard deviation of one.

TABLE 5-4. Selected Background and Achievement Variables: Academic Scientists Receiving Ph.D.'s in 1957-58 (Correlation Coefficients: Men above the Diagonal; Women below the Diagonal)

	I	E	O_1	P_1	Q_1	O_2	P_2	Q_2	O_3	P_3	Q_3
Intelligence (I)	—	.230	.023	.024	.101	.194	.125	.048	.107	.039	.130
Education: Rank of Ph.D. Dept. (E)	.277	—	.300	.093	.169	.300	.164	.188	.177	.173	.213
Occupation at t_1 (1958) (O_1)	.051	.372	—	.175	.166	.602	.219	.160	.434	.225	.081
Productivity at t_1 (1961) (P_1)	-.015	.095	.285	—	.384	.175	.745	.422	.116	.532	.390
Quality of work at t_1 (1961) (Q_1)	.143	.165	.251	.455	—	.034	.310	.785	.001	.240	.532
Occupation at t_2 (1965) (O_2)	.295	.304	.659	.213	.177	—	.238	.100	.721	.279	.146
Productivity at t_2 (1965) (P_2)	.136	.162	.253	.700	.381	.245	—	.390	.228	.695	.414
Quality of work at t_2 (1965) (Q_2)	.144	.143	.297	.482	.817	.268	.465	—	.021	.367	.803
Occupation at t_3 (1970) (O_3)	.309	.318	.471	.191	.078	.846	.233	.256	—	.195	.188
Productivity at t_3 (1970) (P_3)	.171	.151	.274	.357	.275	.219	.396	.353	.131	—	.439
Quality of work at t_3 (1970) (Q_3)	.084	.218	.423	.524	.589	.263	.411	.768	.344	.477	—

VARIABLES

in the occupational structure. In previous work on the reward system of science by Stephen Cole and myself, we found that the quality of work was correlated .28 with department rank for a stratified random sample of 120 physicists.[61] Quantity had a zero-order correlation of .24. The multiple correlation between quantity and quality of research and department rank was only $r/R = .29$. However, quality of research had a greater indedendent influence on appointment than did the sheer size of a scientist's research output. We did find that quality of work and prestige of a scientist's doctoral department have roughly equal influences on the rank of a scientist's academic affiliation some years after receipt of the doctorate. The zero-order correlation between rank of doctoral department and rank of current affiliation is .29. When we controlled for the quality of research, the standardized regression coefficient for doctoral department was .25. The regression coefficient (beta) for quality was .24. These two variables, then, have a roughly equal effect on a scientist's current affiliation. Substantively, the data suggest that both "sponsored" and "contest" mobility operate within science.[62] Although we did not find the site of a scientist's training a strong influence on either visibility or honorific recognition, it was significantly related to mobility between academic positions. In fact, our earlier work on the stratification system of science probably underestimated and did not specify clearly enough the degree and kind of particularism that is often found in appointments to academic departments.

In most previous work in the sociology of science, rank of doctoral department was the only variable used to indicate the quality of individual talent entering the scientific social system. Of course, this was at best a gross indicator of native ability. With the addition of a measure of native ability, such as IQ, we may examine the simultaneous influence of ability and educational origin on recognition.

Let us first consider the aggregate data for five academic disciplines. I begin with a most simple regression model of mobility, which allows rank of doctoral department and IQ to be the sole determinants of the prestige rank of a scientist's first academic job after receiving the Ph.D.[63] The results, reported in Table 5-5, indicate little change from the zero-order correlations. Native ability has virtually no effect on prestige of first appointment ($b* = -.06$) independent of its influence on quality of doctoral department. Quality of doctoral department, however, has a strong influence on first appointment ($b* = .34$). This simple finding tells us a good deal about one feature of life in the academic marketplace. Obtaining a job in academia, whether in the sciences or in the humanities, is not

61. Cole and Cole, *op. cit.*, p. 98.

62. L. Hargens and W. Hagstrom were the first to specify this association in their paper on sponsored and contest mobility (see note 51). Our data corroborated the pattern of findings that they obtained.

63. In this chapter the following subscripts will be used in several tables: i=intelligence; d=prestige of graduate departments; o_1=prestige of first academic job after the doctorate; o_2=prestige of 1965 academic job; o_3=prestige of 1970 academic job; p=scientific productivity; q=quality of scientific work. The reader should note that the subscript for prestige of department is here subdivided into three notations (o_1, o_2, o_3) rather than r for "rank of 1965 department," which was used in Chapters 3 and 4.

TABLE 5-5. The Effect of IQ and Educational Origins on Prestige of Academic Job (Regression Coefficients in Standard Form)

INDEPENDENT VARIABLES	DEPENDENT VARIABLE					
	Prestige of First Job			Prestige of 1965 Job		
	Entire Sample	For Men	For Women	Entire Sample	For Men	For Women
IQ	−.06	−.05	−.06	.21*	.16*	.26*
Rank of Ph.D. Department	.34*	.31*	.39*	.05	.09	.01
First Job				.59*	.56*	.64*

NOTE: Level of statistical significance is .001 for the entries marked by an asterisk. The other coefficients are not statistically significant.

strictly a matter of open competition. Jobs are obtained on the basis of interviews or oral presentations about some aspect of the scientist's work. But in order to get a "foot in the door" of a major department, a scientist has to become known. This is often not done in a strictly universalistic way. Although government affirmative-action requirements have forced many departments to formalize their hiring procedures, the process of hiring still begins most frequently through informal communications between acquaintances at different departments. Naturally almost all departments look to the few superior training centers for candidates. Surely the best departments try to recruit new assistant professors from other similarly evaluated departments. At least in these limited terms, an "old boy network" is still very much alive. Friends try to place their most capable graduate students in departments where other scientist friends are located. Thus, a reciprocal sponsorship system is established, which, some critics claim, closes the door to able young scientists who did not "grow up" in the finest academic departments. The claim that this technique for hiring recent graduates involves a high degree of particularism is not totally accurate. In a social system in which it is difficult in any case to make accurate predictions about the ultimate achievements of any but a rare few young scientists, the use of knowledgeable informants as "truffle dogs" to identify new talent may greatly increase the level of efficiency without much loss in accuracy. At the time they receive the Ph.D. degree, few students have demonstrated their scientific competence either as teachers or as researchers to the point where they have reputations beyond their own graduate departments. Those few who do publish prior to receiving their degrees often do so as collaborators with older, more eminent scientists. Even in such cases it is difficult to assess the relative contribution of the young scientist. Most often it is assumed (sometimes falsely) that the senior collaborator is primarily responsible for the paper.[64] Thus, thoroughly objective criteria for evaluating the competence or potential of young Ph.D.'s are rarely available. In the absence of publications, written recommendations by trusted colleagues at other departments may be the best available indicator of scientific potential, especially when this is coupled with personal interviews and oral colloquia.

This process of identifying talent in junior scientists assumes a high correlation between native ability and the quality of the graduate department attended. The data on later career performance tend to bear this out, although it could be argued that the monopoly of virtually all highly prestigious awards and positions by graduates of top departments is all a result of a self-fulfilling prophecy. Moreover, this process of recruitment assumes that specific forms of specialized training are virtually monopolized by the top graduate departments. Thus, some senior academic scientists might argue that while natively able young scientists can be found outside the top training sites, they represent undeveloped talent

---

64. For a discussion of this process, see Robert K. Merton, "The Matthew Effect in Science," *Science* 199 (5 Jan. 1968):55–63.

which will not fill the current needs of their department. Nonetheless, the data suggest that there is a strong element of sponsorship in obtaining one's first academic job.

The data presented in Table 5-5 indicate one feature of accumulative advantage. Young scientists, both men and women, with roughly equal native ability but trained at graduate departments of varying prestige simply do not have the same opportunities in the job market. Native ability being equal, graduates of top departments are considerably more likely than graduates of lesser departments to obtain prestigious first appointments. This pattern is maintained whether we examine men and woment scientists together or separately.

Stage 2   If IQ has little independent effect on obtaining a first job, does it have any independent influence on positional success at later points in the scientist's career? The zero-order correlations reported above suggest that IQ may have some direct effect. Let me elaborate upon the simple model of mobility. Table 5-5 presents a model of the influence of IQ, rank of doctoral department, and prestige of first job on the 1965 academic affiliation of scientists. IQ and doctorate department remain the only determinants of a scientist's first job. All three variables are presumed to have direct effects upon the prestige rank of a scientist's department in 1965. Intelligence and doctoral training site also are hypothesized to have an indirect effect (however small) on department rank in 1965 through their influence on a scientist's initial job location.

The striking finding reported in this model is the complete reversal of the relative influence of native ability and source of graduate training on 1965 affiliation, compared with their effects on initial placement. In briefest summary, this model indicates that native ability has a definite payoff in the academic marketplace. Controlling for the prestige of first job and doctoral training site, scientists who are natively more able, in terms of IQ, are more likely to be located in superior departments in 1965.

The regression coefficients indicate that intelligence has a significant direct influence on the quality of a scientist's 1965 academic affiliation after accounting for these other antecedent variables ($b^* = .21$). Of course, first-job prestige has by far the strongest effect on 1965 job status ($b^* = .59$). The magnitude of this effect is, in part, simply the result of individuals remaining at the location of their first jobs for an extended time. Nonetheless, it suggests the importance of initial placement within the academic social system. Scientists, once in a given prestige stratum, tend to remain in that stratum. Most of the mobility that does occur is downward—away from the larger and more prestigious departments.

These results are quite logical. Plainly a scientist's educational origin has a determinant effect at the beginning of his or her career. It helps launch the individual. This is particularly so in the absence of personal indicators of ability other than the recommendations of older, established scientists. But once scientists have begun their careers, their social and intellectual background has little direct influence on later positional success. Correlatively, verbal and manipulative skills, as measured by IQ scores, have little effect at the outset of a career, except in the not inconsiderable way that they aid a student in getting admitted to a high-quality graduate school. Other than its appearance on aptitude exam- inations, such as the Graduate Record tests, the manifestations of high IQ are not immediately visible to the scientific audience. Since there has been little time to measure scientific ability among graduate students by what the scientist actually produces, the minimal effect of IQ on quality of first job is not surprising. Once a scientist is initially placed and begins to interact extensively with new col- leagues, a new basis of evaluation emerges. Certainly, whether he or she came from Harvard or Berkeley no longer carries as much weight in the evaluation process. Correlatively, whether the scientist was educated at a second-rate uni- versity but showed enormous potential no longer substantially affects opinion about her or his competence. The collegial evaluation of ability takes on greater significance. Thus, the verbal qualities that are monitored by IQ scores may work to the benefit of the young scientist at the new academic location.

Let us examine the pattern of differences between men and women. These differences are reported in Table 5-5. In general, IQ has a positive effect on the social location of scientists of both sexes. In fact, all of the variables in the model affect academic affiliation in similar ways. IQ has no effect on the prestige of a scientist's first job, whether the scientist is male or female. However, the mag- nitude of the effect of IQ on achievement seven years later, in 1965, does differ for men and women. Women's IQ's have a considerably greater influence than the IQ's of men on positional recognition. The regression coefficient in standard form for IQ among women scientists is about twice that for men. The product- moment correlations are not reduced by controlling for prestige of first job and source of Ph.D. How can we explain this difference?

Consider a *post factum* interpretation. The data to test an interpretation are lacking. The same interpretation probably explains the pattern for both sexes. When rational criteria for promotion or retention in academic positions do not call for a clear choice, other factors affect the decision. For instance, when neither scientist has produced any published papers, or when both scientists have produced roughly an equal number of papers which are perceived as roughly equal in impact or quality, then other variables become more important in appointment decisions. Under such conditions sex, race, or other statuses may affect decisions, and perceived intelligence may also be influential. This may be even more likely for women, because fewer women than men are prolific scien- tists. Thus, at any given level of scientific productivity the women with the higher IQ's are likely to be located in the more prestigious positions.

In attempting to interpret the results presented in Table 5-5, a puzzling set of questions arises. Why does IQ have these effects on positional recognition? Why are scientists with higher IQ's more likely than those with lower scores to be located at higher-ranked academic departments, even after we control for pres- tige of their first job and the quality of their doctoral department?

A logical hypothesis is that scientists of greater native intelligence are placed at the better training sites, can make better use of available resources, and are more likely to work as apprentices to older scientific leaders; by coupling these contextual social advantages with their superior native ability, they can convert these assets into high-quality scientific discoveries. These contributions to scientific knowledge are then rewarded through appointments to outstanding academic departments. Thus, the variable that logically intervenes between native ability and achievement would be quality of scientific role performance. This argument implies that the model presented here is incomplete, since it fails to include any indicator of scientific role performance. If quality of research output is the key intervening variable, then the new logical chain of causation is clear. High native ability should be correlated with high scientific aptitude. In turn, scientific aptitude should be positively related to the production of new discoveries of some import. Finally, these contributions should in due course be recognized by appointment to leading science departments.

Intelligence and Scientific Output

Let us attend to this critical intervening variable—scientific role performance—and see if it really does interpret the relationship between native ability and positional recognition. Recall an earlier point. Virtually all scholars who have examined the relationship between native ability, on-the-job role performance, and recognition have concluded that IQ has little predictive value. Other than Ann Roe's first efforts, there has so far been only one systematic study of the relation between scientific research output and intelligence.[65] Alan Bayer and John Folger examined this relationship for 224 biochemists. They found an insignificant correlation ($R = -.05$) between IQ scores and the total number of citations listed after a scientist's name in the *Science Citation Index*.[66] If citation counts are taken as a rough indicator of research quality, then the finding suggests no relationship between native ability and this form of scientific performance.

My data can be used also to examine this relationship. The results in large part corroborate the findings of Bayer and Folger. Intelligence is almost totally unrelated to scientific productivity as measured by counts of papers (see Table 5-3). There is no correlation to speak of between intelligence or native ability and sheer quantity of research within the first eight years of the scientific career ($r = .05$). The correlation between scientific productivity and intelligence for the first thirteen years of the career is essentially zero, $r = -.03$.[67] Neither figure is

---

65. A number of studies have tried to produce criteria for selecting fellowship recipients that correlate later performance with early indicators of aptitude. Harmon has done one such study, but he does not report correlations between IQ and later productivity. See Lindsey R. Harmon, "The Development of a Criterion of Scientific Competence," in *Scientific Creativity: Its Recognition and Development*, ed. Calvin W. Taylor and Frank Barron (New York: John Wiley & Sons, 1963), pp. 44–53.

66. Alan E. Bayer and John Folger, "Some Correlates of a Citation Measure of Productivity in Science," *Sociology of Education* 39 (fall 1966):389.

67. These product-moment correlation coefficients differ slightly from those presented in Table 5-3 (p. 161). Here

statistically significant. Intelligence is only slightly correlated with citations to scientific output during the shorter period ($r = .09$) and is only slightly lower for the thirteen-year period. I have found no significant change in these correlations when the physical and biological sciences are grouped together and compared with psychology and sociology. Within the more mature sciences, the association between IQ and publication counts is .04 for the shorter period and .02 for the longer one. There is no change in the correlation between intelligence and research quality: $r = .08$ for the first eight years and .09 for the longer period. For the social scientists the correlation is .05 between IQ and productivity, and .12 between IQ and quality for both points in time. We can conclude that among the various fields differences in the relationship are minimal.

However, some variation is obtained in the IQ and quality of research association when the sample is divided into men and women. Over the shorter career span, the correlation between native ability and research performance is twice as strong for men ($r = .20$) as it is for women ($r = .08$), with the difference being slightly less over the longer span ($r = .19$ for men; $r = .12$ for women). The association between IQ and quantity of research is roughly the same for both men ($r = .12$) and women ($r = .16$). For men there does seem to be some positive, albeit not very strong, relationship between native ability and later scientific output. Intelligence is translated into higher-quality work. There is substantially more variance among men in the quality of work, as measured by citation counts. This may account for the stronger correlation for men than for women. There is little variance in the quality of work produced by this sample of academic women, yet a substantial variability in their IQ's.[68] With women we are, in effect, explaining a constant with a variable.

Why isn't there a stronger relationship between native ability as measured in early adolescence and the actual performance of scientists during their careers? Consider several possible explanations. Intelligence scores are based upon manipulations of words and numbers. By design, they are intended to predict performance in school, not the likelihood of translating these skills into later achievement. They certainly are not intended to measure all the variables that interact to produce creative work. There is no special attempt to measure scientific talent in IQ tests. Although there may be close linkages between the skills required to score well on these tests and the skills necessary for scientific creativity, there surely is not a perfect correlation between the two. Furthermore, psychologists have often distinguished clearly between intelligence and creativity. At best intelligence is an indicator of potential. It may be necessary to master certain skills required for creative science, but no one will argue that such mastery is or

the correlations are based upon the total number of years from the doctorate to the point of final measurement. Thus, the total for the years 1957–61 is a subset of the total from 1957–65. In the matrix the productivity scores were mutually exclusive of each other with no overlap. Thus the minor differences in the correlations. This also holds for the correlation in the following paragraph.

68. The variance in citations to the work of women scientists is less than half the level found for men for both the period from 1957 to 1965 and for the period 1957 to 1970.

should be sufficient for producing scientific knowledge. In short, it may be impossible to make significant contributions to science without a fairly high IQ—undoubtedly quite high compared to general population averages. However, having an IQ of 150 or more will certainly not ensure scientific "stardom," or even the discovery of important scientific results. Other social and psychological factors are necessary ingredients for original discoveries. Scientists themselves continually report how serendipity, timing, risk-taking behavior, perseverance, and incredible outputs of energy play an important role in their discoveries.

Another possible reason for a lower than expected correlation between native ability and the quality of scientific research lies in imperfect social arrangements which impede the productivity or creativity of talented young scientists. If such scientists have not attended the best graduate departments they may have difficulty, once engaged in research, in finding the conditions that are conducive for producing important work. We know that the correlation between IQ and quality of doctoral department is only about .30 on the individual level. This suggests, of course, that there are many natively intelligent students who are trained at relatively poorer graduate schools. It is difficult to assess the extent to which creative ideas put forward by natively bright young scientists are resisted—discouraging further research efforts by them. This seems unlikely, however, to be commonplace. Empirical evidence suggests that most papers submitted for publication in the physical sciences are accepted[69] and that there is little resistance to new ideas in the mature sciences.[70] Resistance may be a more significant problem in the social sciences; for these disciplines it is often hypothesized that there is less codification and consensus as to what constitutes acceptable work. If particularistic criteria influence the evaluation of new ideas, then the correlation between ability and achievement is likely to decline.

Finally, IQ tests were developed for use in the entire population. We have noted that there are marked differences, on average, in the IQ's of individuals in occupations of varying prestige. Thus, these measuring instruments simply may not be capable of making the fine distinctions that are necessary to predict scientific potential.

What, then, is the relative influence of intelligence, actual research performance, and social characteristics on scientific achievement? Given the low correlation between the two independent variables, IQ and quality of research output, it is clear that the track record of an individual will not interpret the association between IQ and the prestige of his or her academic position. Let us consider their relative significance in determining a scientist's academic location.

In order to see these relative effects clearly I have at once simplified and expanded the model of career mobility. I have eliminated the influence of first

---

69. Harriet A. Zuckerman and Robert K. Merton, "Patterns of Evaluation in Science: Institutionalization, Structure and Functions of the Referee System," *Minerva* 9, no. 1 (Jan. 1971):66–100; reprinted in Robert K. Merton, *The Sociology of Science* (Chicago: University of Chicago Press, 1973), pp. 460–496.

70. Cole and Cole, *op. cit.*, Chapter 7.

job on 1965 job, and have added an indicator of research performance. In a single regression equation, I regress prestige of a scientist's academic department on intelligence, quality of doctoral department, and productivity (total number of published scientific papers).[71]

The results reported in Table 5-6 indicate almost an equal influence for these three variables on prestige of academic department. After accounting for graduate school prestige and productivity, intelligence has a partial regression coefficient (in standard form) of .17. The independent effect of quantity of research is .19; the independent effect of graduate school quality is .23. Both beta coefficients for intelligence and quantity are very similar to the zero-order correlations. The effect of intellectual background, or training location, is slightly reduced from a zero-order correlation of .26. The same general pattern of results is obtained if we control for quality of research output rather than quantity. Intelligence has a regression coefficient of .16 after controlling for research quality and graduate school quality. Thus, it is reduced one-third from the strength of the zero-order correlation. Here, too, the coefficients are all roughly equal in strength ($b^* = .15$ for quality; $b^* = .23$ for rank of doctoral department). Each of these beta coefficients is statistically significant. Although these variables have roughly equal effects upon prestige of academic department, the coefficient of determination is only about .40, suggesting that other factors not included in these simple models strongly influenced prestige of academic department.

The reduced form models presented in Table 5-6 suggest that IQ has almost as great an influence on prestige of department affiliation in 1965 and 1970 as do both rank of doctoral department and indicators of scientific output. When we examine men and women scientists separately, are the same results obtained? Intelligence has a greater influence on academic location in 1965 for women than for men. When departmental prestige in 1965 is regressed on measured intelligence, quality of research, and educational background, the beta coefficient is .22 for intelligence, .26 for quality of research, and .18 for prestige of doctoral department. For male scientists the strength of the beta for intelligence is just over half that found for women ($b^* = .12$). In these reduced form models, IQ continues to have an effect on prestige of department in 1970 for women. It does not for men. This observed difference can be anticipated, of course, by examining the product-moment correlations between IQ and departmental prestige for men and women scientists in 1970.

What does all this say substantively? It suggests that the prestige of a scientist's academic department is as much a result of her or his measured intelligence as of either research performance or the quality of the graduate school attended. Furthermore, intelligence has a direct influence on this outcome. It is not only indirectly influential through the quality or quantity of research produced. In fact, research performance does not interpret the relationship between IQ and

---

71. This analysis has been done for academic scientists only—both men and women in the natural and social sciences for whom IQ data were available.

positional recognition. Even if we introduce prestige of first job into the regression equation and retain the performance variable, intelligence continues to have a substantial independent effect ($b^* = .20$). With the addition of prestige rank of first job, the coefficient of determination increases significantly to $r/R = .67$.

Stage 3   Does intelligence continue to influence the rank of a scientist's department at still later points in the scientific career? My data on the 1957 and 1958 Ph.D.'s continue through 1970. I reported above that the correlation between IQ and prestige of 1970 department is .20; between IQ and quantity of research for the thirteen-year period, .03; between IQ and quality of research, .09. These figures suggest that the multivariate relationship between these three variables will not reduce significantly the effect of IQ on prestige of department. In fact, there is virtually no reduction. Regressing the prestige of 1970 department on IQ, quantity of work, and rank of Ph.D. department reduces the zero order of .20 to a beta coefficient of .15. Number of research papers and rank of doctoral department have slightly, but not appreciably, stronger effects on the 1970 affiliation. When prestige rank of first job is added to the regression equation, the coefficient of determination increases from .33 to .50. In short, native ability does have a significant influence upon departmental prestige, after accounting for scientific output and intellectual origin.

Measured intelligence has a stronger independent influence on 1970 prestige of department for women than for men (see Table 5-6), just as it did when we considered its effect on 1965 affiliation. For men, the influence of IQ, controlling for scientific output and prestige of doctoral department, is .06; for women, measured intelligence has a beta of .22, reduced from the product-moment level of .30.

If quality and quantity of research performance do not interpret the relationship between native ability and achievement some eight years into the scientist's career, how can the effects of IQ be explained? I cannot "wash out" this relationship with the data I have in hand. I can only speculate on why intelligence does influence appointment and retention at top departments.

There are at least two possible interpretations, and they are compatible with each other. The first might be termed the "house intellectual" explanation. If a science department has two scientists who have produced roughly equal amounts of work but it can rehire only one of them, quite likely it will choose the individual who distinguishes himself or herself in interpersonal interaction—in short, the apparently "brighter" of the two. Within the academic world, there are tangible social rewards for being labeled "brilliant." Moreover, every science department is populated with men and women who have produced discoveries of varying impact. For the most part, even at the most distinguished departments, only one or two persons are truly eminent figures in their fields. Most faculties are made up of very able but less distinguished individuals. Among these scientists it is not unreasonable to find a payoff for being perceived as extraordinarily bright. Almost by definition, colleagues are likely to find articu-

TABLE 5-6. Influence of IQ, Research Performance, and Source of Ph.D. on Prestige of Academic Job (Reduced Form Models; Regression Coefficients in Standard Form)

INDEPENDENT VARIABLES	DEPENDENT VARIABLE					
	Prestige of 1965 Job			Prestige of 1970 Job		
	Entire Sample of Academic Scientists	Men Only	Women Only	Entire Sample of Academic Scientists	Men Only	Women Only
IQ	.17 Δ .17 Δ (.16) Δ	.13* .12* (.12)*	.23 Δ .21 Δ (.22) Δ	.15* .15* (.14)*	.07 .06 (.05)	.24* .22* (.23)*
Rank of Ph.D. Department	.26† .23† (.25)†	.27 Δ .24 Δ (.25) Δ	.24 Δ .22 Δ (.18)*	.19 Δ .16* (.16) Δ	.16* .13* (.12)*	.25* .23* (.24)*
Research Performance	.19 Δ (.14)*	.18 Δ (.11)*	.18* (.26) Δ	.19 Δ (.15)*	.21* (.19)*	.15* (.07)

NOTE: Regression results using citations, rather than total number of papers, as a measure of research performance appear in parentheses. Differences in levels of statistical significance for similar coefficients result from differences in the size of the samples.

Level of Statistical Significance:

* = .05 level
Δ = .01 level
† = .001 level

late, verbal, witty, and well-read academics more interesting people to talk to about almost anything, including their intellectual ideas, than persons who seem duller. Along the same line, every department seems to have several members who have published very few papers of significance, yet are reputed both inside and outside the department to be "brilliant," "fountainheads of knowledge," or extremely imaginative in handling scholarly problems. These individuals, who for one reason or another cannot get down to the business of writing scholarly papers, usually are welcome additions to a department. They are the "house intellectuals" who add a touch of class.

These same individuals, conceded to have enormously fertile minds while being mysteriously unproductive, fulfill other important functions within the department. Often we overlook the fact that academic science departments, the source of much of the basic science in the nation, are also organizations. In order for these organizations to operate effectively, a multitude of needs must be met—needs which in the long run facilitate the production of outstanding work. Many very bright scientists, who are not the stars of the department, are among the best instructors of science, spend hours interacting with undergraduates and graduate students, hours criticizing others' work, and hours performing essential administrative functions. True, they are not placing new scientific ideas in print, but this does not mean that they do not have them and exchange them with colleagues. Only a few members of any department produce significant scientific contributions. Why not fill the other slots or promote to higher rank those who seem to have exceptional minds? Of course, this seems to assume that a department can predict which of its faculty will make fundamental contributions. Actually, the assumption is that departments initially hire individuals on the basis of scientific potential and retain or promote some of them not solely on the basis of their actual publication achievements. This is exactly what the models suggest. Holding productivity constant, natively brighter individuals, who initially get jobs at top departments, are more likely than less natively able scientists to get promoted or have their contracts renewed.

Consider a different kind of explanation for the results presented in Table 5-6. Brighter individuals are better able to understand the norms that operate within the academic community. They understand the reward structure of the community better, and therefore are better able to control their immediate environments in a way that will pay off for them. They are superior in their presentation of self to colleagues and in the use of other skills needed to "make it" in the academic marketplace.

Rewards in science are not simply a function of the actual quality of a scientist's work.[72] Absolute standards are nonexistent. Far more significant is the perceived quality of one's work by colleagues—and, for that matter, the perceived quality of one's mind. There are, of course, many institutional

---

72. For a discussion of the relationship between quality of scientific work as measured by citations and quality of work as perceived by one's colleagues in the field, see Cole and Cole, *op. cit.*, Chapter 4.

mechanisms for conferring the label of "brilliant" on young scholars. The perception of their capabilities results partially from the way young scientists present themselves to others.[73] Some are masters of self-presentation; others fail miserably. Almost everyone who has spent some time around academics knows there is an imperfect correlation between actual talent, as demonstrated in publications, and perceived talent, which is often exhibited orally. The presentation of self is often critical in placing bright young scientists in initially advantageous positions. This can be seen in the making of "subterranean reputations" among the neophytes. Many gain reputations as potential stars on the basis of their quickness, cleverness, and analytic skills transmitted in informal interpersonal association, or by their ability to give distinguished presentations of their ideas at informal colloquia. In fact, in some scientific disciplines there has developed an informal lecture circuit—a university speaking tour on which very bright junior scientists are booked to present some aspect of their current work. Solely on the basis of listening to these talks, offers of outstanding positions frequently are made. Thus rather prestigious, if not lofty, positions may be achieved without putting much into print. Since oral reputations may come more easily than those earned through written work, there may emerge some reluctance among the aspiring stars to commit many of their ideas to print. Once published, the ideas are open to closer review and greater potential criticism. Why risk deflating a growing reputation and the social rewards which are its consequence by publishing? Of course, most of the young, articulate, new stars do publish their ideas at a relatively high rate and do confirm the earlier social judgments. However, in some cases, perhaps in sufficient numbers to produce part of the observed correlation, these extremely bright young scientists do not produce much published work, but do receive recognition through appointment to outstanding science departments.

Consider another aspect of this interpretation. Intelligence can be used in various ways. Although the proposition remains untested, natively more intelligent individuals, on average, may be better able to perceive the organizational structure of authority, power, and influence within a department. They may use this ability to further themselves once they are located in outstanding academic settings. F. Merei,[74] Kalma Chowdhry and T. M. Newcomb,[75] and Alex Bavelas,[76] among many other social psychologists, have found that the ability to quickly comprehend the structure of group norms is associated with ability to become a group leader. Their work suggests a social structural component, beyond personality, which facilitates or hinders an individual who attempts

---

73. On the presentation of self, see Erving Goffman, *The Presentation of Self in Everyday Life* (New York: Doubleday & Co., Anchor Books, 1959).

74. F. Merei, "Group Leadership and Institutionalization," *Human Relations* 2 (1949):23-29.

75. Kalma Chowdhry and Theodore M. Newcomb, "The Relative Abilities of Leaders and Non-Leaders To Estimate Opinions of Their Own Groups," *Journal of Abnormal and Social Psychology* 47 (1952):51-57.

76. Alex Bavelas, "Communications Patterns in Task-Oriented Groups," in *The Policy Sciences*, ed. Daniel Lerner and Harold Lasswell (Stanford: Stanford University Press, 1951), pp. 193-202.

to assume a leadership position. Perhaps intelligence also permits individuals to seize their opportunities by comprehension and manipulation of group dynamics.

Surely variables besides "intelligence" and "track record" affect the process of mobility through the academic hierarchy. Psychological predilections—including such sensibilities as a distaste for predominantly instrumental relationships; a disdain of the hypocrisy often required to manipulate the academic system; a "moral" repugnance toward "jungle tactics" that are sometimes required, in which only the "fittest" survive; and a feeling that the means subvert or defeat the ends—may all come into play in determining whether one or another individual "makes it" in the academy. The question that must go unanswered for now is: To the extent that such requirements exist in the scientific community, do they work against women more than for them? For instance, does participation by women in the game of making it in science become self-defeating insofar as they are then labeled by their male superordinates as being "pushy" or "aggressive" and consequently unworthy colleagues?

We know that there is greater particularism in the structure of appointments and promotions in academic science than in the distribution of other forms of rewards. Although appointments may not involve significant discrimination by sex, race, religion, or age, other forms of particularistic behavior may operate. For example, the structure of friendship surely could influence the offers from and promotions within a department. Most social scientists examine the relationship between status characteristics and recognition for clues to particularistic behavior. An investigation of sociometric patterns of friendship may, in the end, tell us a good deal about how particularism actually operates within academic departments. Conceivably, highly intelligent scientists make superior choices as to which colleagues to impress or befriend in order to gain an advantage over peers with publication records as good as their own. In this way intelligence is linked to the political structure of recognition in academic life. Caplow and McGee catalogue the testimony of academic men and women who admit that the road to success may lie more in one's ability at bridge than in science.[77] Native intelligence may be related to the realization that the management of personal relationships is as important to this form of career advancement as are scholarly publications.

This interpretation is consistent with James S. Coleman's data on adolescent subcultures within high school communities.[78] One of Coleman's most striking findings was that the natively most able students (measured by IQ scores) were rarely the ones who achieved the most scholastically. Rather, the most intelligent youngsters better understood the value system that prevailed in their schools—a system in which membership in the leading crowd was determined by athletic

77. Theodore Caplow and Reece J. McGee, *The Academic Marketplace* (New York: Basic Books, 1958).

78. James S. Coleman, "Adolescent Subculture and Academic Achievement," *American Journal of Sociology* 65 (1960):337–347.

ability and popularity more often than by academic brilliance. Among women, social popularity shaped recognition and social approval. Coleman found that the brightest youngsters, in terms of IQ, gravitated to where the rewards were located—and less natively able youngsters held the top academic ranks in the schools.

> In a school where such [academic] achievement brings few social rewards, those who "go out" for scholarly achievement will be few. The high performers, those who receive good grades, will not be the boys whose ability is greatest but a more mediocre few. Thus the "intellectuals" of such a society, those defined by themselves and others as the best students, will not in fact be those with the most intellectual ability. The latter, knowing where the social rewards lie, will be off cultivating other fields which bring social rewards.[79]

Now, this general pattern surely does not predominate in science. However, we often mistakenly believe that what determines or affects motivation among scientists in the top five or ten academic departments holds for the entire scientific community. Most scientists, bright and otherwise, never are appointed to the few distinguished departments. Most are located at third- and fourth-rank universities or nonuniversity science departments in colleges. In these settings research performance is not as strongly emphasized. In fact, in some settings too much research activity may actually be frowned upon by senior colleagues who are in positions to give promotions and renew or terminate the contracts of young scientists. Within these contexts, the application of intelligence to means of getting ahead other than publication may make the difference between location at a second-level university or at an obscure college.

Within the academic community there are, of course, multiple means which may be used to obtain certain desired ends. Although fraternizing with the most influential members of the department won't help a scientist win a Nobel Prize, it might affect the renewal of his or her contract. Perhaps if intelligence were more frequently directed toward scientific research, the correlation between intelligence and scientific output would be stronger. The application of intelligence to the end of making it within the academic community suggests that it is very difficult to estimate the role of "luck" in the reward process. Outcomes which may appear to be the result of "luck" may be the result of the rational application of certain skills and intelligence to the management of one's immediate environment.

I have been describing and interpreting elements in the career histories of scientists. Data have been presented on the influence of educational background, intelligence, and scientific role performance in achieving departmental affiliations within academic science. I now want to pull these various stages together, by presenting a causal-chain model of the academic-science career.

---

79. *Ibid.*, pp. 340–341

A Causal-Chain Model of Aspects in the Academic-Science Career

As my data are longitudinal, it is possible to follow the same scientists through various career stages, from the point of initial entry into the academic occupational structure to their occupational locations some thirteen years after the doctorate. No synthetic cohorts are needed to simulate longitudinal data. These data have an additional advantage over typical panel data obtained through interviews. They are collected from biographical compilations and publication abstracts, and are therefore less likely to be biased through case losses and processes of self-aggrandizement.

The first fifteen years are unquestionably both the formative and crucial ones in the overall career history of a scientist—especially in terms of academic appointments. Accumulating evidence indicates that if scientists are ever to make significant contributions they must begin when quite young.[80] While many scientists must wait until their fifties and sixties to achieve formal recognition (often for work done when they were much younger), a scientist's position in terms of academic appointments is fairly well set within the first thirteen years. The data already discussed suggest this.

Several causal-chain models of career development have been proposed in the recent stratification literature. Blau and Duncan present one model for the American occupational structure. The model is very simple. It allows the occupational prestige of an individual's first job to be a function of his or her socioeconomic background (father's occupation is the indicator used) and educational achievement. At each successive stage in the individual's career, her or his occupational status is directly dependent upon immediately preceding occupational status, upon educational attainment, and upon father's occupation.[81] One significant feature of this model is that current occupational status is not directly influenced by the status of occupations twice removed from the current position. Unfortunately, Blau and Duncan did not have all of the panel data necessary to make a strong test of their model (correlations of occupational status at different points in time were missing). They constructed synthetic cohorts to simulate panel data and borrowed some correlations from ability studies done in Chicago and Minneapolis. In the Chicago set, present occupational status was correlated .55, .77, and .87 with the occupational status that immediately preceded it.[82] The path estimates presented by Blau and Duncan indicate that present occupational status depends increasingly upon one's immediate prior occupational status and less on family background and education.

There has been a debate within recent sociological literature over the empiri-

---

80. Harriet Zuckerman and Robert K. Merton, "Age, Aging, and Age Structure in Science," in Matilda White Riley, Marilyn Johnson, and Anne Foner, eds., *A Sociology of Age Stratification*, vol. 3 of *Aging and Society* (New York: Russell Sage Foundation, 1972); Stephen Cole, "Age and Scientific Performance," *American Journal of Sociology* 84 (January 1979):958–977.

81. Blau and Duncan, *op. cit.*, pp. 183–184.

82. *Ibid.*, p. 185.

cal accuracy of the Blau-Duncan model. David Featherman, using actual panel data, has questioned aspects of it.[83] Jonathan Kelley, using Featherman's own data, has defended the original model.[84] The key issue in these debates is whether "historical" elements do or do not have a direct effect on the individual's current position.

Actually, all of these mobility models show much the same pattern. Current occupational status is far and away most influenced by one's immediately prior occupational status. Of course, this is hardly surprising. The contour of occupational status over time works much like a ratchet: Once you move into one slot you may not go any farther, but it is very unlikely that you are going to fall backward. Thus the strong temporal correlations reported by Duncan and Hodge,[85] and by Hochbaum.[86] In summary, each of these career models concludes, not surprisingly, that the best way to predict the future is to know the immediate past, and that achieved statuses such as education have a more lasting and direct influence on careers than do socioeconomic characteristics.

This is the pattern for careers in the total occupational structure. Is the pattern the same within a single set of occupations—in particular, academic science? I have constructed a somewhat truncated causal-chain model for scientific careers. The model is applied to both men and women scientists.

I have modified the Blau-Duncan model in order to make it applicable for analysis within the academic-science community. My model brings together the four stages in the careers of scientists that I discussed above: the Ph.D. source; first job; academic position in 1965, roughly eight years into the career; and academic position in 1970. I have two background characteristics of scientists. Blau and Duncan include educational attainment in their model, as do I. But my variable is actually quite different substantively from the Blau-Duncan indicator. Educational achievement for them meant number of years of school completed. For academic scientists this is not really a variable, since each of them received a Ph.D. Thus, my measure of educational achievement must make a different distinction from number of years of schooling. In science, the real distinction lies in the quality of the scientist's educational background. Those who received doctorates from distinguished departments are defined as having achieved more than those graduating from less notable departments. The really important point, however, is that in the context of science, education, as defined here, actually is a social-origins variable much like father's occupation is in the Blau-Duncan model. In their societal model educational achievement represents an achieved rather than an ascribed status. Although dependent on father's education and

83. David Featherman, "Achievement Orientations and Socioeconomic Career Attainments," *American Sociological Review* 37 (April 1972):131–143.

84. Jonathan Kelley, "Causal Chain Models for the Socioeconomic Career," *American Sociological Review* 38 (Aug. 1973):481–493.

85. O. D. Duncan and Robert W. Hodge, "Education and Occupational Mobility," *American Journal of Sociology* 68 (1963):629–644, as reported in Blau and Duncan, *op. cit.*, p. 185.

86. Godfrey Hochbaum et al., "Socioeconomic Variables in a Large City," *American Journal of Sociology* 61 (1955):31–38, as reported in Blau and Duncan, *op. cit.*, p. 185.

father's occupation, in the framework of the Blau-Duncan analysis it represents the "great equalizer" in the American stratification system that operates to overcome inequalities in social origin. Intellectual background and achievement are, in a peculiar sense, transformed in my model into an ascribed characteristic, which regardless of its actual effect is as visible as one's religion. Thus, to have been at Harvard or to have read science at Merton College, Oxford, becomes a permanent part of one's social character. An even more significant aspect of this transformation of achievement into an inextricable part of one's social background can be seen in intellectual apprenticeships. It is not simply being educated at Harvard that becomes part of the social origins of the scientists; it is being a "student of Watson's" or "a student of Parsons' " that one carries throughout a career. To have been at the scene when an intellectual movement got started, or to have been a witness to a great intellectual breakthrough, even if only a student at the time, becomes a badge that one wears—not dissimilar to the marks of gradations in social class. A clue to the social meaning of the educational background of individuals lies in its use. How frequently we find individuals parading their intellectual origins, in hope of eliciting some measure of social deference or esteem, trying to parlay past associations into present gain.

A second background variable in my model is IQ. Both variables are hypothesized to directly affect the scientist's occupational status at three points in the career. I follow Blau and Duncan in allowing occupational status to be influenced only by the occupational status immediately preceding it and not by jobs held at early stages in the career.

The final modification of the societal model lies in the substitution of scientific performance variables for income. In the general society, income clearly is one of the chief forms of social recognition. This is not the case in academic science. While there may be a fair amount of variability in income among scientists at different departments, in different disciplines, or in different academic ranks, science for profit is normatively opprobrious. Recognition in science comes primarily in the form of rewards for original discoveries. Thus, the scientist's honors, visibility, and perceived quality of research can all lead directly to positions of power and prestige. They are the coin of the realm, which is exchanged for other social rewards. However, one should not overlook the enormous personal satisfaction, the intrinsic reward, in publishing high-qaulity work and in doing so regularly. This, after all, is really the ambition or dream of all young scientists. Therefore, I have included in my causal model a second chain—linking several stages of research performance. Quality and quantity of research publications are seen as directly determined by education and prestige of department, or occupation as it is dubbed in this model. Research performance at a given point in the career is seen as a direct influence on prestige of department at the next career stage as well as a direct influence on future research performance. Scientific productivity at each point in time is a completely independent measure of productivity from the previous index. Productivity at Time 1 is the total number of published papers from 1958 to 1961; at Time 2, from 1962 to

TABLE 5–7. Causal-Chain Model of Academic Appointments within the Science Career: Full Sample of All Academic Scientists (Standardized Partial Regression Coefficients)

INDEPENDENT VARIABLES	DEPENDENT VARIABLES								
	Occupation Status*			Scientific Output**					
	t_1 = 1957–58	t_2 = 1965	t_3 = 1970	t_1 = 1957–58		t_2 = 1965		t_3 = 1970	
				P	Q	P	Q	P	Q
Intelligence	−.056	.205†	.016	−.042	(0.57)	.029	(−.078)	—	—
Education (Ph.D. Dept. Rank)	.340†	.052	.001	.041	(.103)	.056	(.040)	.024	(.060)
Occupation at t_1		.585†	−.050	.196†	(.148)†	.012	(.039)	.027	(−.105)
Productivity at t_1		.066		—	—	.719†	(.792)†	.183	(−.101)
Occupation at t_2		—	.781†	—	—	.066	(.101)	.628†	(.805)†
Productivity at t_2			.042	—	—	—	—	−.124	(.264)
Occupation at t_3									
Coefficient of Determination	.330	.664	.766	.214	(.223)	.747	.798	.668	.818
R^2	.109	.441	.587	.046	(.050)	.558	.637	.447	.669

*Occupation at t_1 = Prestige of first job, 1957–58: Cartter ratings.
Occupation at t_2 = Prestige of 1965 job: Cartter ratings.
Occupation at t_3 = Prestige of 1970 job: Roose-Andersen ratings.
Scientific Output: P = Number of scientific papers; Q = Number of citations.

**For regression of occupational status, number of scientific papers is used here as the measure of scientific output.

†F test of significance indicates a statistically significant result at .05 level or greater; standard error of B is less than half the regression coefficient.

TABLE 5-8. Causal-Chain Model of Academic Appointments within the Science Career: Separate Regressions for Men and Women Scientists (Standardized Partial Regression Coefficients)

DEPENDENT VARIABLES

Occupation Status*

	t_1 = 1957–58		t_2 = 1965		t_3 = 1970	
	Men	Women	Men	Women	Men	Women
Intelligence	−.048	−.056	.159	.265	.032	.014
Education (Ph.D. Dept. rank)	.309	.387	.091	.010	.043	.095
Occupation at t_1			.560	.639	−.006	−.184
Productivity at t_1			.065	.034	.067	−.035
Occupation at t_2					.728	.926
Productivity at t_2						
Occupation at t_3						
Coefficient of Determination	.301	.375	.638	.710	.726	.860
R^2	.091	.141	.407	.504	.527	.739

TABLE 5-8. (continued)

Scientific Output

	$t_1 = 1957$–58				$t_2 = 1965$				$t_3 = 1970$			
	Men		Women		Men		Women		Men		Women	
	P	Q	P	Q	P	Q	P	Q	P	Q	P	Q
Intelligence	.010	(.072)	−.029	(.118)	.085	(−.060)	.119	(.006)	.016	(.074)	.030	(.005)
Education (Ph.D. Dept. rank)	.042	(.114)	.003	(.048)	.048	(.055)	.048	(−.038)	−.027	(.154)	.023	(.365)
Occupation at t_1	.162	(.131)	.288	(.227)	.044	(−.036)	−.001	(.038)				
Productivity at t_1					.722	(.785)	.687	(.793)				
Occupation at t_2					.055	(.085)	.049	(.113)	.081	(.047)	.121	(−.562)
Productivity at t_2									.943	(.813)	.891	(.689)
Occupation at t_3									−.028	(.258)	−.099	(−.476)
Coefficient of Determination	.180	(.220)	.287	(.286)	.761	(.791)	.719	(.827)	.955	(.835)	.912	(.836)
R^2	.033	(.048)	.082	(.082)	.579	(.626)	.517	(.684)	.911	(.698)	.833	(.699)

*For occupation regressions productivity, not quality, was used for the figures that appear in this table.

1965; and at Time 3, from 1966 to 1970. Similarly, quality of research is measured by independent counts of citations at three points in time. It is possible, although unlikely, for scientists to be highly productive in the initial career phase and totally unproductive thereafter.

Consider only two among the multiple number of relationships presented in Tables 5-7 and 5-8. First, it is increasingly difficult for scientists to escape their pasts as they move through their careers. A highly prestigious undergraduate college helps an individual gain admission to a superior graduate school. Receiving a doctorate from a graduate department of the first rank significantly increases the chances of getting a prestigious first job. Having been placed in an outstanding first job, chances are increasingly great of maintaining the same position or securing a position in a similarly ranked department. While there are exceptions, the sorting process takes place early on in the scientific career. And there seem to be accumulating advantages, independent of aptitude, for those scientists who initially are located in favorable settings. What we witness is the opening up of opportunity structures for a few and the closing off of opportunities for many. And it takes only about eight to ten years after the doctorate for the process to be more or less complete.

The data presented in Tables 5-7 and 5-8, which summarize the overall results, indicate that academic positions at one point in time are dependent largely upon those which immediately precede them. There is little lingering effect of positions held in the remote past. For my data, scientists' 1970 positions are strongly dependent upon 1965 positions, but not upon first job positions. There is, however, an important indirect effect of the first job on the 1965 and the 1970 jobs.

The tables also show that measured intelligence does have a significant independent influence on academic position, but only at a specific point in the career history. Its initial impact is minimal. It is not an important factor in initial location in the stratification system. Once a scientist is part of the system, after the first job, IQ influences positional recognition directly, and independently of first-job prestige, educational background, and scholarly performance. After twelve years, in 1970, no variables other than current job prestige influence future academic positional recognition directly. The effect of IQ and scientific productivity is indirect through the way they have already influenced previous job status. Of course, some of the strength of the intertemporal effect of past occupational status on future status is due to the relative immobility of scientists after achieving tenured ranks.

Finally, the career histories of men and women scientists seem to be shaped by the same basic factors, which exercise roughly equal influences on the careers of both sexes. The regression coefficients for women are not significantly different from those of men. In science, measured intelligence has a slightly stronger effect on women's occupational destinations than on men's. The data further indicate that for both sexes scientific output depends more upon past productivity than upon social origins, current affiliation, or measured intelligence.

SUMMARY

Similarities may be found in the influence of measured intelligence on occupational careers in science and on occupational status in the total American stratification system. In the larger stratification system, native ability has a greater effect on occupational status than either father's background, father's education, or family size. It is a strong determinant of educational achievement, which is the central link between social background and rank in the hierarchies of occupational prestige and income. Knowledge of native ability increases understanding of the social processes leading to social rankings. In science, the knowledge of native ability increases our understanding of processes of social stratification. Measured intelligence has as strong an influence on occupational achievement as does the source of a scientist's intellectual training. It has a different type of effect at various points in the career history. At some points, its influence is indirect through its association with other variables, such as prestige of doctoral institution. At other points, the influence is direct and independent of other social characteristics. As significant as the influence of IQ is on career development, its effect is not through its relationship to superior scientific publications. The influence of IQ on achievement in science, when we consider academic appointments, suggests the operation of a reward system that reinforces behavior which is not directly linked to scholarly productivity. I have speculated here about the social processes involved in the reward for high IQ, but further detailed study is needed before any firm conclusions can be reached.

Chapter 6

MARGINAL WOMEN:
A HISTORY OF THE
TRIPLE PENALTY

CONTRARY TO SOME OPINION, the twentieth century has been one of extraordinary change for women in American science. Superficially, little of significance appears to have changed for women entering science over the past seventy-five years, at least until the last decade. Imperfectly examined empirical evidence seems to support this belief. Articles and monographs frequently note, for example, that women constituted 15 percent of doctoral recipients in 1920 and only 13 percent fifty years later. Further evidence of the snail's pace toward sexual equality in the sciences is provided by the apparently sorry record of the National Academy of Sciences and other elite scientific societies in opening their doors to women. The small number of women in positions of high rank at top American universities is offered as further evidence for the absence of change. These facts deserve careful attention. As we shall see, this outer appearance of little change obscures significant shifts in the status of women in science during this century. To claim no change on the basis of these data results in what Alfred North Whitehead called ''the fallacy of misplaced concreteness.''

Not that these gross facts are in error although we must take a closer look at them, but they are misleading. By stepping back from them we may obtain a larger perspective on their pointillist quality, identifying actual changes in attitudes and behavior toward women in science, instructive perturbations in the movement toward gender equality, and previously unknown patterns of stability in the relationship between men and women scientists. The purpose of this chapter, then, is to place in historical relief the material I have presented thus far on the current position of women in the academic-science community.

This historical sketch is not intended to be comprehensive, and it will frequently be impossible with the data in hand to move from the level of correlation to causation. The aim is to identify and trace historical trends in the status of female scientists since the turn of the century. First, I shall sketch with very broad brush-strokes some of the cultural values bearing on women's place in the labor force generally and in the academic community particularly, and I shall suggest that these values represented both psychological and social barriers to women who desired careers. For those defying convention, a professional life as a marginal figure was the best that could be expected. Second, I shall outline the shifting position of women in the labor force and in higher education since 1900. Third, I shall present historical data on the changing level of sex discrimination in the academic reward system. Fourth, I shall argue that patterns of change for women have not been linear; that there have been periods of relative improvement and decline, but that there has been progress over the last fifty years toward the greater inclusion of women in the central sectors of the academic-science community.

I shall argue that for the period from 1911 to 1960 there have been a number of significant changes in the degree to which women represent marginal figures in the academic world generally and in science particularly. In attempting to reduce their marginality, women have historically faced three prominent barriers to becoming fully productive members of the scientific community—barriers that Harriet Zuckerman and I have referred to as elsewhere as the triple penalty:

> First, science is culturally defined as an inappropriate career for women; the number of women recruited to science is thereby reduced below the level which would obtain were this definition not prevalent. Second, those women who have surmounted the first barrier and have become scientists, continue to be hampered by the belief that women are less competent than men. Whatever the validity of this belief, it contributes to women's ambivalence towards their work and thereby reduces their motivation and commitment to scientific careers. And third . . . there is some evidence for actual discrimination against women in the scientific community. To the extent that women scientists suffer from these disadvantages, they are victims of one or more components of the triple penalty.[1]

If women face these obstacles today, they were even more formidable in the first half of this century. By tracing over time several indicators of status, I will suggest that the status-sets of men and women scientists have moved increasingly from essentially heterogeneous to homogeneous configurations.

CULTURAL DEFINITIONS OF WOMAN'S PLACE: THE MARGINAL WOMAN

Although the "marginal man" concept originated with Robert E. Park in the 1920s, it was given its first formal and extensive treatment by Everett V. Stonequist in 1937. Park, in his introduction to Stonequist's book, notes that

---

1. Harriet Zuckerman and Jonathan R. Cole, "Women in American Science," *Minerva* 13, no. 1 (spring 1975): 84.

"the marginal man . . . is one whom fate has condemned to live in two societies and in two, not merely different, but antagonistic, cultures."[2] In its early usage the concept was applied to conflicts between races, cultures, and social classes. Thus, in a society that is largely closed to social mobility, the "bourgeois gentilhomme" or parvenu is exposed to severe social stress and becomes a marginal figure, who can neither identify with the social group from which he has come nor be fully accepted by the group to which he aspires. So, too, for the parvenu's logical counterpart, the déclassé, who experiences stress stemming from a rapidly changing social structure. Robert K. Merton, some years after Park, succinctly described the marginal-man pattern as representing

> the special case in a relatively closed social system, in which the members of one group take as a positive frame of reference the norms of a group from which they are excluded in principle. . . . The ineligible aspirant . . . engaging in . . . anticipatory socialization becomes a marginal man, apt to be rejected by his membership group for repudiating its values and unable to find acceptance by the group which he seeks to enter.[3]

Social scientists utilizing the concept of the marginal man have usually referred to the male gender, not to humanity generally. But there is a place for "marginal women" among marginal men. Stonequist, and others before him, were aware that the social and psychic tensions faced by career-oriented women at the turn of the century were analytically akin to those of marginal men:

> The modern transformation in the traditional role of women has produced like dilemmas in ambition and conduct. The declining functions of the home and the growing opportunities in industry, government, and the professions have lured, if they have not compelled, many women to seek careers in the "man-made world" outside the home. But public opinion and moral codes have been slow to sanction the new departures, and the pioneers thus find themselves between two fires: men who resist their encroachment, and women who are outraged by their free and seemingly adventurous conduct. The successfully married stay-at-homes may find time heavy on their hands and look upon the "modern woman" with mingled feelings of envy and dissatisfaction; the successful career-woman in turn may regret the sacrifices in marriage and family life which her ambitions have entailed.[4]

Marginality: Cultural Values as a Context: 1875–1930

Marginality resulted, in part, from several strongly felt values held by Europeans and Americans in the last half of the nineteenth century: specifically that women belonged in the home; that women were innately incapable of creative work; that women should not "usurp" the traditional male role as the head of the

2. Everett V. Stonequist, *The Marginal Man* (New York: Charles Scribner's Sons, 1937), p. xv.

3. Robert K. Merton, *Social Theory and Social Structure,* enlarged ed. (New York: Free Press, 1968), pp. 320, 345.

4. Stonequist, *op. cit.,* pp. 6–7.

family. These beliefs, of course, represented a great barrier to achievement by women. John Stuart Mill's observations in *The Subjection of Women* (1896) were quite exceptional:

> Like the French compared with the English, the Irish with the Swiss, the Greeks or Italians compared with the German races, so women compared with men may be found, on the average, to do the same things with some variety in the particular kind of excellence. But that they would do them fully as well, on the whole, if their education and cultivation were adapted to correcting instead of aggravating the infirmities incident to their temperament, I see not the smallest reason to doubt.[5]

This is the exception. The more common attitudes about women's capacities can be found abundantly in the writings of male scientists and scholars, and in the literature of the last half of the nineteenth century.

Alphonse de Candolle devotes only two pages to the place of women in science in his otherwise extraordinary work *Histoire des Sciences et des Savants depuis Deux Siècles* (1885):

> We do not see the name of any woman on the lists of learned men connected with the principal academies. This is not due entirely to the fact that the customs and regulations have made no provision for their admission, for it is easy to assure one's self that no person of the feminine sex has ever produced an original scientific work which has made its mark in any science and commanded the attention of specialists in science.[6]

Putting aside the question whether such notable nineteenth-century female scientists as Maria Agnesi or Sophie Germain were of the first rank, it is evident that Candolle's position is based not so much on the argument of limited opportunity as on a belief in the limited innate mental capacity of women. As Candolle goes on to remark, the female mind

> takes pleasure in ideas that are readily seized by a kind of intuition; a mind to which the slow method of observation and calculation by which truth is surely arrived at are not pleasing. Truths themselves, independent of their nature and possible consequences—especially general truths which have no relation to a particular person—are of small moment to most women. Add to this a feeble independence of opinion, a reasoning faculty less intense than in man, and finally, the horror of doubt, that is, a state of mind in which all research in the sciences of observation must begin and often end. These reasons are more than sufficient to explain the position of women in scientific pursuits.[7]

Candolle's ideas were not exceptional in their time. Within the scholarly and scientific community in general, a significant research effort was being devoted to establishing the intellectual and physiological inferiority of the female. There

---

5. John Stuart Mill, *The Subjection of Women,* in *John Stuart Mill and Harriet Taylor Mill: Essays on Sex Equality,* ed. Alice Rossi (Chicago: University of Chicago Press, 1970), p. 197.

6. Alphonse de Candolle, *Histoire des Sciences et des Savants Depuis Deux Siècles,* as quoted in H. J. Mozans, *Woman in Science,* 1913 (reprinted, Cambridge, Mass: MIT Press, 1974), p. 392.

7. *Ibid.,* pp. 392-393.

was extensive research on cranial volume and its relation to intellectual capacity.[8] The weight of brains also became a focus of considerable attention, with men of science, otherwise noted for their care in weighing evidence, trying with some difficulty to weigh human brains and to draw the inference that women's smaller capacity was directly related to the lower average weight of their brains. As might be expected, evidence contrary to the "desired" outcome continually confronted these investigators. For instance, the two largest brains on record among distinguished intellectuals had been the former property of the noted zoologist Cuvier and the distinguished novelist Turgenev, the former's totaling a massive 1,830 grams and the latter's weighing in at an even more hefty 2,012 grams. But still larger brains belonged, we are told, to "an ignorant laborer named Rustan," whose lobes totaled 2,222 grams; to a "weak-minded London newsboy," whose brain weighed 2,268 grams; and to "a twenty-one-year-old epileptic idiot," whose brain tipped the scales at more than 2,800 grams.[9]

When the argument about size and weight was apparently falsified, auxiliary hypotheses were formulated. For example, it was suggested that the "development of the frontal lobe exhibited a pronounced difference in the two sexes. It was said to be much greater in man than in woman and was regarded as a distinguishing characteristic of the male sex. This was in keeping with the generally accepted assumption that this portion of the brain is the seat of the higher intellectual processes."[10] These attempts to prove a physiological basis for the lack of female accomplishments in the arts and sciences led, of course, only to dead ends. Yet when at the turn of the century the German scholar Paul Moebius published his *The Physiological Feeble-Mindedness of Women,* there were fewer attacks on the conclusion than on the causal argument.[11]

In 1900 Havelock Ellis, attempting to synthesize available physiological and psychological data on sex differences, devoted the better portion of a chapter in *Man and Woman* to the brain-weight evidence, and two full chapters to the "intellectual impulse of men and women" and the "legal, scientific, and social importance of women's periodicity of function."[12]

Considerable attention was given to the physical frailty of women as well as to their emotional and sexual inferiority to men. Numerous works were published in the late nineteenth century on the relationship between female menstruation, for example, and work capacity.

If the prevailing beliefs embedded in American and European culture about women being innately less well equipped for activities of the mind had deep

---

8. *Ibid.,* pp. 118 ff.

9. *Ibid.,* p. 119.

10. *Ibid.,* p. 122.

11. After locating Moebius' work I discovered an interesting discussion of it in Stephen Kern's *Anatomy and Destiny* (Indianapolis: Bobbs-Merrill Co., 1975): "Their smaller brains, Moebius believed, make women moody, nervous, and animal-like. They are unable to create, and lying and exaggeration come easily to them. Their mental deficiency, however, prepares them for important functions in the family" (p. 96). Moebius believed that female fertility was negatively related to intellect.

12. Havelock Ellis, *Man and Woman* (New York: Charles Scribner's Sons, 1900), Chapters 8, 9.

roots, they were joined with equally strong beliefs about the proper place for women. There was a strong ideology of domesticity. Just as in Germany, where women were to be concerned with "Kirche, Küche und Kinder," a woman's place in America was in the home, or perhaps the kitchen. If domesticity is legitimated today in terms of self-sacrifice, in the late nineteenth century it was a woman's calling. When women worked for pay or profit, it was mostly work within the home.[13] Tocqueville, at midcentury, sang the praises of equality between the sexes in American society, but did so within a framework of "separate but equal." Tocqueville admired the capabilities of American women, but only within the terms of their own domain—the home.[14]

The intellectually restless and independent-minded woman, unwilling to settle for traditional domestic virtues, became, of course, an increasingly familiar character in works by novelists and playwrights of the late nineteenth century. Many works of the late Victorian age thematically examine the consequences of the sharp disjunction between the aspirations of a select group of women and the social standards dominating their societies. In George Eliot's *Middlemarch* (1871-72), for example, Dorothea is portrayed as a modern St. Theresa, but without any culturally acceptable mechanism for making known her intellect or ambitions: "Many who knew her, thought it a pity that so substantive and rare a creature should have been absorbed into the life of another, and be only known in a certain circle as a wife and mother. But not one stated exactly what else that was in her power she ought rather to have done."[15] Dorothea, in the end, chooses the acceptable alternative: She lives through her husband's activity and ambitions. Henrik Ibsen, following the lead of his friend Camilla Collett's novel on the status of women, *The Judge's Daughters* (1853), made female independence a motif in several of his plays—most notably, of course, *A Doll's House* (1879) and *Hedda Gabler* (1890). When Nora forsakes her marriage vows, leaves her children, renounces her duty of wifely obedience, and displays an independent spirit, her behavior is considered scandalous.[16]

Indeed the critical reception of *A Doll's House* was such that Ibsen was reluctantly forced to rewrite the ending for the German premiere: Nora, about to slam the door to be heard around the world, instead leaves it open, looks at her sleeping children, drops her bag, and in despair decides to stay with her husband—once again restoring to its proper place the German sense of masculine supremacy. In preparatory notes to *Hedda Gabler,* Ibsen remarks: "Men and women do not belong to the same century.... The play is to be about 'the insuperable'—the longing and striving to defy convention, to defy what people

13. Robert W. Smuts, *Women and Work in America* (New York: Columbia University Press, 1959), Chapter 1.

14. Alexis de Tocqueville, *Democracy in America,* vol. 2 (New York: Alfred A. Knopf, 1945), pp. 198–214.

15. George Eliot, *Middlemarch* (reprinted Middlesex, England: Penguin Books, 1965), p. 894.

16. Here I have found particularly useful Peter Watts' Introduction to *Ibsen Plays* (Middlesex, England: Penguin Books, 1965). Also helpful was Michael Meyer's Introduction to *Hedda Gabler and Three Other Plays* (New York: Doubleday & Co., Anchor Books, 1961).

accept (including Hedda)."[17] Hedda, of course, does not quite make it, ending it all with suicide. While some critics responded favorably to the play, much of the daily London press after seeing it first performed were of quite another opinion: "Hideous nightmare of pessimism . . . the play is simply a bad escape of moral sewage gas. . . . Hedda's soul is a-crawl with the foulest passions of humanity."[18]

Many other works by major authors also developed the theme of females striving for an independent identity, only to be beaten back by contrary values and conventions. Among them are George Gissing's *Odd Women* (1893) and Henry James's *The Bostonians* (1886).

One consequence of the frustrations surrounding the quest for some independent identity was that late-nineteenth-century women with increasing frequency defined themselves in terms of illness. There was a proliferation of "women's diseases," mostly focusing on the female reproductive organs. Middle-class American women and the larger society began to maximize rather than minimize the idea of female frailty.[19] For the woman, sickness was a method of legitimating her perceived "failure"; for the society, it helped create a self-fulfilling prophecy about women's ability to handle energy-consuming jobs.

If we think that the forefathers of modern sociology and psychology held substantially different views from their contemporaries in other fields when it came to assessing the capacities and potentials of women, consider the following observation by Emile Durkheim in *Suicide:*

> It is said that woman's affective faculties, being very intense, are easily employed outside the domestic circle, while her devotion is indispensable to man to help him endure life. Actually, if this is her privilege it is because her sensibility is rudimentary rather than highly developed. As she lives outside of community existence more than man, she is less penetrated by it; society is less necessary to her because she is less impregnated with sociability. She has few needs in this direction and satisfies them easily. With a few devotional practices and some animals to care for, the old unmarried woman's life is full. If she remains faithfully attached to religious traditions and thus finds ready protection against suicide, it is because these very simple social forms satisfy all her needs. Man, on the contrary, is hard beset in this respect. As his thought and activity develop, they increasingly overflow these antiquated forms. But then he needs others. Because he is a more complex social being, he can maintain his equilibrium only by finding more points of support outside himself and it is because his moral balance depends on a larger number of conditions that it is more easily disturbed.[20]

---

17. As quoted in *Ibsen Plays (op. cit.)*, pp. 265–266.

18. As quoted in Meyer, *op. cit.*, p. 263.

19. On this problem, see Ann Douglas Wood, "The Fashionable Diseases: Women's Complaints and Their Treatment in Nineteenth-Century America," *The Journal of Interdisciplinary History* 4, no. 1 (summer 1973): 25–52.

20. Emile Durkheim, *Suicide*, 1897 (English translation from the French, New York: Free Press, 1951), pp. 215–216.

Herbert Spencer, whose work had an extraordinary impact around the turn of the century—far more extensive, for example, than Durkheim's—held even more dramatic biologically determinist views. Spencer believed that women represent "a somewhat earlier arrest of individual evolutions," which was nonetheless necessary and allowed them to conserve their energy for reproduction, and for bearing and nursing children.[21]

When we remember that up from the lowest savagery, civilization has, among other results, caused an increasing exemption of women from bread-winning labour, and that in the highest societies they have become most restricted to domestic duties and the rearing of children; we may be struck by the anomaly that in our own days restriction to indoor occupations has come to be regarded as a grievance, and a claim is made to free competition with men in all outdoor occupations. This anomaly is traceable in part to the abnormal excess of women; and obviously a state of things which excludes women from those natural careers in which they are dependent upon men for subsistence, justifies the demand for freedom to pursue independent careers. That hindrances standing in their way should be, and will be, abolished must be admitted. At the same time it must be concluded that no considerable alteration in the careers of women in general, can be, or should be, so produced; and further, that any extensive change in the education of women, made with the view of fitting them for businesses and professions, would be mischievous. If women comprehended all that is contained in the domestic sphere, they would ask no other. If they could see everything which is implied in the right education of children, to a full conception of which no man has yet risen, much less any woman, they would seek no higher function.[22]

In the middle of the nineteenth century Auguste Comte, the putative father of sociology, worked out an elaborate rationalization of the intellectual inferiority of women, at the same time putting them on a moral pedestal in order to keep them in their place.

Sociology will prove that the equality of the sexes, of which so much is said, is incompatible with all social existence, by showing that each sex has special and permanent functions that it must fulfill in the natural economy of the human family, and that concur in a common end by different ways, the welfare that results being in no degree injured by the necessary subordination, since the happiness of every being depends on the wise development of its proper nature. . . . The social mission of woman in the positive system follows as a natural consequence from the qualities peculiar to her nature.

In the most essential attribute of the human race, the tendency to place social above personal feeling, she is undoubtedly superior to man. Morally, therefore, and apart from all material considerations, she merits always our loving veneration, as the purest and simplest impersonation of humanity, who can never be adequately represented in any masculine form. But these qualities do not involve the possession of political power, which some visionaries have claimed for women, though without their own consent. In that which is the great object of

21. Spencer quoted in Kern, *op. cit.*, p. 96.
22. Herbert Spencer, *Principles of Sociology*, vol. 1, 3d ed. (New York: D. Appleton, 1896), pp. 768–769.

human life, they are superior to men, but in the various means of obtaining that object they are undoubtedly inferior. In all kinds of force, whether physical, intellectual, or practical, it is certain that man surpasses woman, in accordance with the general law prevailing throughout the animal kingdom. Now, practical life is necessarily governed by force rather than by affection, because it requires unremitting and laborious activity. If there were nothing else to do but to love, as in the Christian utopia of a future life in which there are no material wants, woman would be supreme. But we have above everything else to think and to act, in order to carry on the struggle against a rigorous destiny; therefore, man takes the command, notwithstanding his inferiority in goodness. Success in all great undertakings depends more upon energy and talent than upon goodwill, although this last condition reacts strongly upon the others.

    . . . Hence we find it the case in every phase of human society that women's life is essentially domestic, public life being confined to men.[23]

In 1896, W. I. Thomas published a paper in the *American Journal of Sociology,* "On a Difference in the Metabolism of the Sexes," in which he described a set of biological factors that produced in women a distinctive quality of passivity that rooted them to a lower position than men on the evolutionary scale. Although Thomas would later give up these biologistic causes in favor of social explanations, he represented at the turn of the century serious sociological thinking and research on sex-based differences in capacities.[24] Coming to sex-linked differences from another angle of vision, Freud found women wanting nonetheless:

The sexual behaviour of a human being often lays down the pattern for all other modes of reacting to life. A special application of this proposition . . . can easily be recognized in the female sex as a whole. Their upbringing forbids their concerning themselves intellectually with sexual problems though they nevertheless feel extremely curious about them and frightens them by condemning such curiosity as unwomanly and a sign of sinful disposition. In this way they are scared away from any form of thinking, and knowledge loses its value for them. . . . I do not believe that women's "physiological feeble-mindedness" is to be explained by a biological opposition between intellectual work and sexual activity, as Moebius has asserted in a work which has been widely disputed. I think the undoubted intellectual inferiority of so many women can rather be traced back to the inhibition of thought necessitated by sexual repression.[25]

Perhaps the leading feminist social critic in the early twentieth century was Charlotte Perkins Gilman. While most male observers at the time attributed sexual differences to physiological and biological causes, Gilman attacked the basic social fabric of American society, particularly the family, as the fundamental source of sex inequality. The root of female problems, she asserted, could be found in the singular focus of women on marriage and the family. The traditional

23. Quoted from *Auguste Comte and Positivism: The Essential Writings,* edited and with an Introduction by Gertrud Lenzer (New York: Harper & Row, Torchbooks, 1975), pp. 268–269, 373–374.

24. W. I. Thomas, "On a Difference in the Metabolism of the Sexes," *American Journal of Sociology* 3, no. 11 (July 1897).

25. Sigmund Freud, " 'Civilized' Sexual Morality and Modern Nervous Illness," in *Collected Works,* vol. 9 (London: Hogarth Press, 1959), pp. 198–199.

division of labor produced in women dulled minds and restricted opportunities.[26] To continue the sexual division of labor meant only continued subjugation of the female sex. Women had to move out of the home and into industry if they were to achieve equality: "[T]he highest emotions of humanity arise and live outside the home and apart from it. . . . science, art, government, education, industry—the home is the cradle of them all, and their grave, if they stay in it."[27]

In addition to being restricted by social values and attitudes, women had few legal rights in either Europe or America until the late nineteenth century. For example, legal existence was basically suspended during marriage. Women could not enter contractual relations, could not separately will away real or personal property, and of course could not exercise such fundamental rights of citizenship as voting. The formidable British justice, Blackstone, summed up in his *Commentaries* a married woman's place under the common law: "Husband and wife are one, and the husband is the one."

In *Muller* v. *Oregon* (1908) the Supreme Court for the first time addressed the issue of "protective legislation" for women only.[28] The case involved an Oregon statute that restricted the number of hours women could work in industries. Louis Brandeis, in a 113-page *amicus curiae* brief on behalf of the National Consumers' League, compiled large quantities of empirical evidence

> about (1) the physical differences between men and women, (2) changed industrial conditions due to the introduction of machinery, (3) the "bad effects" of long hours on women workers' health and morals, on job safety, and on the health and welfare of future generations, (4) the conditions of laundries [Muller had been convicted of requiring a female employee to work more than ten hours a day in his laundry], (5) the reasonableness of the limit chosen [ten hours], and (6) the need for uniform restrictions (i.e., without provisions for exceptions such as overtime at premium rates of pay).[29]

The Court upheld the Oregon law, and in its opinion Justice Brewer stated:

> History discloses the fact that woman has always been dependent upon man. . . . Doubtless there are individual exceptions, and there are many respects in which she has an advantage over him; but looking at it from the viewpoint of the effort to maintain an independent position in life, she is not upon an equality. Differentiated by these matters from the other sex, she is properly placed in a class by herself, and legislation designed for her protection may be sustained, even when like legislation is not necessary for men and could not be sustained.[30]

---

26. Charlotte Perkins Gillman, *Women and Economics* (Boston: Small, Maynard & Co., 1898).

27. *Ibid.,* p. 222.

28. *Muller v. Oregon* (208 U:.S. 412, 28 S. Ct. 324, 52 L. Ed. 551 [1908]) is a classic case in the history of protective legislation. It should be related to the earlier and critical decision by the Court in *Lochner* v. *New York* (198 U.S. 45, 25 S. Ct. 539. 49 L. Ed. 937 [1905]), which also dealt with limitations on the work week. The Court found unconstitutional a New York statute that set a ten-hour-a-day, sixty-hour-a-week limit on the employees of a bakery. For discussion of how the *Lochner* and *Muller* cases can be distinguished, see Paul Brest, *Processes of Constitutional Decisionmaking: Cases and Materials* (Boston: Little, Brown & Co., 1975), pp. 725–754; Barbara Allen Babcock et al., *Sex Discrimination and the Law: Causes and Remedies* (Boston: Little, Brown and Co., 1975), pp. 19–41; Leo Kanowitz, *Sex Roles in Law and Society: Cases and Materials* (Albuquerque, N.M.: University of New Mexico Press, 1973), pp. 46–48.

29. Babcock, *op. cit.,* p.29.

30. *Muller (op. cit.).*

Muller was only the first in a series of cases in which the Court over the next two decades upheld "protective legislation."[31] In 1924 Felix Frankfurter wrote: "Nature made men and women different . . . the law must accommodate itself to the immutable differences of Nature."[32]

This, then, was the context in which women with a mind for a career were placed. But how did these cultural values affect women entering science and why were they marginal figures during this period? From the last quarter of the nineteenth century until the middle 1920s, women determined to pursue academic-science careers were penalized for violating norms of appropriate female behavior. The relationship between careers and marital status contains the ingredients that led to one aspect of this marginality. In the early part of the century, more than today, American women who did not marry were considered failures, even if M. Carey Thomas, the pioneering president of Bryn Mawr College, tried to turn opprobrium into distinction by declaring, "Our failures only marry!" Women who decided for science in effect decided against marriage. If they were being asked to emulate men in some respects, they were also being asked to abandon marriage, family, and children. Indeed, in some quarters the sacrifice was taken as a measure of their professional commitment. As we shall see, the number of unmarried female academics in the first three decades of the century was extraordinarily high, but it remains moot what portion of this proportion is attributable to women's choice or to the stigma attached to female professionals.

Whatever the cause, women thinking of careers as scientists were cut off from traditional marital and family statuses, and were, in this sense, pushed into a marginal position.[33] To be sure, the hopeless spinster type sketched by Charlotte Brontë in *Shirley* was succeeded by the ambitious, independent, and self-sufficient single woman, who was well educated and acting to further her personal development as well as a variety of other social causes. But in departing from the "proper role" for women, she was apt to be victimized by a society that held fast to its ideal of female domesticity.

Women wanting to enter academia had to do so through the back door. Most were educated at distinguished private women's colleges, and when they later sought jobs after earning the doctorate, they found them, if at all, most often at women's colleges or at newly founded state schools, rather than at centers of

---

31. *Muller* and a similar case, *Bunting* v. *Oregon* (243 U.S. 426 [1917]), marked the first cases in which the Court was confronted with large amounts of empirical evidence in support of a position. Brandeis pioneered in the use of such data. In *Bunting* the Court was presented with a brief in two volumes, totaling 1,020 pages, and it included tables, graphs, and analysis of the harmful effects of long hours. For a pivotal discussion of the impact of the use of such statistical evidence, see Felix Frankfurter, "Hours of Labor and Realism in Constitutional Law," *Harvard Law Review* 29 (1916).

32. As quoted in William H. Chafe, *The American Woman: Her Changing Social, Economic and Political Role* (London: Oxford University Press, 1972), p. 129. Chafe's book on the changing role of the American woman has been an extremely valuable source for information on this period.

33. Willystine Goodsell, *The Education of Women: Its Social Background and Its Problems* (New York: Macmillan Co., 1924), esp. pp. 35–45.

scientific and scholarly activity. Thus, women scientists faced much the same dilemma of contradictory statuses that Everett C. Hughes suggests the black physician has faced in American society.[34] The black physician, confronted with prejudice and restricted opportunity, became marginal to the professional community by limiting his practice to black patients in predominantly black residential neighborhoods. Structurally similar adaptations were available for women Ph.D's in science during the first third of the century. They clustered at teaching institutions that were usually women's colleges; they worked alone on their research, or were segregated from male researchers; and they were largely excluded from membership in scientific societies and professional associations.

At those institutions that would hire them, women faced grosser inequities than they do today in terms of salary, professorial rank, and prestige of their affiliations. The men who controlled these institutions set women apart by forcing them to accept essentially second-rate jobs. Furthermore, there continued to prevail throughout the period a strong ideology that women simply were not capable of producing good research, although they might make excellent teachers at the secondary-school and college level. There is evidence that the beliefs about differential capacity for scholarship were accepted by many women scientists, and this could only have impeded their attempts to produce new discoveries. Even where motivation for such work persisted, women scientists faced markedly different opportunities than male scientists for achieving their research goals. These forces placed women scientists in the position of being "outsiders" inside the scientific community.

MARGINAL WOMEN IN AN ECONOMIC
DEPRESSION: PUBLIC ATTITUDES TOWARD
WORKING WOMEN

By the end of the 1920s the energy in the feminist movement had been exhausted. Hopes and aspirations about social gains for women, which reached an apex with the passage of the Nineteenth Amendment, were slowly replaced by despair. Visions of a leap forward toward sexual equality quickly faded with the onset of the Great Depression. And with economic decline came a set of significant reversals for women striving to reduce economic and occupational inequality. Among the general public and within the academic community, attitudes about the value of women's work, which had never exactly undergone an enormous transformation, were reinforced by economic "necessity" and translated into greater resistance to women's participation in the labor force. Opportunities in the marketplace for women contracted, and in the midst of an economic depression few Americans, men or women, believed that married women

34. Everett C. Hughes, "Dilemmas and Contradictions in Status," *American Journal of Sociology* 50 (March 1945).

should work and thus "take jobs away" from men with families. Not until the American entry into World War II did women regain a prominent position in the labor force. And the war itself ironically provided the impetus for new gains for working women. Let us examine several indicators of the shifting cultural beliefs about woman's place in American occupational life from the 1930s to the early 1950s.

A new barometer of social attitudes was widely used in the 1930s—the public opinion survey. From the late 1930s through the forties, numerous polls were taken to determine Americans' attitudes toward married women working. Clear trends of these attitudes cannot be established because of the wide variety of questions. Without assigning significance to small percentage differences, we can identify some distinct fluctuations in public attitudes toward women working between 1930 and the 1950s.

A 1937 Gallup poll indicated, for example, that 82 percent of the American public were opposed to a married woman "earning money in business or industry if she has a husband capable of supporting her."[35] A year later, it was much the same: 75 percent of the women and 81 percent of the men were against married women working. Among those who felt that women should not work, the three major reasons given for such an expressed attitude, in a 1936 *Fortune* poll, were that women would take jobs that otherwise would be filled by men (36 percent); that woman's place was in the home (35 percent); and that children would be healthier and home life happier if women didn't work (21 percent).[36]

In the late 1930s bills were introduced into twenty-six state legislatures prohibiting married women from working in business or industry if their husbands earned more than, for example, $1,600 a year in Illinois or $1,000 a year in Massachusetts.[37] A 1939 Gallup poll found that a substantial majority in a national sample of the American public supported these pieces of legislation: Some 74 percent approved of the Illinois statute and 66 percent supported the one in Massachusetts. As the historian William Chafe has noted, "No massive outcry greeted the passage of legislation restricting women's rights to gainful employment in the 1930s."[38]

The family was the basic unit for analysis of income equity and for formulating public policy throughout the period from 1930 to the late 1950s. Husband and wife were not separate subjects for analysis. Since the early 1960s, however, the unit of analysis has shifted, especially among legislative decision makers. When legislators propose laws to prevent job and salary discrimination, such as the

---

35. Hadley Cantril, *Public Opinion 1935–46* (Princeton, N.J.: Princeton University Press, 1951), p. 1044. This section has drawn particularly on the provocative book by William Chafe, *The American Woman* (*op. cit.*).

36. Cantril, *op. cit.*

37. On restrictions on women's working, see Ruth Shallcross, *Should Married Women Work?* Public Affairs Pamphlet no. 49, 1940; Valerie Kincade Oppenheimer, *The Female Labor Force in the United States*, Population Monograph Series, no. 5 (Berkeley, Calif.: University of California Press, 1970).

38. Chafe, *op. cit.*, p. 111.

various civil rights acts of the 1960s, they do not act in terms of family units, although the family remains a key analytic unit for economists. Rather, salary discrimination is largely considered in terms of individuals, regardless of their marital or family statuses. Throughout the first half of this century, the obvious consequence of using family income as the basic frame for analysis was to systematically and purposefully discriminate against married women, who were not treated as primary breadwinners. They were passed over for jobs that they qualified for, and were paid less for those they did obtain. By shifting the standard of equality to individuals in the 1960s and 1970s, legislatures have been able to directly address sex-based salary discrimination. The consequence in many occupations, including academic ones, has been a reduction in salary differences between men and women, and an increase in individual equality.

But this policy may be having the added and perhaps unintended consequence of increased income inequality in the overall society, in which families are still used as the economic unit of analysis. This would result, naturally, from occupational assortative mating. To the extent that working professionals marry each other, and sales and clerical workers marry each other, the couple that has the relatively higher income to begin with will have an even greater comparative income when sex differentials in salaries are eliminated, assuming that the extent of salary discrimination is more or less the same in the various occupational classifications. For example, assume that women in all occupations earned 60 percent of men's salaries in 1975, and that this differential was entirely a result of sex discrimination. Consider two couples: one a professional couple—the husband earns $20,000 a year, his spouse $12,000; the other a sales couple—the man earns $10,000, the woman $6,000. By 1980 all sex discrimination in individual salaried income will be eliminated. The family incomes, assuming no change in the men's salaries, will be $40,000 for the "professional family" and $20,000 for the "sales family." The elimination of sex differences will have produced a greater *absolute* difference in the incomes of the two families ($4,000) in 1980 than existed in 1975. The final outcome is increased inequality in the incomes of the two families.[39]

Negative attitudes toward women's work in the 1930s cannot be directly linked, however, to any explicit social policy designed to equalize family incomes in American society, even if public attitudes toward married women's work were tied closely to social and economic conditions. To the extent that the public believed that necessity rather than volition guided decisions of married women to work, these decisions were palatable. But the suggestion that work was intrinsically important for the welfare of women continued to be resisted. This is evidenced by varying results obtained in the opinion surveys depending on the wording of the questions. When questions were phrased to suggest a basic

---

39. For further discussion of this idea, see Lester C. Thurow, *Generating Inequality* (New York: Basic Books, 1975).

structural change in family relationships, the public attitudes were consistently negative. When they emphasized economic necessity, they were more apt to be answered positively.

The importance of perceived necessity is dramatically seen in the change in public attitude toward married women's work from that in 1937, at the height of the Depression, to what it was five years later, after the American entry into the war. By 1942 some 60 percent of Americans believed that married women *should* work in war industries, while only 13 percent were opposed; and in a second National Opinion Research Center (NORC) poll taken in the same year, 71 percent favored married women without children working in war industries.[40] But when the 71 percent who approved were asked if they also believed that married women with children were needed in war industries, the overwhelming majority said no.[41]

The more egalitarian public attitudes toward married women's work did not outlast the war. In 1945 a Gallup poll found 62 percent of the public disapproving work by married women, which, while a reduction from the prewar highs of 80 percent, was not exactly indicative of extraordinary change. A study by Richard Centers in 1946 found that most men in both the middle and working classes continued to feel that women's place was in the home, although middle-class men were more apt to have permissive attitudes.[42] In the same year, another poll indicated that 80 percent of workingwomen in New York and 75 percent in Detroit wanted to continue working.[43] But the measure of how conservative public opinion remained about women entering the male occupational world is reflected in two 1946 *Fortune* polls in which the overwhelming majority of both men and women believed that men "should nearly always" hold the position of mayor of a city or of head of a school board. It was quite all right, on the other hand, for women to head the local PTA.[44] In all of these polls there is surprisingly little difference between men and women in the expressed attitudes. So if men were suffering from false consciousness, so were women.

FEMALE PARTICIPATION IN THE LABOR FORCE

As we might expect, the changing degree of marginality of women in the occupational world parallels their increased participation in the American labor force. In 1900 women represented roughly one-fifth of the labor force; today they are about 45 percent. Examined by decade, the increase has been consistent but uneven. Between 1900 and 1950 the proportion rose steadily but slowly from 18

---

40. Cantril, *op. cit.*, p. 1045.

41. *Ibid.*

42. As cited in Oppenheimer, *op. cit.*, pp. 48–49.

43. These data are drawn from a survey reported in Chafe (*op. cit.*), p. 178.

44. Cantril, *op. cit.*, p. 1054.

to 28 percent; it then jumped to 37 percent in the decade of the 1950s, and to 43 percent by 1970.[45] In the single decade between 1950 and 1960 the proportion of the total labor force that was female rose from 28 to 37 percent, about equal to the total rise that was witnessed in the first half-century. There was, contrary to common belief, a far greater expansion of female participation in the fifties than during the war years of the forties. Today fully half of all adult women are active if intermittent participants in the American labor force.

Valerie Kincade Oppenheimer demonstrates that "from 1900 to 1960 the great majority of female workers were concentrated in occupations that were disproportionately female . . . during the 1900–1960 period, between 60 and 73 percent of the female labor force were in occupations where the majority of workers were women, and between 30 and 48 percent were in occupations which were *80 percent or more female.*" [46] Since 1940 women have been concentrated in a small set of occupations: "About half of all working women were in only 20 occupations and no less than 30 percent were either elementary school teachers, retail sales clerks, bookkeepers, waitresses, or 'stenographers, typists and secretaries.' "[47] Furthermore, the distribution of women among the various standard occupational categories has remained remarkably constant over time, in contrast to the situation for men and racial minorities. The national trend toward an increased proportion of professional and technical workers, from 7.5 to 14.4 percent of the work force from 1940 to 1970, results entirely from shifts in the job structure of male workers.[48]

---

45. Data on the historical participation of women in the American labor force are reported in W. Elliot Brownlee and Mary M. Brownlee, *Women in the American Economy: A Documentary History, 1675 to 1929* (New Haven: Yale University Press, 1976). From official sources for 1900–70, the data by decade are as follows:

Participation of Women in American Labor Force, 1900–70

	WOMEN'S SHARE OF LABOR FORCE (%)	WORKING WOMEN IN THE FEMALE POPULATION (%)
1900	18.3	21.2
1910	20.0	24.8
1920	20.4	23.9
1930	22.1	24.4
1940	24.3	25.4
1950	27.9	29.1
1960	37.1	34.8
1970	42.8	42.6

Sources: U.S. Bureau of the Census, *Historical Statistics of the United States, Colonial Times to 1957* (Washington, D.C.: Government Printing Office, 1960), pp. 67–72; U.S. Bureau of the Census, *Occupational Trends in the United States: 1900–1950,* Bureau of the Census Working Paper no. 5 (Washington, D.C.: Government Printing Office, 1958); and U.S. Department of Commerce, *Statistical Abstract of the United States, 1971* (Washington, D.C.: Government Printing Office, 1971), pp. 211–212.

46. Oppenheimer, *op. cit.,* pp. 68, 70.

47. Donald J. Treiman and Kermit Terrell, "Women, Work and Wages," in *Social Indicator Models,* eds. Kenneth Land and Seymour Spilerman (New York: Russell Sage Foundation, 1975), pp. 157–199.

48. Among white male workers the proportion occupied as professional and technical workers increased from 6.0 to 14.8 percent in the thirty-year period. For white females, the increase was only from 14.6 to 15.4 percent during the same three decades. In 1940 there was a slightly higher proportion of females in professional and technical occupations than in 1970, although this is surely confounded by their extraordinary representation among nurses and schoolteachers, and their absence from high-ranking positions within these professions.

Oppenheimer has persuasively argued that expansion of female labor force participation from 1940 to 1960 did not result from basic shifts in attitudes about women's place or from increased supply of female labor, but from increased demand for it. In the war years this need was self-evident, and after the war, because of inflation and rising consumer demands, the need continued to expand, contrary to the expectations of many. But the demand continued to be selective in terms of occupations, and segmentation of the labor force into predominantly male and female jobs continued. It was in this period that demand for workers in traditionally female jobs expanded rapidly. After the war, when women left jobs in heavy industry, they reassumed positions that had for decades been sex-labeled.[49]

The basic change that did occur in the forties was not in the percentage of workingwomen or in the level of equality achieved by women, but in the altered social composition of the female labor force. "At the turn of the century, the young, the single, and the poor had dominated the female labor force. Fifty years later, the majority of women workers were married and middle-aged, and a substantial minority came from the middle class.'"[50] In terms of equality, little had been gained in the 1940s beyond the opening of work opportunities for new sectors of the female population. Nationally, by 1948 the proportion of women entering the professions continued to decline; there was a dip in the number of female lawyers and school superintendents; medical schools continued to impose strict quotas on female admissions at roughly 5 percent (70 percent of all hospitals refused to admit female interns, and women were barred from numerous medical societies); and although women held 25 percent of all government jobs, almost none held high-level positions.[51] Finally, women's wages in manufacturing industries were roughly two-thirds of men's, according to the National Industrial Conference Board. In short, those changes that did obtain were not translated into employment or salary equality.

What can we conclude? First, between 1930 and the 1950s there were few fundamental changes in public attitudes toward women's work as a means for personal gratification or self-expression. But there was increased flexibility in these attitudes, especially when women's work was perceived as an economic exigency rather than a function of a threatening ideological movement.

The importance of this new attitudinal flexibility and these compositional changes should not be underestimated. Although women were not expected to pursue careers, by the mid-1950s a woman who did want to combine a career with marriage and a family was not a statistical anomaly. The striking change in the standards relating to marriage and families among female academics was noted in 1954 by Virginia Gildersleeve, dean of Barnard College, who at an

49. Oppenheimer, *op. cit.*, pp. 42–52.

50. Chafe, *op. cit.*, p. 195.

51. *Ibid.*, pp. 184–185; see also Cynthia Fuchs Epstein, *Woman's Place: Options and Limits in Professional Careers* (Berkeley, Calif.: University of California Press, 1971), p. 7, Table 1.

earlier time had had to defend married academic women and who now had to defend the unmarried:

> It was right to do what we could at Barnard to aid married teachers; but of recent years another aspect of the question has troubled me. I have occasionally thought that in schools and colleges there has arisen a particularly cruel and unwholesome discrimination against *unmarried* women for some teaching and administrative posts. This is due in part to the attitude toward the unmarried of certain of the less responsible psychologists and psychiatrists of the day, which tends to voice disrespect for spinsters in the teaching profession as "inhibited" and "frustrated."[52]

Another important effect of the demand for workingwomen during the war was the entrance of married women, older women, and middle-class women into the labor force in large numbers and the acceptability of their presence under specific conditions. By 1950 it was commonplace for these types of women to work; twenty-five years earlier they had been anomalies.

Also, although the value placed on women remaining at home did not undergo radical change, there was a fundamental change between 1900 and 1950 in the rationales and explanations for woman's place. In the early part of the century, much of the argument was based on a "natural order of things," on biological and physiological factors, which supposedly made women less capable of performing specific occupational roles involving physical strength or mental creativity. But after the Depression and the war, after women had clearly demonstrated their ability to handle even physically taxing jobs, the argument took a dramatic turn. The shift was to predominantly social and economic explanations. Now workingwomen presumably threatened men's jobs; threatened the structural integrity of the family and the value placed on the male as primary wage earner; and interfered with the socialization of children—leading, inevitably, to increases in juvenile delinquency.

In some ways, however, the public attitudes toward women's careers in the late 1940s and 1950s were remarkably like those faced by independent-minded women at the turn of the century. The dominant themes of the fifties, which can be seen in a casual perusal of random issues of the *Ladies' Home Journal, McCall's, Good Housekeeping,* or the *New York Times* Sunday magazine section, were "togetherness," "homemaking," male emasculation, juvenile delinquency, and domestic tranquillity. Betty Friedan summed up the prevailing attitudes: "The feminine mystique says that the highest value and the only commitment for women is the fulfillment of their own femininity."[53] The normatively approved, if not prescribed, role for women was again what it had been for Dorothea in *Middlemarch*: A woman's success is measured by her husband's achievements.

---

52. As quoted in Jessie Bernard, *Academic Women* (University Park, Pa.: Pennsylvania State University Press, 1964), p. 210.

53. Betty Friedan, *The Feminine Mystique* (New York: Dell Publishing Co., 1963), Chapter 2.

The values and beliefs expressed in the public opinion surveys created a curious paradox. At the same time that women were being enjoined in the media to move back to the home, relinquish career aspirations, and be content with domestic roles, there was simultaneously the greatest expansion in the twentieth century of female participation in the labor force. And despite the enormous expansion in higher education in the fifties, it is firmly believed today that women's position relative to men's deteriorated in the academic community.[54] How can this disjunction between ideology and behavior be explained? By examining several forces operating simultaneously to influence the numbers of women entering careers and the conditions they faced.

In fact, the cultural attitudes toward careers did significantly influence women's behavior, but selectively across different social classes and educational strata. Among middle-class women—especially those educated at elite colleges, who represented a large portion of the pool for potential graduate training—the new domesticity had a significant impact. Getting married and raising families rather than obtaining a master's degree or a doctorate became an option that was increasingly taken. Many young women were married or engaged upon graduation. And these women did not *have* to work for a living. They were marrying men who either were professionals or had sufficient income to support their wives and families comfortably. Women from the other social classes were "forced" into the ranks of labor; it was the most effective way for them to expand their family's purchasing power. Thus, the growth in female participation did not come from the expansion of female professionals.

Recent evidence supports the idea of differential labor force participation. Linda J. Waite, in a trend analysis of working wives from 1940 to 1960, notes that in the early 1950s a wife's participation in the labor force was negatively related to educational attainment.[55] While this pattern was particularly true in the 1950s, it apparently has continued to the present time. The *Economic Report of the President* said in 1973: "Thus, the rearing of children of preschool age causes all women, regardless of education, to curtail their work outside the home. However, the drop in participation during this childbearing period is most pronounced for highly educated women who in other circumstances have higher participation rates."[56]

---

54. Bernard, *op. cit.*, pp. 29–40.

55. Linda J. Waite, "Working Wives: 1940–1960," *American Sociological Review* 41, no. 1 (Feb. 1976): 65–80.

56. Council of Economic Advisers, *The Economic Report of the President, 1973* (Washington, D.C.: U.S. Government Printing Office, 1973), as quoted in Waite, *op. cit.*, p. 77. For extensive discussions on the conditions under which there is differential labor force participation of women, see among others, Jacob Mincer, "Labor Force Participation of Married Women: A Study of Labor Supply," in *Aspects of Labor Economics*, ed. National Bureau of Economic Research (Princeton, N.J.: Princeton University Press, 1962); Glen G. Cain, *Married Women in the Labor Force* (Chicago: University of Chicago Press, 1966); William Bowen and T. Addich Finegan, *The Economics of Labor Force Participation* (Princeton, N.J.: Princeton University Press, 1969); Waite, *op. cit.*

WOMEN IN HIGHER EDUCATION

The Academic Marketplace: 1940-60

Much of the confusion over the believed deterioration of the position of women entering the academic marketplace arises from an incomplete analysis of the data on female doctorates. To be sure, statistics on female doctorates reflect a sharply declining interest in graduate education. In every subject area the proportion of all doctorates awarded to women is considerably lower in the 1950s than in the forties. This is most true for the disciplines in which women had been best represented. For example, in the humanities the 22 percent of all Ph.D.'s who were female in the 1940–44 period falls off to roughly 15 percent in the 1950s (see Table 6–1). Declines of this proportion also appear in the physical, biological, and social sciences.[57] Since the overall production of Ph.D.'s was expanding virtually exponentially during this period, and the production of female doctorates was simultaneously declining, the growth rate was accounted for entirely by men. The important distinction lies in the difference between relative and absolute numbers. The proportion of female Ph.D.'s was on the wane in the 1950s, but the absolute number of women earning doctorates and beginning academic careers was rapidly growing. There were 755 women who earned doctorates in the arts and professions between 1950 and 1954, compared to 501 for the years 1945–49, more than a 50 percent increase in absolute numbers.[58]

Simultaneously there was extraordinary growth in the American system of higher education. Bernard Berelson remarked in 1960: "More doctorates were granted in this country in the past decade than in all the years up to then."[59] This did not result simply from growth in preexisting doctoral programs. "In 1940, about 100 institutions gave the doctorate and about 300 the master's; in 1958 the figures were 175 and 569."[60] Actually the expansion came on all fronts. Among the fastest-growing institutions were the junior colleges and former teachers' colleges, which were being reorganized as parts of enlarged state university systems.

In the 1950s and early 1960s, there was a rapid expansion in the academic community of institutions which for decades had had a favorable record in hiring women, yet the proportion of their faculties that was female was declining. How was it possible, then, for women's status to improve, since their representation declined? The demand for faculty, especially less expensive junior faculty, was high in this period. But the culture was simultaneously leading women who

57. Source: *Doctorate Production in United States Universities: 1920–1962*, Publication 1142, Table 26 (Washington, D.C.: National Academy of Sciences, National Research Council).

58. *Ibid.*

59. Bernard Berelson, *Graduate Education in the United States* (New York: McGraw-Hill Book Co., 1960), p. 32.

60. *Ibid.*, p. 35.

TABLE 6-1.   Percentage of All Doctorates Awarded to Women, 1920–64, 1972

FIVE-YEAR PERIOD	PHYSICAL[1] SCIENCES	BIOLOGICAL SCIENCES	SOCIAL SCIENCES	ARTS-PROFESSIONS	EDUCATION
1920–1924	7.6	15.4	20.2	21.4	15.2
1930–1934	6.8	15.5	15.8	22.6	17.6
1940–1944	4.1	11.3	15.7	21.7	23.3
1950–1954	2.8	9.2	9.7	14.7	17.0
1960–1964	2.8	9.4	12.9	17.3	19.6
1972	4.4	18.4	18.7[2]	24.4	23.2

NOTES:
1. Includes engineering.
2. 1972 includes history.

Sources: National Academy of Sciences–National Research Council, *Doctorate Production in United States Universities, 1920–1962,* publication 1142 (Washington, D.C.: NAS-NRC, 1963), pp. 52–63; National Research Council, Office of Scientific Personnel, *Doctorate Recipients from United States Universities,* Summary Report (Washington, D.C.: NRC, 1972), Table 2.

might normally enter the pool of eligible professors away from the training that would make them candidates for positions. So the supply of female academics relative to males was actually declining in the 1950s, just at the time when growth of the academic industry was increasing opportunities for women. With the number of faculty positions expanding and proportionately fewer women available to fill them, it follows, of course, that the overall percentage of faculty positions held by women would decline. And this is what happened in the fifties in women's colleges and in state college and university systems.[61]

But the critical point is not to equate this proportional decline with sex discrimination in the academic marketplace, as is so often done. The proportional decline does not mean that opportunities were closing for those women who did obtain the Ph.D. That inference depends on the selection of an inappropriate indicator of opportunities—the proportion of faculties that are female. Actually the marketplace for jobs was expanding for women who earned doctorates in the middle and late 1950s. And the consequence of this expansion is underscored by the virtual absence of an association between sex status and the prestige of first academic job for the 1958 cohort of men and women Ph.D.'s (discussed in Chapter 3).

In fact, there are only two historical points in this century when the zero-order correlation between gender and prestige of academic deportment essentially disappears: in the 1920s and the late 1950s and 1960s. These were the two periods of extraordinary growth in the size of the academic marketplace. Historically, conditions for academic women have improved when the demand for professors has expanded rapidly. In fact, growth in the academic community has probably played a more significant role than ideology in increasing appointment opportunities for women. The hiring of females may not have been due to universalistic principles, but their chances of obtaining jobs in colleges and universities became roughly equal to those of male graduates when the need for professors increased markedly. This does not mean, of course, that other forms of sex discrimination did not continue to exist. But what discrimination there was began to center more on promotions and salaries than on hiring.

The hostile climate of opinion about professional women in the early and middle 1950s created a thick barrier to the selection of scientific and other professional careers among the talented young women in American colleges, and thereby radically reduced the number of women who could take advantage of structural changes in American higher education. In the 1950s the supply-and-demand curves for women Ph.D.'s were poorly equilibrated, but this did not result in better jobs for the relatively small group of women prepared, since the positions could easily be filled by men. Only ten years later, in the 1960s, when government affirmative-action policy began to pressure colleges and universities to hire women, would the relative scarcity of women in the academic marketplace begin to increase their bargaining power.

---

61. For a statement of the inference drawn from these figures, see Bernard, *op. cit.,* p. 43.

A Closer Look at Graduate Training

But let us step backward and examine more closely the changing position of women in American higher education. After all, some women did seek higher degrees leading to careers in academia, and we want to look at their opportunities after they entered the tracking system.

The results reported here are based primarily on historical sources and on data I have collected for five cohorts of male and female Ph.D.'s. Five samples of Ph.D.'s were selected from cohorts separated by ten-year intervals, beginning in 1911. For each decade one sample was selected. The names of scientists and scholars were obtained from *Dissertation Abstracts,* which in 1911 began publishing fairly complete lists of doctoral recipients in the United States. In each of the periods I searched for female Ph.D.'s and then for males, each matched to a woman in terms of discipline, specialty (where possible), and source of doctorate. For the 1911 and 1921 cohorts there were few female doctoral recipients. In order to increase the sample size, I extended the search to 1913 and to 1923 for these two cohorts. In effect, for the first two cohorts I have the complete listing of females and a random sample of males within the several sampling criteria. For the three later cohorts I selected stratified random samples of both men and women. This results in data for a total of 1,178 men and women recipients of the Ph.D. Roughly 65 percent of the cases are in the cohorts from the 1940s and 1950s.

Marital status, number of children, awards received, job histories, and other background information were obtained from early editions of *American Men of Science* (AMS). The careers of the men and women were traced for roughly twenty years from the year of their degree. Information on scientific productivity was obtained from the various science abstracts. Citation indices only appeared in 1961, and no citation data were collected for the preceding period.

There are several significant limitations to these data. First, unlike the data presented in the rest of this book, those presented here include some non-scientists.[62] Information was collected on women and their male matches in the

---

62. The disciplines included in my historical data, and their distribution, are as follows:

For All Five Time Periods

FIELD	NUMBER	PERCENTAGE
Math, Physics, Astronomy	61	5
Chemistry	132	11
Biomedical, Biological	436	37
Psychology	208	18
Other Social Sciences	129	11
Arts and Humanities	195	17
Education and Teaching	1	0
	1162	99
No Field Code	16	
	1178	

Thus, roughly a fifth of the sample were drawn from the humanities rather than the physical, natural, or social sciences.

humanities as well as in the physical, natural, and social sciences. However, productivity data are limited to the sciences.[63] Second, there is an abnormally large amount of missing data for individuals included in these samples. Some degree recipients were not included in *AMS*.[64] Consequently, the analysis is based upon a sample of men and women who are probably slightly more prominent than the entire cohort of Ph.D.'s in the period. Third, a great deal of specific information about the work experiences of these men and women cannot be adequately captured in short biographical summaries. Therefore, it will be impossible to empirically test a number of interpretations of the relationships obtained. It is important, for each of these reasons, to accept the data presented here with some skepticism and to treat the findings as tentative. Wherever possible I have tried to corroborate the statistical findings with historical materials.

The history of graduate education in the United States is, of course, a short one, and in its infancy, before the turn of the twentieth century, it did not cut much of a figure in the education of either men or women. Yale began to offer graduate courses in philosophy and arts in 1846. Harvard (1872), Johns Hopkins (1876), the University of Pennsylvania (1882), and Columbia (1890) opened graduate schools between 1872 and 1890.[65] The first American university to award a Ph.D. to a woman was Boston University in 1877, but in the next twenty-five years few women received doctorates from any of the major universities. Bryn Mawr College, perhaps the leader in women's education at the time under President M. Carey Thomas, offered graduate instruction from its beginning in 1885, and required as much preparation for the doctorate as any of the larger universities, if not more.[66]

---

63. No productivity data were collected for the humanities Ph.D.'s. There are no reliable abstracts of papers produced by these scholars, but more importantly the primary mode of publication is books rather than articles. All correlations that include productivity scores exclude these humanists. The statements and conclusions about productivity differences between men and women apply only to those in the sciences.

64. Following are the percentages of male and female Ph.D.'s in each cohort who were included in *American Men of Science:*

	Cohort				
	1912	1922	1932	1942	1952
Men	63	70	86	71	73
Women	48	53	65	47	46
Percentage Difference	15	17	21	24	27

Note that the difference in male-female inclusion rate increases over time. Although a greater absolute number of women are included in the compilation over time, there are changes in the relative proportions of men and women. Interestingly, the difference is least for the earliest cohort. Why these changes occurred is an open question, but the point here is that there is a significantly greater amount of missing background information for female Ph.D.'s than for males. This low inclusion rate, limiting the amount of complete data, is a further limit to this data set, and argues even more forcefully for a cautionary position vis-à-vis the data.

65. Thomas Woody, *A History of Women's Education in the United States,* vol. 4, bk. 2 (New York: Science Press, 1929), ch. 7. This book has been a particularly valuable source for material on higher education from 1900 to 1925.

66. The University of Pennsylvania announced in 1888 that "women are admitted to any course for the Ph.D. degree on the same conditions as men." From 1890 to 1892 Yale and Columbia formally acknowledged the admission of women to their graduate programs. Yale's President Dwight "recommended that in the academic year 1891–92 and afterwards, the courses of graduate study leading to the degree of doctor of philosophy should be open to the graduates of all colleges and universities without distinction of sex." (Both quotes from Woody, *op. cit.*, p. 335.) Brown University opened its doors formally to women in 1891.

The University of Chicago, under President William Rainey Harper's leadership, championed women's education from its start in 1892. Harvard, however, remained reluctant to offer equal opportunity to women in higher education. While it admitted women to courses designed for graduate students, they were not allowed to earn a Harvard degree. They received degrees from Radcliffe "with the approval of the President and Fellows of Harvard College."[67] This approval was not always forthcoming. Mary Whiton Calkins, who wrote her thesis under the guidance of William James and Hugo Münsterberg, was refused her degree by the Harvard Corporation in 1896. In 1902 she was offered a Radcliffe degree, but she in turn refused it, insisting on a degree from Harvard.[68] This *de jure* separatism remained in effect at Harvard throughout the first quarter of the twentieth century, although beginning in 1921 women were offered degrees of Master and Doctor of Education on the same conditions as men.[69]

Until the last decade of the nineteenth century, there was a striking paucity of women graduate students at major American universities, although these campuses were not swarming with men either. After that, however, their proportion began to increase steadily. In 1892 women represented 14 percent of the graduate students in public and private institutions; by 1900 this had increased to 29 percent; and it stood at 37 percent in 1920. Of the women receiving advanced training at major universities President Harper of Chicago asserted in 1901: "The women now being graduated, with the Doctor's degree, from our strongest institutions, are, in almost every particular, as able and as strong as the men. If opportunity were offered, these women would show that they possess the qualifications demanded."[70]

The institutions that first admitted women without special qualification were also, for the most part, the leading universities in the nation. The policy did not extend to all universities, most notably those in the South, which throughout the period from 1900 to 1925 placed extraordinary restrictions on the admission of women to graduate programs.[71]

The ratio of male to female recipients of college and advanced degrees has been altered significantly since 1900. Today women have about 40 percent of the B.A. and B.S. degrees, a third of the master's degrees, and a fifth of the Ph.D.'s. As the data in Table 6-2 suggest, the pattern of change has differed for the

---

67. Woody, *op. cit.*, p. 336.

68. Margaret W. Rossiter, "Women Scientists in America Before 1920," *American Scientist* 62, no. 3 (May-June 1974): 312–323. This paper presents interesting materials, both quantitative and qualitative, on the career patterns of over 500 women scientists. It is particularly strong in its discussion of honorific recognition.

69. Even by 1925–26 no degrees for graduate courses, other than education, were being granted by Harvard to women. In the same year Yale awarded graduate degrees to twenty women. Among the large eastern universities, Columbia and Pennsylvania were the clear leaders. For example, in 1925–26 Columbia awarded a total of 2,017 graduate degrees of all kinds; 56 percent of these were conferred upon women. See Willystine Goodsell, "The Educational Opportunities of American Women—Theoretical and Actual," *The Annals* 143 (May 1929): 1–13, esp. p. 4. Pennsylvania advanced beyond Columbia in one respect: It admitted women to its undergraduate program in 1925.

70. As quoted in Woody, *op. cit.*, p. 327.

71. Willystine Goodsell, "The Educational Opportunities of American Women," *op. cit.*, p. 3.

TABLE 6-2. Degrees Conferred in United States Colleges and Universities, 1900–72

YEAR	BACHELOR'S DEGREE[1]			MASTER'S DEGREE			DOCTORATES		
	% Female	Number Female	Total	% Female	Number Female	Total	% Female	Number Female	Total
1900	19	5,237	27,410	19	303	1,583	6	23	382
1910	23	8,437	37,199	26	558	2,113	10	44	443
1920	34	16,642	48,622	30	1,294	4,279	15	93	615
1930	40	48,869	122,484	40	6,044	14,969	15	353	2,299
1940	41	76,954	186,500	38	10,223	26,731	13	429	3,290
1950	24	103,217	432,058	29	16,963	58,183	10	643	6,633
1960	35	136,187	389,183	33	25,727	77,692	10	1,028	9,829
1970	42	346,373	833,322	40	82,667	208,291	13	3,976	29,866
1972	42	388,931	926,866	41	101,973	250,075	16	5,270	33,330

NOTE: 1. Includes first professional degrees requiring at least six years.

Source: U.S. Bureau of the Census, *Historical Statistics of the United States: Colonial Times to 1970*, Bicentennial Edition, Series H 751–765 (Washington, D.C.: Government Printing Office, 1976), pp. 20–21.

several degrees. The largest increase in the sex ratio of B.A. and M.A. recipients occurred between 1910 and 1930, with a notable decline between 1940 and 1950, and a rise again by the 1960s to the prewar level. In short, up to 1940 there were significant gains made by women at the predoctorate level, with little advance since then. The pattern of female Ph.D.'s is more variable: The proportion of women rises steadily from 1900 to 1930, moving from 6 to 15 percent, and dips back to only 10 percent by 1950, only to rise significantly from roughly 11 to 18 percent between 1965 and 1975.

The data here also provide additional clues to the shifting size of the academic social system. I have noted already that the proportion of female Ph.D. recipients has remained roughly constant over time, but that the absolute number of female Ph.D. recipients has increased sharply. Put concretely, in 1910, 10 percent of doctorates represented only 44 persons; by 1950, 10 percent represented 643 persons. The increase in absolute numbers of B.A. and M.A. recipients is much larger still. Of course, as these numbers grow so does the size of the academic social system that has until recently absorbed them, and a casual glance at the absolute numbers suggests that the elite universities had an increasingly larger pool of females with doctorates to consider for jobs as the century moved on.

The absolute number of women holding higher degrees is of some import, because rather early on these degrees became necessary conditions for admission to the academic-science community. And it was not until late in the nineteenth century that women were admitted at all to most graduate programs in the United States. Without a doctorate, even by 1920 a person's chances of obtaining a quality position within the academic-science community was small. In 1925, for instance, 85 percent of all Columbia and 87 percent of all Chicago professors held the doctorate.[72] The rising proportion and absolute number of women holding the union card made it increasingly difficult for employers in the first quarter of the century to point to the absence of a degree as the reason for not hiring women.

The similar proportion of female Ph.D.'s in 1920 and 1960, referred to at the beginning of this chapter, is often used as an indicator that there has been little change since the early part of the century in the status of women in the academic community. By inference it is considered an indicator of the persistent exclusion of women from academia. Before accepting this inference, let us examine the data more carefully. Even if the percentage of female doctorates has remained invariant, the institutional origins of training for men and women, which are so influential in their careers, may have changed significantly over time. Is there evidence that women in the earlier period were systematically excluded from excellent centers of training?

Explaining the limited educational opportunities of men and women is,

---

72. Harry Dexter Kitson, "Relation between Age and Promotion of University Professors," *School and Society* 24 (July-Dec. 1926): 400–404, esp. p. 402.

perhaps, as crucial to understanding sex discrimination as is explaining the reasons for the small proportion of advanced degrees earned by women. As I have suggested, the historical paucity of female Ph.D.'s may result in part from discriminatory exclusion, but it also surely results from a set of forces leading women to turn away from graduate school. If women are, on average, trained at poorer universities, the pattern could result from nondiscriminatory factors involving self-selection, but if men and women are trained at the leading graduate departments in equal proportion, we can tentatively conclude that little discrimination obtains in the admissions process.[73]

The recent situation regarding graduate school origins is relatively clear. Many people think that women are less apt to have been trained at distinguished universities than men, since it is believed that they are discriminated against in admissions and that the range of their alternatives in choosing graduate schools is more restricted. In fact, the best evidence suggests that women are just as likely as men to receive their training from the most distinguished universities. "Among 1961 doctorate recipients, 52 percent of the men were trained in either 'distinguished' or 'strong' departments, as compared with 51 percent of the women."[74]

The educational origins of male and female Ph.D.'s in the earlier period, 1920 to 1924, can be compared with these relatively recent data. To uncover the relationship between gender and the quality of doctoral institution, data for male Ph.D.'s were obtained from *Doctorate Production in United States Universities: 1920–1962,* which contains detailed information on the production of Ph.D.'s by American universities, subdivided into several time periods.[75] Data for women Ph.D.'s in the period were obtained from a survey of over 1,000 female doctorate recipients published in 1929, which provides a yearly breakdown of the sources of graduate training.[76]

To determine whether men and women in 1920–24 received their training from universities of similar rank, we need some measure of the quality of American universities during the early 1920s. In 1925 Raymond M. Hughes published such an assessment, based upon peer evaluations, i.e., appraisal of quality made by scholars in the field.[77] The rankings obtained by Hughes are used here as the indicator of the quality of graduate schools.

73. To draw this conclusion, it is necessary to assume that the men and women are roughly the same in all relevant respects except sex. Data for the contemporary period on IQ of Ph.D.'s, for example, has been presented in Chapter 3 and Chapter 5, but no comparable data exist to my knowledge for this earlier period.

74. John K. Folger, Helen S. Astin, and Alan E. Bayer, *Human Resources and Higher Education* (New York: Russell Sage Foundation, 1970), p. 285.

75. Source: *Doctorate Production in United States Universities: 1920–1962 (op. cit.),* Table 7, p. 19; Appendix 2, pp. 63–73.

76. Emilie J. Hutchinson, *Women and the Ph.D.,* Bulletin no. 2 (Greensboro, N.C.: The North Carolina College for Women, 1929). This study of extraordinary interest on the careers of female Ph.D.'s presents both results of a quantitative survey of 1,025 women who have taken the doctorate and qualitative material on their perceptions of career barriers. I have reanalyzed some of Hutchinson's data that were presented in raw form. The study has been an extremely useful source of data for the early period.

77. Raymond M. Hughes, *A Study of the Graduate Schools of America* (Oxford, Ohio: Miami University Press, 1925).

214   FAIR SCIENCE: WOMEN IN THE SCIENTIFIC COMMUNITY

A scan of the leading producers of male and female Ph.D.'s in the period reveals several significant features. Bryn Mawr College, which ranked seventh in terms of number of female Ph.D.'s, is not among the overall national leaders in the period. While women who attended Bryn Mawr undoubtedly were required to meet standards as rigorous as those required at the leading universities in the nation, there clearly was less opportunity for Bryn Mawr graduate students to work with leading scientists and scholars. In fact, several schools that were leading producers of doctorates did not offer degrees to women, the most notable of these being Harvard and Princeton. Radcliffe College, which offered doctorates only to women, was the third largest producer of female Ph.D.'s.

So much for particulars. Is there any solid evidence that women were systematically excluded by the best universities? I have found little evidence of systematic bias in the awarding of degrees. Table 6-3 presents in rank order the

TABLE 6–3. Comparison of Quality of Graduate Departments Attended between 1920 and 1924[1]

QUALITY OF SCHOOL		MEN*			WOMEN		
		No.	*%*	*Cum. %*	*No.*	*%*	*Cum. %*
1.	Chicago	436	.114	.114	37	.100	.100
2.	Harvard	338	.089	.203	0	.000	.100
3.	Columbia	377	.099	.302	72	.194	.294
4.	Wisconsin	256	.067	.369	16	.043	.337
5.	Yale	201	.053	.422	27	.043	.410
6.	Princeton	94	.025	.447	0	.000	.410
7.	Johns Hopkins	261	.068	.515	19	.051	.461
8.	Michigan	102	.027	.542	11	.030	.491
9.	Berkeley	175	.046	.588	9	.024	.515
10.	Cornell	233	.061	.649	25	.067	.582
11.	Illinois	160	.042	.691	16	.043	.625
12.	Pennsylvania	108	.028	.719	15	.040	.665
13.	Minnesota	94	.025	.744	11	.030	.695
14.	Stanford	57	.015	.759	9	.024	.719
15.	Ohio State	83	.022	.781	5	.013	.732
16.	Iowa	112	.029	.810	10	.027	.759
**17.	Northwestern	27	.007	.817	2	.005	.764
**18.	North Carolina	26	.007	.824	1	.003	.767
**19.	Indiana	26	.007	.831	4	.011	.778
20.	Other	651	.171	1.002	82	.221	.999
	Total	3815			371		

NOTE: 1. Quality of graduate departments based upon ratings by Raymond M. Hughes, *A Study of the Graduate Schools of America* (Oxford, Ohio: Miami University Press, 1925).

*Male totals = Total Ph.D.'s listed in NAS–NRC Publication 1149 (Washington, D.C.: 1965) minus number of women shown in this table.

**Information available only for 1920–29. Figures shown here are estimated, based on the assumption that of all Ph.D.'s awarded between 1920–29, .35 were awarded in 1920–24. These estimates are .35 of the 1920–29 totals.

nineteen most distinguished American universities offering graduate programs in 1925. It also presents in separate columns the cumulative percentage of men and women receiving degrees from these universities, moving from Chicago, rated first, to the residual category of universities not rated among the top nineteen.

In examining the data, first note that women represent only about 9 percent of the total—that is, less than the national proportion for this period, suggesting that the survey data for women are incomplete. In fact, responses were not obtained from all women who had received degrees. Second, the number reported as male Ph.D.'s is obtained by subtracting from the NRC-NAS total for each school the number of females receiving doctorates from it. Third, a small number of schools account for a very high proportion of doctorates. In fact, the leading five schools in this period accounted for roughly 42 percent of the total; the leading ten produced about 65 percent of all doctorates. This great concentration in production by a relatively few distinguished schools has changed dramatically since 1920. There has been a marked diffusion in the educational origins of American Ph.D.'s. In 1960–61 the five most distinguished graduate universities, according to the 1964 Cartter report, produced only about 12 percent of the doctorates; the top ten produced roughly 28 percent.[78]

The most interesting feature of Table 6–3 is the comparison of the cumulative percentages of doctorates awarded to men and women according to the rank of the graduate school. Focus initially on the proportion awarded by the top five schools: 42 percent of doctorates earned by males and 41 percent of those earned by females were awarded by these institutions, even assuming that Radcliffe degrees were not de facto Harvard degrees. Among the top ten schools in terms of their assessed quality, there remains little difference in the proportion of males and females granted doctorates: 65 percent of the males compared with 58 percent of the females. Finally, the top nineteen graduate schools accounted for 83 percent of the male and 78 percent of the female Ph.D.'s. In sum, there is no statistical difference of any consequence in the 1920–24 period in the quality of graduate schools attended by men and women. If over the forty-year period from 1920 to 1960 women and men were equally likely to have received their doctoral training from the better universities, it is not because women were doing as poorly as they had ever done, but because they were doing as well as they had done in the past. However, to leap from these statistics to inferences of comparable experience at the major American universities in this period would be premature indeed.

Although the statistical association between the quality of graduate departments and sex status of doctorate recipients is insignificant, the quality of the educational experience at these schools for men and women was quite different. It is doubtful that women attending Radcliffe between 1900 and 1925, for example, received the same educational opportunities as their male counterparts at

78. Allan M. Cartter, *An Assessment of Quality in Graduate Education* (Washington, D.C.: American Council on Education, 1966); H. W. Magoun, "The Cartter Report on Quality in Graduate Education," *Journal of Higher Education* 37, no. 9 (Dec. 1966): 481–492.

Harvard. There is some evidence that even though women were allowed to follow the same graduate course as men at the major universities, their positions were marginal, even within the hallowed Ivy League walls. In her autobiography Mary Whiton Calkins describes restrictions on the activities of women at Harvard at the turn of the century:

> I shall not let this opportunity pass by to record my gratitude for the friendly, comradely, and refreshingly matter-of-fact welcome which I received from the men working in Laboratory as assistants and students, by whom the unprecedented incursion of a woman might well have been resented. . . . I interrupt myself to interpolate a frivolous reminiscence, of a much later date, which sets off in bold relief the friendly tolerance of my Harvard fellow-students. I was a member in 1905 of the Executive Committee of the American Psychological Association. Dr. Münsterberg had planned a lunch-meeting of the Committee at the Harvard Union, but the burly head-waiter stoutly protested our entrance. No woman, he correctly insisted, might set foot in the main hall; nor was it possible to admit so many men, balanced by one woman only, to the ladies' dining-room. It was almost by main force that Professor Münsterberg gained his point and the Committee its lunch.[79]

One wonders whether at the time interaction between men and women students at these major universities was not highly restricted by formal and informal rules of association.[80]

Job Location and Academic Rank

Thus far I have sketched the representation of women in the labor force, and in academic graduate programs. Let us now consider the historical evidence of sex discrimination within the academic community.

By the 1920s extensive empirical research was being conducted on the status of women in the academic community. In conjunction with the data I have collected, these empirical studies allow us to examine the opportunities that women Ph.D.'s had in terms of five aspects of scientific careers: job location, academic rank, salaries, honorific recognition, and scientific research performance.

The Hutchinson survey recorded job histories of female Ph.D.'s. Data were obtained from a questionnaire returned by 1,025 women who had received doctorates from American colleges and universities between 1877 and 1924. Although the survey accounts for only two-thirds of all doctorates awarded to females in

79. Mary Whiton Calkins, *Autobiography,* in *A History of Psychology in Autobiography,* ed. Carl Murchison, vol. 1 (Worcester, Mass.: Clark University Press, 1930).

80. Virginia C. Gildersleeve, former dean of Barnard College, discussed the politics of furthering the cause of women's education in the context of a larger and far more powerful brother institution. At one point she says: "Men were opposed to letting women in some courses and professional schools largely because they thought the women would cause trouble, would probably weep and faint at inconvenient moments, expect special consideration and privileges, perhaps lower the standards, and in general be a nuisance . . . when I took office . . . [t]here were still a good many graduate courses not open to women." Virginia C. Gildersleeve, *Many a Good Crusade* (New York: Macmillan Co., 1954), pp. 97, 98.

this period, and does not present comparative data for male Ph.D.'s, it is a rich source of material on the job opportunities open to women. Most of the women Ph.D.'s had completed their course of study in the early part of the century and were entering the job market between 1915 and 1924: Roughly 60 percent of the women surveyed received their degrees between 1915 and 1924; another 27 percent, between 1905 and 1914.[81] What did they do after obtaining the Ph.D.?

Teaching was their predominant activity. Of the 1,025 women, 58 percent became teachers, and 11 percent administrators or executives; 8 percent worked at research jobs, and another 8 percent at a variety of other jobs; 16 percent were not employed. There was very little variation among the fields in which the degree was earned. Although females in the humanities were more apt to enter teaching than those in the natural sciences and mathematics, the difference is only about 5 percent. Social scientists were least likely to enter teaching, primarily because of the number of psychology Ph.D.'s who entered private practice.

Few of these women worked in secondary schools (10 percent). Thirty-one percent of all Ph.D.'s held jobs in universities, about a third of them in the natural and social sciences and a fourth in the humanities. Roughly half the teachers were affiliated with colleges.[82] A high proportion of the women who worked at colleges were employed either at women's colleges or, less frequently, at new state colleges. Forty years later, in 1965, Helen Astin found that women Ph.D.'s had employment settings very similar to those found by Hutchinson. In 1965 roughly 70 percent of female Ph.D.'s worked in colleges and universities, another 10 percent in junior colleges or secondary school systems, and 14 percent in industry, nonprofit organizations, and government.[83]

In 1901 President Harper of Chicago summed up the existing opportunities for women in the academic community:

> In colleges and universities for men only, women may not find a place upon the faculty. In a certain great State university, in which there are as many women students as men students, women are represented in the faculty by a single individual, and she has been appointed in the last three years. . . . In some of the women's colleges women find a place. In others, second-rate and third-rate men are preferred to women of first-rate ability. The number of faculties of colleges and universities on which women have appointments in any number is very small, and even in certain institutions in which women have gained secure footing there is often greater or less distress among the men of the various departments if even one or two women are appointed. . . . So far as I can ascertain, during the past year the appointments of women, east and west, even in coeducational institu-

---

81. The median age for women scientists who took the doctorate was 31.3; for men of science, the age was lower. One study reporting on 9,000 doctorates showed that the degree was obtained when the recipients were between the ages of 28 and 31. See Ching-Ju Ho, "Personnel Studies of Scientists in the United States," master's thesis, Teachers College, Columbia University (New York, 1928), pp. 23, 24.

82. Hutchinson, *op. cit.,* p. 55.

83. Helen S. Astin, *The Woman Doctorate in America* (New York: Russell Sage Foundation, 1969), Table 21. Lindsey Harmon found similar distributions for samples of female Ph.D.'s of 1955 and 1960. See *Profile of Ph.D.'s in the Sciences,* Career Patterns Report no. 1 (Washington, D.C.: National Academy of Sciences, National Research Council, 1965).

tions, have numbered very few—fewer, perhaps, than ever before. Is this progress? Or is it rather a concession to prejudices which, instead of growing weaker, are growing stronger?[84]

Not much changed in the following two decades. Female marginality was maintained in practice partly by creating opportunities in colleges segregated by sex. Women had some opportunities to teach at women's colleges, but few at larger coeducational universities and state colleges. In 1910, 9 percent of faculties at state colleges and universities were women, but this conceals a skewed distribution. For example, 56 percent of the faculty at the Florida State College for Women were women, but of the combined teaching staffs of 1,258 at the University of California, MIT, University of Maine, and Cornell, only 25, or about 2 percent, were women.[85]

In 1921 the status of women in colleges and universities was assessed by a committee of the American Association of University Professors. Results were based upon a questionnaire sent to "all of the 176 institutions that are represented in the membership of the American Association of University Professors." Data were obtained from 145 of these institutions.[86]

The findings on the academic affiliation of female Ph.D.'s lend additional support for the idea of their marginal status. First, at colleges and universities exclusively for men, which included

> nearly all of the more noted eastern universities, . . . until quite recently no woman held any grade of professorship in these institutions. At present only two women are found among the nearly two thousand professors in these [29] colleges and universities. One woman was given a professorship of the third rank in the Harvard Medical School about two years ago and another a professorship of the second rank in the Yale School of Education this fall.[87]

One dean at a prestigious eastern university remarked on opportunities for women at his school:

> When we discover a woman who can handle some subject in our course of study *better* than a man could handle it, we shall not hesitate to urge the appointment of the woman and we shall, in all probability, be successful in getting it confirmed. . . . There is, of course, a general prejudice at————against the appointment of women, but it is a prejudice that arises out of the traditional masculinity of the institution. It is neither a violent antagonism nor a judgment based on study and experience [emphasis added].[88]

---

84. Woody, *op. cit.*, p. 328.

85. *Ibid.*; see also Preliminary Report of Committee W, "Status of Women in College and University Faculties," *Bulletin of the American Association of University Professors* 7, no. 6 (Oct. 1921): 21–32. This report and its follow-up provide extraordinarily useful data for the period 1920–25. The members of the committee were: A. Coswell Ellis, Texas (chairman); Florence Bascom, Bryn Mawr; Cora J. Beckwith, Vassar; Harriet W. Gigelow, Smith; Isabelle Bronk, Swarthmore; Carleton Brown, Minnesota; Caroline Colvin, Maine; John Dewey, Columbia; Anna J. McKeage, Wellesley; D. C. Munro, Princeton; Helen M. Searles, Mount Holyoke; W. F. Willcox, Cornell.

86. Committee W report, *op. cit.*, p. 21.

87. *Ibid.*, p. 21.

88. *Ibid.*, p. 23.

The hypothetical woman candidate could achieve appointment only by being "better" than her male competition.

Second, at the fourteen women's colleges sampled, which included most of those top-ranked in the nation, women were extraordinarily well represented in all professorial ranks. Of the 294 full professors, 55 percent were women; of the 167 associate professors and the 152 assistant professors, 80 and 78 percent respectively were women. Fully 86 percent of the instructors at these schools were female. Yet even in these women's colleges 52 percent of the men as compared with 22 percent of the women were full professors.

Third, at the 104 coeducational colleges and universities surveyed, women held 4 percent of the 4,560 full professorships. If we exclude those women with full professorships in either home economics or physical education, the proportion drops by a quarter to roughly 3 percent. Twenty-six percent of these institutions had no women at all on their arts and sciences faculty; 12 percent had only one.[89] Forty-seven percent of the schools did not have a single female full professor on their arts and sciences faculty. A more detailed breakdown indicates that women represented 4 percent of the full professors in arts and sciences departments and in education, and one-half of one percent of the full professors at medical schools; they were totally unrepresented at the full-professor levels at schools of law, journalism, theology, agriculture, and engineering. Women were more apt to be represented in lower ranks, of course, comprising roughly a tenth of the associate and assistant professors and almost a quarter of the instructors. These percentages, however, are somewhat misleading because of the vastly different absolute numbers of men and women among the professoriate.

When the data are recomputed, we find that significant differences in rank lie at the extremes: 43 percent of the men as compared to 12 percent of the women were full professors, and 26 percent of the men and 67 percent of the women were instructors. Nonetheless, the data compiled by the AAUP committee, by Woody, and by Hutchinson clearly suggest that women had little opportunity to obtain full-time professorial positions at major universities and colleges. Their best bets were at women's colleges.

Thirty years later, in the 1950s, women professors were still far more likely than men to be located at sex-segregated colleges. In a national study of social science faculties, 49 percent of the women and 7 percent of the men held jobs at women's colleges. Correlatively, only 3 percent of the women and 19 percent of the men were located at men's colleges. Fully 75 percent of the men but only 48 percent of the women worked at coeducational institutions.[90] Of course, many of the women at the women's colleges had been there for many years; we have no access to age-specific distributions.

The employment opportunities for women Ph.D.'s in effect served as a

89. *Ibid.*

90. Stanley Budner and John W. Meyer, "Women Professors," Bureau of Applied Social Research, Columbia University, 1961, mimeo. These data were drawn from Paul F. Lazarsfeld and Wagner Thielens, *The Academic Mind* (Glencoe, Ill.: Free Press, 1958). I thank John Meyer for an original copy of this unpublished paper.

structural mechanism for isolating women scholars and reducing to a minumum the contact between women faculty and the most productive and creative male professors. While women with admirable skills could easily be found at small women's colleges, few of the most able and productive men were located there. Thus, female Ph.D.'s rarely interacted as colleagues or collaborators with the established men in their fields; they were in effect cut off from the centers of scientific and scholarly activity as a result of their job location.

A woman of exceptional ability could reduce her marginality by a simple exchange mechanism. She could, in effect, give up her chances for a professorship in order to be situated at a better academic location. Data drawn from the early empirical studies indicate that women with doctorates working at universities were largely in the lower professorial ranks, while those at colleges were more apt to hold high rank. Only 12 percent of the women at universities as compared with 45 percent at the colleges were full professors; 52 percent at universities and 41 percent at colleges were either assistant or associate professors;[91] and 29 percent at universities and 5 percent at colleges were instructors.[92]

Women at top universities like Columbia and Chicago also were promoted more slowly.[93] For male professors at Columbia, twelve years passed, on average, between the Ph.D. and full-professor status. It took women fifteen years to travel the same route. The same three-year differential is observed for Chicago faculty, although it took both men and women somewhat longer to achieve full professorial rank than at Columbia. By the time a woman did manage to become a full professor, she was already well into middle age.

These figures speak to the limited opportunities for positional rewards. Further evidence comes from the female Ph.D.'s of this period themselves, who recount stories of obstacles to achievement. A great number of women Ph.D.'s in all fields commented on the price women had to pay in the academic community. Consider several selected accounts of their perception of job opportunities:[94]

> The place of women in science is very doubtful. It is better than it used to be. More openings are appearing; nevertheless, they are few [botany—asst. professor].

> The field for women with a Ph.D. in economics is decidedly limited. Women may be given minor positions, instructorships, etc., in first class colleges and universities, but would rarely be considered for one of major rank. A good college embarrassed for funds may unwillingly take a woman [economics—professor].

> When every agency in the country tells you that in English particularly, every president and head of department insist on having only men in higher positions, it seems to me idiotic to encourage young women to take the higher degree with the

---

91. Unfortunately, the data cannot be disaggregated any further.
92. These data were recomputed from those appearing in Hutchinson, *op. cit.*, p. 55. Seven percent at universities and 9 percent at colleges held assorted lower-ranking positions.
93. Kitson, *op. cit.*, p. 401. His data on age of men and women Ph.D.'s in the 1920s are very similar to those on age of Ph.D.'s in the 1970s.
94. These quotations are drawn from Hutchinson, *op. cit.*, pp. 175, 178, 181, 186, 189, 190.

thought of getting anything like a fair deal. I could write a volume on the effects of this situation on one's whole viewpoint [English—assoc. professor].

I think that a woman ought to realize in advance of working for the degree that in many fields promotion is difficult except in women's colleges. She ought to understand the limitations as well as the opportunities [Latin—instructor].

The factor of sex makes a noticeable difference in securing university teaching positions, and unless a woman intends to make practical use of her psychological training in clinical, vocational, or other lines I would not advise her to get a Ph.D. for the purpose of teaching in a university [psychology—assoc. professor].

I am each year confronted with the problem of advising girls to work for their Ph.D. degrees. For the most part they want to do college teaching, but I realize that in my field there are few openings except in women's colleges. They resent the discrimination. On the other hand, I am confident that the boys I send away will receive offers to teach in universities [sociology—asst. professor].

I find myself apologizing for being a woman. The men who are heads of departments are so insistent upon having men, and they tell me that women ought to be satisfied to go into high schools and junior colleges. . . . During the past 10 years I have stood by and helped recommend men for at least six positions that I wanted desperately badly, but being a woman I wasn't even considered [zoology—asst. professor].

These data suggest that from 1900 to the early 1920s major American universities had few women on their faculties and almost none at the full-professor level. Academic men and women remained virtually sex-segregated until the period of the rapid growth of state university systems, which were more open to women faculty members.

Let us turn to the longer perspective and examine the trend in affiliation and academic rank of men and women. This perspective is gained by examining the historical data I have collected for the five Ph.D. cohorts.

In Chapter 3 I presented data for the cohort of 1958 Ph.D.'s, which suggested that women were as apt as men to be affiliated with high-ranking departments seven and twelve years after receiving their doctorates ($r = -.02$, $r = .01$ respectively). They were less likely, without accounting for research performance and labor force participation, to hold high rank ($r = -.34$, $r = -.28$) for the two periods. This is where matters stood by 1970; the historical data suggest where they have come from. Tables 6-4 and 6-5 present the time-series and intra-cohort data for the five historical points.

Let us first consider the relationship between gender and the prestige of academic affiliation over time. The argument rests on one important assumption—that the relative standings of academic departments remain roughly constant over time. Rather than rely on the very limited scope of the 1924 and 1934 surveys of departmental prestige by Hughes or on the equally limiting 1960 Berelson study, I have used the 1966 Cartter study evaluations as the measure of departmental ranking at each point in the historical time-series. This decision to use the Cartter data is supported by analysis by H. W. Magoun, who provided

TABLE 6–4A.   The Relationship between Sex Status[1] and Prestige of Academic Department[2] for Five Historical Cohorts[3] (Zero-Order Correlation Coefficients)

PRESTIGE OF DEPT.	COHORT (YEAR OF PH.D.)				
	1911–13	*1922–23*	*1932*	*1942*	*1952*
First Job	−.38	−.05	−.03	−.08	−.13
5 Years After Ph.D.	−.16	−.08	−.20	−.15	−.21
10 Years After Ph.D.	−.22	−.03	−.19	−.17	−.17
15 Years After Ph.D.	−.39	−.14	−.15	−.12	−.23
Last Job or 20 Yrs.					
After Ph.D.	−.47	−.07	−.22	−.13	—
Range of N:	(48–52)	(31–39)	(84–97)	(127–140)	(136–142)

TABLE 6–4B.   The Relationship between Sex Status[1] and Prestige of Academic Department[2]: Level of Success at Different Points in Women's Careers, by Year (Zero-Order Correlation Coefficients)

YRS. SINCE PH.D.	IN 1922	IN 1932	IN 1942	IN 1952	IN 1962
New Ph.D.	−.05	−.03	−.08	−.13	—
10 Yrs. Out	−.22	−.03	−.19	−.17	−.17
20 Yrs. Out	—	−.47	−.07	−.22	−.13

NOTES:
1. Sex status was coded: 1 = Male; 2 = Female.
2. Several different editions of *American Men of Science* were used to obtain data on the affiliations of each scientist over the course of his or her career.
3. Pairwise deletion of missing values accounts for the varying number of cases on which the correlations for a cohort are based.

comparative ranking data for the 1925, 1957, and 1964 studies.[95] Substantial evidence suggests that departments, at least in terms of their relative faculty standing, do not vary significantly in their prestige position over time. There is a .81 rank-order correlation coefficient between the first Hughes survey in 1925 and the department ranks produced by Kenniston in 1957. An equally high coefficient exists between the Kenniston and Cartter ranks (.83). Before moving to the data let me emphasize again that the men and women being compared in each of the five cohorts have been matched in terms of discipline, professional age, and institutional source of the doctorate.

In matters of appointments, the data in Table 6–4 corroborate the Hutchinson and Committee W findings. First, looking at the 1911 cohort, women are far less likely than men to be located at prestigious university departments. The correlation between gender and departmental prestige is −.38, and although it seems to diminish slightly within the first ten years of the career, by our last reading some

95. Magoun, *op. cit.* (note 78).

twenty years later there is a very large disadvantage for the women Ph.D.'s ($r = -.47$). What is more interesting, however, is the trend that emerges when we examine the first three cohorts. Considering only first appointments, the substantial association between sex status and prestige of department found in 1911 is reduced to insignificance by 1922, only ten years later: The association is only $-.05$ in 1922 and remains the same in 1932. Although these results are based upon a distressingly small number of cases, the pattern for the 1920s is consistent: Men and women who received degrees in the early 1920s tended to be located in roughly similar departments in terms of prestige. Note that there is only a small difference ($r = -.07$) in the departmental ranks of the 1922 men and women roughly twenty years later. Contrasting the 1911 and 1922 cohorts, there is a strong suggestion of greater equality of the sexes in this feature of the academic reward system, which might be expected to continue in later years.

But data for the 1930s cohort, and in fact, for the 1940s and 1950s cohorts as well, suggest a trend away from equality. Although the first appointments of male and female Ph.D.'s are roughly similar, within the career cycle of each cohort there is a steady tendency for women to move into less prestigious positions than the men in their cohort. Thus, by our last measurement for the 1932 cohort, some twenty years after the degree, women are significantly less apt than men to hold appointments at the better universities and colleges ($r = -.22$). This basic pattern holds for the 1942 and 1952 cohorts as well. Thus, historically there is a reduction of the importance of gender in first appointments, but a curvilinear relationship in the relation of sex status to the prestige of affiliations within careers.[96]

These perturbations make sense within the historical context. The sharpest decline in the relationship between sex status and prestige of departments comes immediately after a decade of extraordinary feminist activism in American society generally and in the American academic community particularly, plus a rapid expansion in the size of the community. Granted, most of the force of the feminist and equal rights movement was spent by the early 1920s, but the consequences of that activism and the expanded demand for professors continued to have an impact throughout the 1920s. Among academics, there was clearly concern in the early 1920s about the status of female Ph.D.'s and their limited opportunities, as can be seen in the many surveys and statistical reports on the status of women. Distinguished members of the educational community—men such as John Dewey and women such as Florence Bascom and Cora J. Beckwith—not only were trying to better understand the position of women but also were working to facilitate social change where sex discrimination was identified. Furthermore, there was rapid growth in both the absolute and relative numbers of women entering academic careers just before and during the early 1920s. Academic women were not such uncommon specimens after 1920 as they

---

96. For each of the cohorts from 1911 to 1942, the final-appointment data are taken after roughly twenty years; for the 1952 cohort, after roughly fifteen years.

had been previously. They represented more of an interest group and they could effect social reform by their common action. Indeed, the Committee W report, in analyzing salary differences between men and women Ph.D.'s, concluded:

> There is evidence of a tendency to increase women's salaries in the institutions in which the women of the faculty or of the state demand it. . . . From [one] state university a male professor writes: "Standard full professor's salary is $4,000. Probably $3,000 is about the limit for women at present. One woman professor was kept much below that until another one, who was highly valued, started a row over what she regarded as a general policy of discrimination against women. It ended in a scale suggested by the faculty which (wrongly, I think) took no account of sex."[97]

Women as a collectivity were rapidly becoming a force for academic administrators to contend with.

The reversal in the trend toward equality in appointments appearing in the 1930s was a result *inter alia* of the dissipation of female activism and a declining market for female Ph.D.'s. The next period after the twenties that had insignificant correlations between gender and positional recognition was, not surprisingly, the activist period of the sixties. These data suggest an important role for activism in reducing job discrimination. Of course, many other forces were at work in these two periods to produce significant gains for women, but the combination of general economic expansion and the pressure placed on university and college administrators probably influenced a reduction in prima facie sex discrimination during the 1920s and 1960s.

The data in Table 6-4 also suggest that each Ph.D. cohort has its own history, which is largely independent of previous or subsequent groups. Although these histories are often quite similar, in periods of social change considerable variability may be found. For example, within the 1932 cohort there is a modest association between gender and affiliation ($r = -.19$) some ten years after the degree, that is, in roughly 1942. But the association for the new cohort of Ph.D.'s in 1942 is less than half the strength of that found for the 1932 cohort ($r = -.08$). But after its first ten years the association between sex status and departmental rank for the 1942 cohort reaches $-.17$. Correlatively, if we compare the 1922 cohort twenty years after the Ph.D. with the 1932 cohort ten years after the degree—that is, at the same point in time—we find significantly different correlations between sex status and this form of positional recognition. The 1920s cohort, which enters the community under more favorable conditions, maintains its advantages; the 1930s group continues to be disadvantaged, indicating, of course, that positions stabilize early on in the academic career. These patterns further imply that significant social gains are cohort-specific, and that they are not likely to improve the lot of those women who have previously been discriminated against.

Let us shift focus to academic promotions. Recall that there was a moderately

---

97. Committee W report, *op. cit.*, p. 27.

strong association for the contemporary cohort (1957–58) between gender and
academic rank ($r = -.34$ after seven years; $-.28$ after twelve). The historical
pattern is strikingly clear and remarkably invariant over time (see Table 6–5).
Women start off in roughly the same ranks as men. This is, of course, expected.
But within five years of the degree, a significant gap in rank exists between men
and women; after ten years the discrepancy is even greater. Data for the first two
cohorts corroborate quantitatively the perceptions by faculty members at the time
that promotions to high rank were a rarity for women scholars and scientists. For
those men and women who received their degrees around 1911, there exists a
notable association between sex status and rank ($r = -.36$) by the early twen-
ties. An even stronger association obtains for the 1922 cohort after ten years ($r = -.49$). Furthermore, since these data are drawn from early editions of *American
Men of Science,* in which the women represented a relatively more select group
than the men, the association between gender and academic rank is, if anything,
underestimated.

Examining the trend after the career path is all but settled, we still find
significant differences in rank. And how remarkably close the data parallel those
presented in Chapter 3 for the contemporary period. With the notable exception
of the 1922 cohort, the associations after twenty years are roughly equal: $-.27$
for the first cohort; $-.34$ for the 1930s cohort; and $-.33$ for the 1940s cohort.
Although I have not yet attempted to explain these associations, it is worth
emphasis that even after twenty years these zero-order associations remain
strong. After all, it is a well-established fact of academic life that if a scientist or
scholar sticks around long enough, he will eventually be promoted to full professorial rank. Not so for women, apparently. Even after twenty years many women
still have not been promoted to higher professorial rank.

Consider the deviant case, the 1920s cohort, in terms of the relationship
between departmental affiliation and academic rank. While the associations between gender and departmental prestige are lower for this cohort than for any of
the others, the associations between sex status and promotion are strongest. This
is wholly consistent with the history of the period. It reinforces the idea that

**TABLE 6–5.   The Relationship between Sex Status and Academic Rank for
Five Historical Cohorts (Zero-Order Correlation Coefficients)**

ACADEMIC RANK	COHORT (YEAR OF PH.D.)				
	1911–13	*1922–23*	*1932*	*1942*	*1952*
First Job	.10	−.07	.03	.02	−.04
5 Yrs. After Ph.D.	−.21	−.33	−.19	−.03	−.15
10 Yrs. After Ph.D.	−.36	−.49	−.20	−.12	−.37
15 Yrs. After Ph.D.	−.29	−.51	−.30	−.31	−.32
Last Job or 20 Yrs.					
After Ph.D.	−.27	−.65	−.34	−.33	—
Range of N:	(41–57)	(31–44)	(87–104)	(128–145)	(129–153)

women exchanged the possibilities of promotion for more prestigious affiliations. There were more women, proportionately, at major universities than before, but they may have accepted lower-ranking positions with limited opportunities for positional advancement. So after twenty years the correlation
between gender and departmental prestige was only − .07, but it was − .65 between
gender and academic rank.

Male and Female Salaries

It is widely assumed that there is a great deal of sex discrimination in
academic salaries. I have not personally investigated this assumption, first, because I did not generate my own salary data, and second, because there is much
outstanding research on the subject. But in discussing historical trends I must
turn briefly to the question of salary differences.

Simply put, there has probably been as much overt sex discrimination in
matters of salary as in any other aspect of the American occupational structure.
The record is clear and requires little elaboration. For example, if we look
backward only to 1959, "the working female earned, on the average, 53 percent
as much as her male counterpart. Even on an hourly basis, the wage differential
was over 30 percent. . . . Further analysis reveals that for single males and single
females of the same age and level of education, hourly wage differentials averaged
only 18 percent, while being over 50 percent for married males and females."[98]
Even after a great deal of anti-bias legislation in the early 1960s, including the
Civil Rights Act of 1964, in the economy as a whole women earned about 60
percent of what men earned. On the basis of data extrapolated from the 1960
census, Victor Fuchs reported that women earned 60 percent as much as men in
hourly earnings, and that this figure was increased only to 61 percent after taking
into account race, schooling, age, and city size, and to 66 percent when further
controls were placed on marital status, class of worker, and the length of the
work experience.[99]

The sex differential in hourly wages has remained rather indifferent to historical change.[100] There was little change in the ratios between men's and women's

98. Solomon W. Polachek, "Discontinuous Labor Force Participation and Its Effects on Women's Market Earnings," in *Sex, Discrimination, and the Division of Labor,* ed. Cynthia Lloyd (New York: Columbia University
Press, 1975), pp. 90–91.

99. Victor Fuchs, "Differences in Hourly Earnings between Men and Women," *Monthly Labor Review,* vol. 44,
no. 5 (May 1971): 9–15; see also Larry E. Suter and Herman P. Miller, "Income Differences between Men and
Career Women," *American Journal of Sociology* 78, no. 4 (Jan. 1973): 962–974.

100. Of course, the differential value assigned not only to work but to the value of male and female persons is
ancient. Curiously, the ratio of .6 seems to be a long-term historical constant. To go back only as far as the writing
of the Scriptures, note the following evaluations drawn from Leviticus 27: 1–4:

And the Lord spake unto Moses, saying, Speak unto the children of Israel, and say unto them, When a man
shall make a singular vow, the persons *shall be* for the Lord by thy estimation.

And thy estimation shall be of the male from twenty years old even unto sixty years old, even thy estimation
shall be fifty shekels of silver, after the shekel of the sanctuary.

And if it *be* a female, then thy estimation shall be thirty shekels.

The Scripture goes on to evaluate the worth of males and females in other age categories, coming to much the same
ratio for each.

salaries reported in manufacturing industries between 1885 and 1948. For example, one study of average daily wages of men and women employed in seventeen industries in 1885 showed that women earned roughly 60 percent of what the men did. (Similar findings emerged from a comparative study of wages completed in 1890.)[101] Data on average hourly wages in twenty-five manufacturing industries between 1914 and 1948 also suggest remarkable stability to the pattern: Women earned 59 percent of the hourly wage of men in 1914; 66 percent in 1924; 70 percent in 1934; 65 percent in 1944; and 73 percent by 1948.[102] Studies that have tried to explain these differences between the sexes in terms of occupational rewards have done a fair job in explaining occupational and educational achievement, but have not done well in explaining in rational terms the wage differential. They conclude that there is more evidence for sex-based salary discrimination than for almost any other form of sex discrimination in the American stratification system.[103]

I have repeatedly emphasized that the attitudes, beliefs, and social structures which generate income and prestige differences between the sexes on the societal level may not operate similarly within various institutions, and particularly in the academic community. Today there is little evidence that significant differences obtain in the salaries of men and women in academia, especially below the level of full professor. This was not always so, and I turn now to the trend in academic salaries since the early twentieth century.

The meaning of the slogan "Equal pay for equal work" is by no means clear, simple, or straightforward when applied to the academic marketplace. "Equal pay" is probably easier to define than "equal work," although its meaning is hardly self-evident. To note only a few of the problems, some universities and colleges pay professors on a nine-month basis, others on the basis of twelve months. Should we examine salary or total income? Do we want to examine total annual wages or some sub-unit of the total, such as hourly wages? These issues need to be resolved before comparing the wages of men and women.

The meaning of equal work is even more problematic. What is to be included or excluded from a definition of equal work in academic jobs? Frequently this concept remains at best implicitly defined. Most often it is limited to course teaching-load. But of course some professors spend hours preparing for their courses and seeing students while others do not prepare at all; some teach very large classes, others only small ones. Beyond the teaching role, other social roles might enter discussions of equal work. The amount and duration of administrative work; the supervising of undergraduate theses or doctoral dissertations; the extra work that brings credit and esteem to the college or university, such as serving on national fellowship panels or government advisory committees—all

101. Clarence D. Long, *Wages and Earnings in the United States, 1860–1890* (Princeton: Princeton University Press, 1960), pp. 105–106.

102. Source: U.S. Department of Commerce, Bureau of the Census, *Historical Statistics of the United States, Colonial Times to 1957*, Series D (Washington, D.C.: U.S. Government Printing Office, 1975) 654–668, p. 94.

103. Donald J. Treiman and Kermit Terrell, "Sex and the Process of Status Attainment: A Comparison of Working Women and Men," *American Sociological Review* 40, no. 2 (April 1975); 174–200.

this could fairly enter into the calculation. Finally, and perhaps most important, part of "equal work" in major academic institutions is the production of new knowledge through original research.

Historically, and to the present day, discussions of salary discrimination have taken into account only one or a few of these roles, and they rarely measure adequately those that are examined. This makes estimating the salary differences attributable to gender difficult. But there is a further difficulty in establishing the price women pay by way of lower salaries: No study of salary differences has attempted to measure the multiplicative and cumulating effects of one form of sex bias on another. The forms of recognition under review *are* clearly interrelated. We can hardly examine the professional rank attained by women without determining the quality of the institution at which the rank has been achieved. The problem of salary differentials between men and women poses similar problems of adequate specification. It is therefore unacceptable simply to compare the salaries of men and women in the same academic ranks, since one method of holding down salaries is by not giving promotions. If women are discriminated against in promotion decisions, a true estimate of salary discrimination requires us to estimate the direct influence of gender on salary differences within a particular rank, plus the indirect influence that gender has on salaries by reason of differing possibilities of being promoted to a specific rank. One woman Ph.D. interviewed in the 1920s was quite aware of the problem: "The A.A.U.W. has done much to bring about equal pay so that in most of the colleges men and women of the same status on the faculty are paid on the same basis. The inevitable result is, however, that most of the headships of departments and full professorships with the attendant high salaries are going to men."[104]

Economic analyses of discrimination have tended to cluster into one or another of three types. Most previous research, particularly Gary Becker's, has concentrated on "wage discrimination," which results from differentials that are not attributable to productivity. This approach examines the wages of individuals performing the same tasks with equal efficiency. If wage differences still obtain, there is wage discrimination.[105] This type of analysis generally fixes on a single point in time. A second orientation examines "occupational discrimination," which obtains when the proportion of women in particular occupations is based on criteria other than productivity. A third approach, one articulated in the work of Janice Madden, looks at "cumulative discrimination," which examines current wages in terms of current productivity and the effects of past discrimination on the level of current productivity.[106] Work in the sociology of discrimination,

104. Hutchinson, *op. cit.*, p. 18.

105. Gary Becker, *The Economics of Discrimination* (Chicago: University of Chicago Press, 1957); see also, among others, John P. Formby, "The Extent of Wage and Salary Discrimination Against Negro-White Labor," *Southern Economic Journal* 35 (Oct. 1968): 140–150; Nancy Gordon, Thomas Morton, and Ina Braden, "Faculty Salaries: Is There Discrimination by Sex, Race, and Discipline?" *American Economic Review* 64 (June 1974): 419–427; Barbara Reagan and Betty Maynard, "Sex Discrimination in Universities: An Approach Through Internal Labor Market Analysis," *AAUP Bulletin* 60 (spring 1974): 11–21.

106. Janice Fanning Madden, *The Economics of Sex Discrimination* (Lexington, Mass.: D. C. Heath & Co., Lexington Books, 1973).

with rare exception, has failed to consider the effects of cumulative discrimination on current rewards, especially on salaries. Difficult as it is to measure these cumulative effects, not measuring them leads invariably to weak estimates of the actual level of discrimination. How do we adequately measure the effects of past discrimination against an individual on current financial rewards for that person's scholarly productivity? How do we effectively distinguish between discrimination and other social or biological causes of differences in productivity? Almost all studies on sex discrimination in academic salaries are flawed by their failure to address the problem of cumulative discrimination. Nevertheless, many of the findings are instructive and suggest the extent to which sex status has influenced academic salaries over the past seventy-five years.

Through the 1920s, women faculty members were systematically denied salary equity with men in colleges and universities. The AAUP committee, investigating salary differentials, found that even in women's colleges women were not paid as well as men in similarly ranked positions. In only half the coeducational colleges surveyed were salaries roughly on a par. "In 47 percent of the coeducational institutions and 27 percent of women's colleges in which it is frankly admitted that women are given less salary and lower rank than men for the same work, the difference stated ranges from 10 percent to 50 percent, averaging 18 percent."[107] In another inquiry into the relationship between gender and salaries, at 104 coeducational institutions "only 17.8 percent of all the women reported that they shared equally with their male colleagues in matters of salary."[108] For the women who reported their salaries, it is noteworthy that there is a strong correlation ($r = .64$) between academic rank and salary. The exclusion of women from high rank effectively cut them off from higher salaries.[109] And even in those institutions which reported equal salaries within the same ranks, it was generally admitted that "the promotion of women is much slower even when they do equally as good work as the men."[110]

Women Ph.D.'s responding to Hutchinson's questionnaire repeatedly complained of salary inequities. Here are three typical examples:[111]

Women who expect to engage in college or university teaching should be aware of the reluctance of administrations to promote women or to pay them salaries equal to those paid men of similar attainments for similar work. In many such cases the women are to blame for submitting to such discrimination [botany—asst. professor].

Nothing but the most earnest conviction that she could never be satisfied without a Ph.D. in mathematics would justify a woman's setting herself that end. It is a long, hard road and when the degree is obtained, she finds that all the calls for mathematics teachers are for men, and that when a woman is employed in one of the large universities she is practically always given long hours and freshman

---

107. Committee W report, *op. cit.*, p. 26.
108. Marion O. Hawthorne, "Women as College Teachers," *The Annals* 143 (May 1929): 146–153, esp. p. 149.
109. *Ibid.*; see Table 3, p. 149.
110. Committee W report, *op. cit.*, p. 27.
111. These quotations are drawn from Hutchinson, *op. cit.*, pp. 174–175, 185–186, 188.

work for *years*, with less pay than a man would receive for the same service [mathematics—asst. professor].

I should never advise any woman to enter this line of work, as women are not wanted in medical colleges. The competition with men with the same training is very acute, and women are not nearly as well paid as men with the same or less qualifications [physiology—instructor, medical school].

Since salary discrimination was often a matter of public policy at colleges and universities, administrators in the 1920s were called upon to justify it. They most frequently did so by calling on supply-and-demand arguments and the quite independent sociological claim that men had to support their wives and families. "One president of a well-known western college justifies maximum salaries in his institution of $2,500 for women and $4,000 for men on the ground that men have families to support. No evidence was offered that he practiced the same discrimination against unmarried men."[112] A third form of justification for salary discrimination was that without sex differentials men would not enter college teaching. In effect, college and university administrators were asking women to pay for fulfilling the social need that men continue to teach at the college level. Even if the social need were beyond question, it never was made clear why women rather than society should foot the bill for this socially desirable outcome.

As noted above, the family was the main unit for economic analysis in the late 1920s and throughout the Depression. Individual income was related to support of the family unit, and women's income was seen as largely supplemental. Employed male and female professionals, for the most part, had two different pay scales, and this sex difference was justified by the greater value placed on the male as the primary breadwinner. This policy completely overlooked the low rate of marriage among female professionals during this period. As we shall see, married female academics had great difficulty finding jobs within colleges and universities, which produced a strong disincentive to marry. Women were thus being paid less than men under the assumption that their salaries supplemented that of nonexisting husbands. The gross inequality in salary served only to exacerbate their marginal status.

As we have glimpsed, during the Depression the argument for salary differentials between men and women gained additional support throughout the American occupational structure. Indeed, many of the advances that were made just before and during the early 1920s were quietly rolled back in the Depression. Sentiment against women entering the ranks of the employed was strong within specific institutions and even stronger among the public at large. William Chafe summarizes this sentiment:

The advent of the Depression provided the final blow to feminist hopes for economic equality. During a time of massive unemployment many people believed that women should sacrifice personal ambitions and accept a life of economic inactivity. Congresswoman Florence Kahn spoke for most of her col-

---

112. Committee W report, *op. cit.*, pp. 27–28.

leagues when she declared that "woman's place is not out in the business world competing with men who have families to support," but in the home. Dean Eugenia Leonard of Syracuse University urged women college graduates to enter volunteer work rather than accept a salary. . . . Liberal women reformers voiced similar sentiments. Francis Perkins [the future secretary of labor] denounced the rich "pin-money worker" as a "menace to society," [and] a selfish short-sighted creature, who ought to be ashamed of herself.[113]

Employers, following public opinion, increasingly refused to hire women. Of 31, fully "77 percent refused to hire wives and 63 percent dismissed women teachers if they subsequently married. . . . From 1932 to 1937 federal legislation prohibited more than one member of the same family from working in the civil service."[114] And as we have seen, the public overwhelmingly supported this policy. Indeed George Gallup said he had "discovered an issue on which voters are about as solidly united as on any subject imaginable—including sin and hay fever."[115]

Although public opinion was to fluctuate sharply between the late 1930s and the early 1940s, salary differentials between women and men continued to obtain in most occupations. By 1945 women in manufacturing industries earned roughly two-thirds of what men were paid; women in public administration, about three-quarters.[116] Dual salary scales were maintained for male and female teachers in almost all secondary schools. Clear sex differentials in salary schedules persisted in both college and university settings, although the level of discrimination was beginning to abate and did not approach those in other institutional spheres.

Until recently even the best work on salary differences did not attempt to explain with empirical data relationships between gender and salary level. The figures do not partition the influence upon observed salary differences of employment patterns, academic setting, discipline, rank, and scientific and scholarly productivity. The data in those studies, nevertheless, provide useful clues. In 1954 Sylvia Fava, reporting on the median salaries of men and women sociologists, found that the salaries of women with Ph.D.'s at colleges and universities were 89 percent of those received by men; it was 84 percent for all faculty regardless of highest earned degree.[117]

In 1959-60 the salaries of men and women typically varied most among full professors: At nonpublic universities women earned 81 percent of the salaries of men; at municipal universities women actually earned 113 percent of the men's salaries; at state colleges female full professors earned 94 percent of the salaries

113. As quoted in Chafe, *op. cit.*, p. 107. Chafe's book makes extensive use of public opinion surveys drawn from the collection by Hadley Cantril.

114. *Ibid.*, p. 109.

115. *Ibid.*, p. 108. Original source: "America Speaks, the National Weekly Poll of Public Opinion," Nov. 15, 1936.

116. Chafe, *op. cit.*, p. 185.

117. Sylvia Fava, "The Status of Women in Professional Sociology," *American Sociological Review* 25, no. 71 (1960): 273. For a more detailed and comprehensive report see United States Government, Department of Labor, Bulletin no. 1169 (1954), pp. 119-121, esp. Tables A-27 to A-29.

of their male peers. Among assistant professors, salary differences were minimal: Women earned 96 percent of the men's salaries at both state universities and state colleges. Only at nonpublic universities did the ratio dip to 91 percent.[118]

By 1960 the salary differentials had stabilized. Jessie Bernard reports that female full professors earned median salaries that averaged 86 percent of those of their male peers, but that female instructors' median salaries were 94 percent of those for similarly ranked males.[119] Greater comparability in the lower ranks is a result of more restricted variances in salaries at these ranks, where salaries tend to be pegged to seniority rather than to market processes. The trend toward greater salary equity is further evidenced by data reported by Alan Bayer and Helen Astin for a cohort of male and female academics in 1964.[120] Rough controls for length of employment, setting, and academic rank produced reduced differences in male and female salaries. For the entire sample, the mean salary of women was 92 percent of men's. At universities, however, men and women in lower ranks earned roughly the same salaries (99 percent); within higher ranks the ratio was 91 percent. Similar results obtained when the focus shifted to colleges.[121]

Income data compiled for the years 1970–75 continue to show significant differences between men and women, when pertinent social characteristics of academics are included in the analysis. In 1973 the *Chronicle of Higher Education* reported a 17 percent difference in salaries, but when work experience and location were taken into account the figure varied somewhat.[122] Among men and women Ph.D.'s with twenty-two or twenty-three years of experience, and who were located at universities, women earned 83 percent of the salaries of men; there was a 12 percent difference between the men and women with thirteen or fourteen years' experience; and 11 percent among those with five or six years' experience. The disparities were slightly lower for men and women with comparable experience who were located at colleges rather than universities.[123]

John Centra has found that controlling for academic setting and rank affects the sex and salary correlation. While female professors at universities earn 85 percent as much as male peers, female assistant professors at universities receive roughly 93 percent as much as comparably situated males.[124] Furthermore, controlling for length of employment also affects the results. Male and female full professors who have been employed at universities for only five or six years have

---

118. Source: National Education Association, Research Report, 1960–R3, Tables 5, 9, pp. 12, 16. The figure of 113 percent at municipal universities is striking and is not explained in the original source. A number of factors could account for the parity: state laws, seniority differentials, etc.

119. Bernard, *op. cit.*, p. 180.

120. Alan E. Bayer and Helen S. Astin, "Sex Differences in Academic Rank and Salary among Science Doctorates in Teaching," *Journal of Human Resources* 3, no. 2 (1968): 191–200.

121. *Ibid.*, Table 2.

122. As reported in John A. Centra, *Women, Men and the Doctorate* (Princeton, N.J.: Educational Testing Service, 1974), pp. 78–92.

123. *Ibid.*, p. 83.

124. *Ibid.*, p. 87, computed from Table 5–6.

virtually identical incomes (women earn 98.5 percent as much as men); for those employed from thirteen to fourteen years, women's salaries are 95 percent of men's. Many women clearly perceive current sex differences in salary as resulting from discrimination. One woman, responding to Centra's 1973 inquiry, said: "Employers still pay females less for equal services rendered at this university. Some even go so far as to say—you have a husband so you don't need as much. Nonsense—a person should be paid for his ability and competency regardless of his situation. Being a female does not stop people from asking me to serve on committees *et al.*, but they expect me to be satisfied with less pay."[125] It must be emphasized again that these salary differences, which range roughly from 17 percent to 1, do not include controls for the quality of any form of role performance, particularly as registered in published research.

George E. Johnson and Frank P. Stafford examined the 1964 and 1970 salaries of male and female academics who received their doctorates in four disciplines: economics, sociology, mathematics, and biology.[126] The data are drawn from the National Science Foundation Register. Their study focused on two explanations for salary differences: (1) sex discrimination within the academic community, and (2) differences "generated by the market's reaction to choices by females with regard to lifetime labor force participation and on-the-job training," that is, differences in acquired skills and productivity between men and women.[127] In each of the four fields there was only a small difference in salaries of men and women immediately after receiving their doctorates, ranging in 1970 from 11 percent in biology to 4 percent in sociology. It will be noted that these differences are essentially the zero-order associations between sex status and salaries, without controlling for other possible explanatory variables. Johnson and Stafford did find, however, that sex-based differences increased as males and females moved on in their careers. For example, among men and women who had received their degrees around 1950, they found the 1970 salary ratio ranged from 72 percent in mathematics to 85 percent in economics and sociology. Of course, the increased differences resulted in part from emerging rank differences and market processes, but this is exactly what the authors wanted to test. They tentatively concluded that over the entire academic career about 40 percent of the wage differential between the sexes is attributable to discrimination and the other 60 percent to "human capital differences," that is, to on-the-job training, productivity, and number of years employed. They did not reject the discrimination hypothesis, but implied, without testing the implication, that discrimination alone does not account for most of the salary differences found among male and female academics.

---

125. *Ibid.*, p. 197. All of these data raise the question of the visibility of salaries and of sex-based inequities. In short, how do men or women know whether they are receiving less for equal work? How good are perceptions of salary differences?

126. George E. Johnson and Frank P. Stafford, "Women and the Academic Labor Market," in Lloyd, *op. cit.*, pp. 201–222.

127. *Ibid.*, p. 202. It should be noted that Johnson and Stafford's findings have been criticized in the economics literature for the absence of using more sophisticated models.

A 1972 National Education Association study of salary differences for roughly 210,000 faculty showed that for all ranks combined "the median salary of women faculty is 82.5 percent of the median salary of men faculty."[128] More interestingly the report notes that "in 1959-60 it was 84.9 percent and six years ago it was 83.4 percent." These data, which include salary information for all faculty in four-year institutions, many of whom do not hold professorial positions, are quite consistent with what Johnson and Stafford found for four disciplines. In 1971-72, among women full professors, their salaries were 90 percent as much as those of men full professors; among associate and assistant professors, 94 percent; and among instructors and lecturers, 95 percent.[129]

These data on salary differences over the past twenty years indicate not gross discrimination, as we might have expected, but only slight discrimination. This inference about the level of salary discrimination is plausible, since most of the studies have not considered possible legitimate, nondiscriminatory reasons for differentials in salaries. For example, none of the studies took into account the relative level of scientific or scholarly productivity of men and women scientists. A recent extensive study on academic salary differences by Alan E. Bayer and Helen Astin corroborates these conclusions. Their study strongly suggests small salary differences between men and women.[130] They begin by noting that in 1972-73 the salaries of male academics exceeded those of females, on average, by roughly $3,000. But in attempting to explain the difference they go on to do an extensive multiple regression analysis (unfortunately using a stepwise procedure that automatically introduces variables into the equation in terms of the amount of variance they explain). Among the variables entering the regression are: professional age, self-reported scientific or scholarly productivity, academic rank, type of institution, and so on. Most of the variance in salary obtains among full professors. Bayer and Astin find negligible associations between sex status and salaries among instructors, lecturers, and assistant professors.[131] The relationship is barely significant among associate professors. This skewed distribution of inequality has been the pattern for the past twenty-five years and it continues to be the basic story. Although Bayer and Astin do not present the zero-order associations between gender and salaries, they note that the partial correlation between sex status and salaries after controlling for nine variables is $-.04$. This is lower than the partial correlation of $-.16$ they obtained in a 1968-69 survey.

Academic rank was included in the initial regression, a questionable decision since discrimination in promotion could indirectly influence these salary differentials. Aware of this problem, Bayer and Astin produced a separate regres-

---

128. Source: National Education Association, *Salaries Paid and Salary-Related Practices in Higher Education,* 1971, Research Report 1972-R5, p. 11.

129. *Ibid.*

130. Alan E. Bayer and Helen S. Astin, "Sex Differentials in the Academic Reward System," *Science* 188 (23 May 1975): 796-802, esp. pp. 799-800.

131. *Ibid.,* Table 5, p. 799.

sion omitting academic rank. They still found a partial correlation of $-.05$ between sex status and salary, regardless of rank. Although these partial correlations continue to be statistically significant, do they have any substantive meaning? Gender explains very little of the variance on salaries, and despite the improvement on earlier studies, which failed to control for appropriate explanatory variables, there remain many additional factors that could further reduce, if not entirely eliminate, these apparent residual salary differences. For instance, other features of the academic marketplace (which, of course, may also be related to prejudice and discrimination), such as geographic mobility or the option to pursue alternative offers that can be translated into salary increases in a competitive market, remain unaccounted for in the Bayer-Astin study.

Let us place these salary differences between men and women in another perspective. There are greater differences in the salaries paid to scientists and scholars working in different fields than there are differences between the sexes. The 1952 department of labor analysis of sex differences in academic salaries within several fields allows us to compare the relative influence of discipline and gender. Mathematical statistics had the highest median salary of the nine fields reviewed. Let us relate salaries in other fields to it. Economists earned .96 as much as statisticians in 1952, but anthropologists, geographers, historians, sociologists, and academics in the humanities earned about .75 as much as statisticians. Similar reanalysis of Centra's 1973 data corroborates the Department of Labor findings. Since among females, social scientists had the highest salaries on average, they serve as the reference point. While female biological scientists with five or six years of experience earned .85 as much as their female social science peers, women in the physical sciences earned only .83 as much, and female humanities professors earned only .75 as much. In sum, the field differences are greater than the sex differences reported in the same studies. Does this mean that we have a prima facie case of field discrimination? Not necessarily. There are, of course, various explanations other than discrimination that might account for these field differences, many of which would rest on market conditions confronting humanists as compared with social and physical scientists. And if we finally conclude that salary differences between fields signal a form of discrimination, then we at least know that field "discrimination" accounts for more of the differences in academic salaries than does sex discrimination.

All of this cannot for a moment condone salary discrimination based on gender. There can be no doubt that such discrimination does exist in individual cases, but in the aggregate the trend is clear. Early in the century significant differences reflected a conscious social policy based on a set of widely held values about the relative importance of men and women in the labor foce. By the end of World War II some of the ideological fervor about the value of women's work and about equal pay was diminishing, but salary differentials, probably based largely on discrimination, continued. Since the mid-1950s, however, there has been movement toward salary equity. One gets a clear sense from historical materials that cumulative discrimination played a major role in the justification of salary inequality during the first half of this century. Since women were systematically denied access to research facilities and to resources that would aid their research productivity, it follows directly that they rarely performed as well at research as men. To close the circle, to the extent that salary, especially within the upper ranks, was dependent on research productivity, women were bound to lose out in the race with men.

Honorific Recognition

Shifting now to honorific recognition, we see a record of exclusion of women from the prestigious scientific societies and academies, and from the lists of most honored scientists. We have already noted the absence of women from the lists of Nobelists and Academy of Sciences members. But what is the more extensive record of their achievements?

For the social system of science examined in its entirety, we know from data presented in Chapter 3 that today there is almost no statistical association between sex status and the number of honorific awards scientists receive ($r = -.07$; $b* = .00$ after controlling for scientific productivity). But consider this relationship some fifty to seventy-five years ago.

Until the first quarter of this century, women scientists throughout the world were excluded from most prestigious scientific societies. In fact, the constitutions of the Royal Society and the French Academy of Sciences excluded women from membership. The issue of female membership became a question of some debate in the European press of the early 1900s, especially after Marie Curie was twice awarded the Nobel Prize. A writer for *Nature* made the following cautious observations in 1911 about Marie Curie's claim to membership in the French Academy:

> There may be room for difference of opinion as to the wisdom or expediency of permitting women to embark on the troubled sea of politics, or of allowing them a determinate voice in the settlement of questions which may affect the existence or the destiny of a nation; but surely there ought to be no question that in the peaceful walks of art, literature and science, there should be the finest possible scope extended to them, and that, as human beings, every avenue to distinction and success should unreservedly be open to them.
>
> All academies tend to be conservative and to move slowly; they are the homes of privilege and of vested interest. Some of them incline to be reactionary. They were created by men for men and for the most part at a time when women played little or no part in those occupations which such societies were intended to foster and develop. But the times have changed. Women have gradually won for themselves their rightful positions as human beings. We have now to recognize that academies as seats of learning were made for humanity and that, as members of the human race, women have the right to look upon their heritage and property no less than men. This consummation may not at once be reached, but, as it is based upon reason and justice, it is certain to be attained eventually.[132]

132. As quoted in Mozans, *op. cit.*, pp. 394–95.

Writing only a few weeks later, another author, confronting the same issue, invoked basic scientific creed for justifying female membership in honored societies:

> As scientific work must ultimately be judged by its merits, and not by the nationality or sex of its author, we believe that the opposition to the election of women into scientific societies will soon be seen to be unjust and detrimental to the progress of natural knowledge. By no pedantic reasoning can the rejection of a candidate for membership of a scientific society be justified, if the work done places the candidate in the leading position among other competitors. Science knows no nationality and should recognize no distinction of sex, color or creed among those who are contributing to its advancement. Believing that this is the conclusion to which consideration of the question must inevitably lead, we have confidence that the doors of all scientific societies will eventually be open to women on equal terms with men.[133]

If this typifies the emerging ethos in science, it did not receive expression in actual behavior. For few women were actually honored, even when their work was clearly outstanding. In the most comprehensive study to date on the subject, Margaret Rossiter analyzed the career patterns of the 504 women scientists listed in the first three editions (1906–21) of *American Men of Science*.[134] Women represented from 3.5 to 4.8 percent of all biographies in these editions. These women were probably somewhat atypical in being more eminent than the average women in academic life at the time; their inclusion in the compilation suggests this is so. Rossiter compares the careers of the 504 women with 502 men selected from the third edition of *American Men of Science* (1921). The samples of men and women look remarkably alike. In particular fields, roughly the same percentage of males and females held doctorates; about the same proportion were located in academic institutions (68 percent of the women; 63 percent of the men); and although Rossiter considers the difference significant, similar proportions worked in museums and research institutions (10 percent of the women to 5 percent of the men).[135]

The men and women scientists did, however, differ in their academic location. Women were almost never professors at men's colleges and universities, and were largely concentrated at women's colleges. Women were also far less likely than men to be married and more apt to be unemployed. But what of the eminence of these women? Taking inclusion itself as an indicator of distinction, we find in this period that these women of science were largely excluded from the best institutions. In fact, in the years 1900–1920 only 9 percent of women physicists were employed outside of women's colleges.[136] Nevertheless, the employment pattern for women scientists who were entered in *AMS* significantly changed in the first two decades of the century:

---

133. *Ibid.*, p. 395.
134. Rossiter, *op. cit.*
135. *Ibid.*, p. 316.
136. *Ibid.*, p. 318.

The number of jobs at the women's colleges more than doubled between 1906 and 1921 (from 43 to 96), as the older colleges expanded and some newer ones . . . were established. The overall importance of women's colleges, however, began to decline after 1906, as more and more opportunities for women opened up in state universities and other schools. [There was] a steady decline in the percentage of women employed in the women's colleges, from 57.3% in 1906 to 36.5% in 1921. The state universities showed the largest overall increase, jumping from 2 jobs in 1906 (2.7%) to 53 in 1921 (20.2%).[137]

Even the women who were granted greater recognition did not fare too well. James Cattell, who holds paternity rights to *AMS*, tried to measure the eminence of scientists appearing in his compilation by having a sample of people in the same field evaluate one another's work. By this ranking procedure Cattell identified the top 1,000 scientists, and he described the chosen ones as "starred scientists." This allows us to compare the proportion of men and women who were recognized by their peers as leading contributors to their disciplines. Within the first three editions of *AMS*, 30 of the 504 women, or about 6 percent, were starred; 73 of the 502 men, or almost 15 percent, were marked as distinguished. Although the absolute number of starred women scientists increased from edition one to three, the proportion of starred women declined, from 13 to 5 percent. "The generally low level of accomplishment among women scientists and their virtual segregation raises the question of whether they really were a part of the profession."[138]

Some prestigious scientific societies in America did offer membership to women early on. The American Academy of Arts and Sciences opened its doors to Vassar's astronomer, Maria Mitchell, in 1848; the California Academy of Sciences invited women to join in 1853; the American Mathematical and American Chemical societies offered membership in the last quarter of the nineteenth century. But the most prestigious organization of scientists was more resistant to admission of women. It was not until 1925 that the National Academy of Sciences voted in a woman, the anatomist Florence Sabin, and not until 1931 that a second woman, the psychologist Margaret Washburn, was added to the Academy rolls.[139] Other accomplished female scientists, among them the mathematician Charlotte Scott and the astronomers Annie Jump Cannon and Williamina P. Fleming, were not similarly honored.[140]

Despite the variety of obstacles facing women scientists in this period, some did make important discoveries. To note only a few, Nettie Stevens had an independent multiple discovery with E. B. Wilson in 1905 about the relation between X and Y chromosomes and sex determination. Margaret Washburn and Mary Whiton Calkins made significant contributions to the development of experimental psychology; Florence Sabin did significant work on the lymphatic and

---

137. *Ibid.*
138. *Ibid.*, p. 320.
139. *Ibid.*, p. 321.
140. *Ibid.*

TABLE 6–6.   The Relationship between Sex Status and Honorific Recognition[1] for Five Historical Cohorts (Zero-Order Correlation Coefficients)[2]

1911–13	1922–23	1932	1942	1952
−.26 (62)	−.16 (52)	−.09 (137)	−.12 (212)	−.12 (238)

NOTES:
1. Honorific recognition is measured by the total number of awards, prizes, and postdoctoral fellowships listed in *American Men of Science; Dictionary of American Scholars; Who's Who.*
2. Pairwise deletion of missing values is used.

vascular systems; and Annie Cannon and Williamina Fleming did significant work on classifications of stellar spectra.[141]

Of course, most scientists and scholars, regardless of sex, fail to receive honorific awards. Using the number of honorific awards listed after the person's name in *AMS* as the indicator of eminence, we can estimate only very crudely the influence of gender on the level of honorific recognition for each of the five historical cohorts of scientists.[142] Recognition includes postdoctoral fellowships as well as truly prestigious prizes and awards. The indicator is a weak one not only because it fails to take into account the differing prestige of awards but because the listing of specific awards was made by the selected scientists and scholars themselves. With these caveats, we can examine the trend data presented in Table 6–6. The correlation coefficients in the 1911 and 1922 periods are based upon relatively small samples of men and women, but the data suggest that there were significantly greater differences in the honorific recognition of men and women in the first decade of the century than in the following two decades. The association between gender and awards falls by the 1930s to one-third of its 1910 level. It is noteworthy, however, that the association stabilizes at roughly −.10 between 1930 and 1940 and remains at that level as of 1970, as the data reported earlier in Chapter 3 suggest.

Research Productivity over Time

A major theme of this book maintains that sex differences in scientific productivity are a key interpretative factor in understanding the set of relationships between gender and forms of recognition. We are still unable to explain the sex differences in productivity, and this remains a major future task. The associations between sex status and the quality and quantity of output vary somewhat within

---

141. Here I follow Rossiter closely. She selects several women who made particularly notable contributions and discusses briefly barriers to their honorific achievement, despite their record of scientific performance.

142. Data on the number of honorific awards received are less complete for the female than the male scientists because of women's lower rate of inclusion in biographical compilations. See note 64 for the percentage of males and females included in *AMS*.

scientific fields and within different samples. In the period 1958–70 the zero-order correlations between gender and the quantity of output vary from about $-.20$ to $-.35$. Somewhat lower but equally consistent associations obtain when we look at the relationship between gender and citation counts as a measure of quality. This is the current situation, but has it been so throughout the twentieth century?

There is a widespread perception today that male and female scientists have differential opportunities to pursue their research interests successfully. But difficult as it is, the current situation bears no comparison with the situation of academic women in the first quarter of this century. Until recently, the ideology that women were incapable of creative research activity, which limited their access to quality research laboratories and other research facilities, resulted in a sharp discrepancy between the research behavior and output of men and women in academic science.

In 1920 the Committee W of the AAUP, as part of its larger study on the status of women, sent a questionnaire to the dean and to one male and one female professor in each of its member institutions in order to find out whether they felt that the lower salaries, slower promotion, and lower rank of women were a consequence of less innate ability to do scientific work. The results indicate that an overwhelming majority of both male and female professors felt there were no sex differences in teaching ability. Similar beliefs existed about the comparative physical ability of men and women to handle their jobs. But when it came to evaluating ability to contribute to productive scholarship, the majority thought women were inferior to men: 61 percent thought men were more able than women; the remaining 39 percent saw no sex differences; no one thought women were superior. One woman physicist typified widespread opinion: "In my own experience I have known only one woman who has shown much initiative or creative ability."[143] This opinion was voiced frequently by women as well as by men.

Leaders in higher education, all of them men, at the turn of the century publicly voiced this belief. President Hyde of Bowdoin College put it succinctly: "Productive scholarship should remain and will remain in men's hands."[144] Eliot of Harvard felt that women had not shown ability to do more than "learn from teachers and practice what they have been taught."[145] E. E. Alosson said that "it is the prevailing opinion that they [women] are as a rule inferior to men in work requiring initiative and originality."[146]

Widespread authoritative sentiments like these create conditions that lead to self-fulfilling prophecies. The familiar argument went like this: If women are incapable of originality, why waste resources on a futile effort? And impedi-

---

143. Hutchinson, *op. cit.,* p. 188.
144. As quoted in Woody, *op. cit.,* p. 339.
145. *Ibid.*
146. *Ibid.,* p. 340.

ments were placed in the paths of women interested in research. The experience in Germany of Lisa Meitner, the noted physicist, resembles what happened in the United States. Otto Hahn, who collaborated with Meitner for twenty-five years, recalls the difficulties she faced in pursuing her research goals:

> The beginning was difficult for her. Emil Fischer, the director of the Chemical Institute [at Berlin], did not then accept women, but he did make a concession in her favor. With the condition that she was not to enter the laboratories where male students were working, she was permitted to work with me in the wood shop. In 1907 this was a really large concession . . . and in time he also developed an attitude of fatherly friendship toward Lisa Meitner. But the rule that she had to stay in the wood shop (which was later extended to include another basement room) remained in force.[147]

Lisa Meitner managed to surmount some of these obstacles and to produce many scientific papers, but we cannot estimate how much more she might have produced had she at least been allowed out of the basement.

Women with less energy and fortitude would probably have withdrawn entirely from research. Many women who received their doctorates in the period 1900–1920 reported that they were not allowed to pursue certain dissertation topics that were deemed "irregular" or "not a nice problem for a lady."[148] It remains unknown whether men during these years were also restricted, albeit in different ways, in selecting their topics. Most of the women surveyed by Hutchinson in the 1920s continued to pursue some research after their degree, but generally at a reduced level of effort. They claimed that heavy teaching commitments were a major obstacle to research productivity. One woman put the problem tersely: "Lack of material and money to get it." Another said: "I am inclined to think that the greatest difficulty with regard to research occurs after a woman has begun her first teaching. An unfavorable atmosphere at that time is apt to stop a research spirit that is not well fixed."[149]

Few women in the Hutchinson sample produced much published research beyond the dissertation, which at that time traditionally had to be published. Forty percent of the women professors in language and literature, a third of those in the social sciences, and about a quarter in the natural sciences were "silent" scientists, never having published a single paper after their dissertations. Of course, most male professors were not particularly productive scientists or scholars either. Whether they faced similar obstacles in pursuing their research is questionable.

I have collected productivity data from scientific abstracting journals for each of the scientists and scholars in the historical study. These data are limited in several important ways. First, there are significant amounts of missing data. Many of the abstracting services did not begin to produce their compilations until

---

147. *Otto Hahn: A Scientific Autobiography* (New York: Charles Scribner's Sons, 1966), pp. 51–52.

148. Hutchinson, *op. cit.*, p. 41.

149. *Ibid.*, p. 85.

TABLE 6–7. The Relationship between Sex Status and Research Productivity[1] for Five Historical Cohorts (Zero-Order Correlation Coefficients)

RESEARCH PRODUCTIVITY	1911–13	1922	1932	1942	1952
First 5 Years Following Degree	−.46 (24)	−.13 (35)	−.19 (190)	−.19 (313)	−.15 (383)
First 10 Years	−.20 (53)	−.24 (71)	−.28 (190)	−.22 (345)	−.27 (383)
First 15 Years	−.26 (53)	−.24 (71)	−.30 (190)	−.25 (345)	−.27 (383)

NOTE: 1. Research productivity is measured by the number of published papers per year after receipt of the doctorate.

well into this century. Second, there is a strong bias toward inclusion of data for the social, physical, and natural sciences as opposed to the humanities, since the abstracts emerged first in the sciences. Third, there is the purely technical problem of errors made in attributing particular papers to particular scientists. This occasionally involves misallocations of credit. But the substantial data that have been collected have two basic features: They are for the five Ph.D. cohorts, and they include annual productivity counts for the first fifteen years of the career after the Ph.D.

Table 6–7 presents for each of the five cohorts a set of correlations between sex status and research output at three points in the first fifteen years of the academic career. Let us first look at the historical trend, taking as our point of comparison the total productivity over the fifteen years. The correlations are remarkably consistent in both direction and magnitude. Throughout, male scientists are, on average, more prolific than females. In the 1911 cohort the association was −.26. Almost fifty years later for the 1958 cohort, as reported in Chapter 3, the association was −.30. In the intervening years there was great stability to this association: For the 1922 cohort it was −.24; for the 1932 cohort, −.30; for 1942, −.25; and for 1952, −.27. It is remarkable that such a constant pattern is found, considering the historical evidence of increased access to facilities and resources open to women since the beginning of the century. It may well turn out, of course, that there are historically different causes for the similar correlations.[150]

_____

150. Virtually all studies of the productivity of men and women scientists that have been reasonably designed have found that male scientists tend to publish more than females. Budner and Meyer, *op. cit.*, present data for 264 women social scientists teaching at universities and colleges in the middle 1950s which show that women are less productive than men of comparable age. Their analysis is particularly interesting because it corroborates the finding that productivity differentials between men and women increase over time. Budner and Meyer found an 11-percentage-point difference in the proportion of men and women under forty who were "highly productive." For the age group forty-one to fifty the difference increased to 21 percent, and for professors over fifty it was 24 percent. There are, of course, many not mutually exclusive hypotheses to account for this pattern. Consider only three hypotheses. First, women tend to work in institutional settings which do not expect or reinforce research productivity. In predominantly teaching institutions, research performance has a lower payoff in promotion and salary

Examining the relationship between gender and scientific or scholarly productivity over the first fifteen years of the academic career, we find a similar pattern within all the cohorts except the earliest one. There tends to be a steadily increasing correlation between sex status and scientific and scholarly productivity as the career develops. The pattern identified for the 1958 cohort—that is, the increasingly strong association between productivity and gender from 1958 to 1970—appears to be similar in four of the five historical groups.[151] The lower associations in the first five years compared to those in the fifteen-year history are at least consistent with the hypothesis that the motivation to continue working on research problems becomes dampened for most women.

In the first several decades of this century only a few major universities had relatively good facilities for scientific and scholarly research. These centers of excellence attracted the most talented research-oriented scientists and reinforced their activity by creating conditions necessary for productive scholarship. It should not surprise us, therefore, to find that productivity is more highly correlated with departmental prestige in the early than the later decades. By the forties and fifties there had been some diffusion of quality among American universities. The great state university systems began to provide exceptionally fine research facilities and resources for following through on research plans. But in the first third of the century women were almost totally excluded from the very places that stimulated research activity. Whether association with leading researchers at major universities or location at universities that had facilities and libraries of quality would have significantly altered the level of women's productivity must, strictly speaking, remain conjectural, but we can surely say that it would not have hurt their productivity. Lack of such access represented yet another way in which women academics were cut off from the mainstream of their profession.

Thus far, I have presented historical data on the relationships between sex status and various forms of recognition. Are associations reduced by controlling

---

than it does at large research-oriented universities. So women, on average, do less research. Second, until recently the reward system did not reinforce productive women, so they tended then to discontinue doing research. In the absence of immediate recognition for their early research efforts, they have less motivation to continue to work on hard problems and on the difficult tasks of writing papers and research monographs. Third, women are less interested in doing research. One study showed that a smaller proportion of women than men are interested primarily in research: 39 percent of men and 15 percent of women who hold academic appointments in universities state that they are primarily interested in research. Of course, interest in research probably determines rate of publication and may be reinforced by it.

151. Budner and Meyer, *op. cit.,* also find increasing differences in male and female productivity over time for their 1950s sample. They present, on p. 12 of the paper, the following Table (10):

Productivity of Men and Women Professors in Each Age Group (% Who Are Highly Productive)

		WOMEN	MEN
	Under 40	19 (85)	30 (1031)
Age	41–50	39 (70)	60 (544)
	Over 50	44 (107)	68 (548)

Their measure of "high productivity" distorts the actual proportions in the older groups who are prolific, but that is not as interesting as the percentage difference between men and women, which increases from 11 to 24.

for the research productivity of men and women? To examine the independent effect of sex status on departmental affiliation, academic rank, and honorific recognition, I regressed these features of positional and reputational success on sex status and average yearly productivity. The first two historical cohorts have been excluded because there are too few cases for multivariate analysis. The results are presented in Tables 6–8 through 6–10.

For each of the three cohorts there is a substantial reduction in the effect of sex status on departmental affiliation, and the reduction tends to be more pronounced as careers move on. Taking the thirties group as an example, after research performance is accounted for, there is only minimal reduction in the influence of gender after five years, but a more substantial reduction after ten. Similar patterns are found for the two later cohorts. Of course, the precise meaning of these interpretations is somewhat obscured by the absence of a direct measure of sex discrimination surrounding research opportunities. If women are systematically denied these opportunities, and rewards are, at the same time, based largely on research output, we can expect a reduced influence of sex status. But this finding simply adds a dimension to the discrimination argument, suggesting that differences in recognition must be examined in terms both of the direct effects of gender on rewards and of the effects gender has on access to facilities and resources which are rationally linked to rewards. There is no available direct quantitative measure of access to research facilities, but there is historical evidence that research productivity is associated with rewards. Table 6–11 presents for each cohort zero-order correlations between total published productivity and several forms of academic recognition. The pattern is consistent. Research performance is significantly associated with departmental pres-

TABLE 6–8.   **Regression of Prestige of Academic Department on Sex Status and Research Productivity for Three Historical Cohorts[1] (Correlations[2] and Regression Coefficients[3])**

	1932 COHORT		1942 COHORT		1952 COHORT	
	r	beta coefficient	r	beta coefficient	r	beta coefficient
1st Job	−.03	.00	−.08	−.06	−.13	−.11
5 Yrs.	−.20	−.17	−.15	−.10	−.21	−.17
10 Yrs.	−.19	−.09	−.17	−.11	−.17	−.08
15 Yrs	−.15	−.03	−.12	−.05	−.23	−.18
20 Yrs. or Last	−.22	−.08	−.13	−.06	—	—

Notes:
1. In all cases, the betas for research productivity substantially exceeded the betas for sex status.
2. Correlations between sex status and prestige of department.
3. Regression coefficients in standard form of prestige of department on sex status, controlling for research productivity.

TABLE 6–9. Regression of Academic Rank on Sex Status and Research Productivity for Three Historical Cohorts[1] (Correlations[2] and Regression Coefficients[3])

	1932 COHORT		1942 COHORT		1952 COHORT	
	r	beta coefficient	r	beta coefficient	r	beta coefficient
1st Job	.03	−.01	.02	.00	−.04	−.04
5 Yrs.	−.19	−.24	−.03	−.02	−.15	−.15
10 Yrs.	−.20	−.22	−.12	−.08	−.37	−.33
15 Yrs.	−.30	−.28	−.31	−.25	−.32	−.26
20 Yrs. or Last	−.34	−.30	−.33	−.30	—	—

NOTES:
1. Early in the career, productivity has a stronger independent effect on academic rank than does sex status. As the career moves on, the strength of the two variables are roughly equal.
2. Correlations between sex status and academic rank. Academic rank was coded: 0 = Post-doctoral Fellow; 1 = Lecturer; 2 = Assistant Professor; 3 = Associate Professor; 4 = Full Professor.
3. Regression coefficients in standard form of academic rank on sex status, controlling for research productivity.

TABLE 6–10. Regression of Number of Honorific Awards on Sex Status and Research Productivity for Three Historical Cohorts (Correlations and Regression Coefficients[1])

COHORT (YEAR OF PH.D.)	r_{sa}	$b^*_{sa.p}$	r_{ap}	$b^*_{ap.s}$	N
1932	−.09	.05	.48	.49	137
1942	−.12	−.07	.21	.19	203
1952	−.12	−.07	.21	.20	238

NOTE: 1. In this table, I provide the correlations and betas for both sex status and research productivity. The subscripts are: s = Sex Status; a = Honorific Awards; p = Research Productivity.

TABLE 6–11. The Relationship between Research Productivity and Academic Rewards for Both Sexes—Historical Trend for Final Three Cohorts: 1932, 1942, 1952 (Zero-Order Correlation Coefficients)

ACADEMIC REWARDS	1932	1942	1952
Prestige of First Job	.16 (97)	.21 (120)	.29 (136)
Prestige of Job 15 Yrs. After Ph.D.	.46 (84)	.27 (131)	.24 (148)
Age at Promotion to Associate Professor	−.12 (72)	−.21 (117)	−.25 (126)
Number of Honorific Awards	.48 (137)	.21 (203)	.21 (238)

tige, age at promotion to associate professor, and number of honorific awards. Data for the Depression cohort suggest an unusually high payoff in departmental prestige and honorific awards for high productivity. But were the payoffs for men and women similar? The sex-specific patterns are presented in Table 6–12. They indicate that women gain almost as much return for high productivity as men, except in terms of age at promotion. For men, high productivity consistently leads to early promotion. For women, the pattern has changed historically. In the 1930s and 1940s high productivity did not increase chances for early promotion, but by the 1950s there was a small return for research performance.

If interpreting the influence of gender on departmental prestige is problematic, no such difficulty exists in matters of promotion. Role performance does not reduce the moderate associations between gender and academic rank in any period. The standardized regression coefficients are almost identical to the bivariate correlations. *Historically, productivity patterns simply will not explain the sex differences in academic rank.* Other social and economic variables might explain these associations, of course, but in the absence of adequate data to test alternative hypotheses I tentatively conclude that there has been extensive sex discrimination in promotion opportunities over the past forty years.

In analyzing current data on honorific recognition, I found that scientific productivity completely accounted for the sex differences. Although the results must be viewed with caution until more precise indicators of both eminence and research performance are available, the regression of number of honorific awards on scientific or scholarly output, as well as on gender, does partially explain the original association in each historical period. The modest associations between sex status and honorific recognition are reduced to insignificance. For the 1930s cohort, after controlling for published productivity, the women are actually slightly more apt to receive awards. For the two later cohorts, the initial relationship is reduced roughly in half. This should not be taken to mean that during these years there were no individual cases of overt and even blatant sex discrimination in the allocation of honorific awards and prizes. It does suggest, however, that for academe treated as a single community there were few honored men or women, and that gender fails to explain those differences in rewards that did obtain between men and women scientists and scholars.

Until now I have focused on how the professional statuses of academic women resulted in their marginal position in the academic-science community. The historical data suggest a considerable reduction in that marginality. The belief that women should not work at all has changed, and the assumption that they are incapable of generating important scientific ideas is no longer widespread. Women can increasingly be found in positions at top universities and colleges, and at work in superior research laboratories. They are increasingly incorporated into the informal aspects of the scientific community; they are seldom excluded from meeting places, dining halls, laboratories, or professional societies. Many of the formal and explicit forms of bias have disappeared. But we have not considered the changes in nonacademic statuses, changes which

TABLE 6-12.   The Relationship between Research Productivity and Academic Rewards for Men and Women Separately—Historical Trend for Final Three Cohorts: 1932, 1942, 1952 (Zero-Order Correlation Coefficients)

ACADEMIC REWARDS	1932		1942		1952	
	Men	*Women*	*Men*	*Women*	*Men*	*Women*
Prestige of First Job	.10 (51)	.38 (46)	.20 (73)	.20 (47)	.33 (83)	.13 (53)
Prestige of Last Job after Ph.D.	.53 (44)	.30 (40)	.24 (78)	.28 (53)	.22 (88)	.17 (60)
Age at Promotion to Associate Professor	−.19 (40)	.19 (32)	−.23 (66)	−.02 (51)	−.27 (79)	−.10 (47)
Number of Honorific Awards	.51 (76)	.36 (61)	.20 (121)	.14 (82)	.18 (145)	.25 (93)

have produced male and female status-sets that look more alike today than ever before. I shall focus on only one of these, marital status.

A NONACADEMIC STATUS: MARRIAGE

Until recently, female college graduates remained unmarried far more often than other American women. Willystine Goodsell, in 1924, reported extensive marital data for female college graduates. The marriage rates varied by region and type of college. Looking at graduates of elite female colleges, only 53 percent of Vassar graduates from 1867 to 1892 and only 50 percent of the Wellesley women who earned degrees from 1875 to 1889 were married as of 1915.[152] A 1918 study of 16,739 female graduates of eight elite women's colleges and Cornell indicated that 39 percent had married.[153] At coeducational schools and at western universities the marriage rate was roughly 50 percent.[154] These data are significant, of course, only when compared to national averages. The 1910 United States census shows that 79 percent of women between twenty-five and seventy-four years of age were married, and 89 percent of those between thirty-five and forty-four.

The historical data are better than the contemporaneous explanations. A typical interpretation concluded:

> It is probable that the woman who has enjoyed the benefits of higher education in the liberal arts unconsciously forms for herself a lofty conception of the marriage relation and refuses to compromise with her ideals when she discovers that her suitors fall short of them. Not driven by the sex drive to the same extent as men, women find it more possible to hold steadily before them an ideal of marriage as a close spiritual comradeship, a community of interests and aims, extending to every vital aspect of life. Rather than fail in attaining the best many a college woman cheerfully accepts the lot of the spinster worker whose economic and personal independence is her chief compensation for turning her back on home and motherhood.[155]

There are, of course, multiple alternative and more plausible explanations for this abnormally low marriage rate. Granted that for highly educated women the pool of men from roughly similar backgrounds was small, and that the cultural image of the educated woman may have turned off eligible men. But perhaps more important, especially for women with graduate degrees, were the structural barriers to marriage. Formal and customary rules barred married women from many occupations, and academic science certainly was no exception.[156] There

---

152. These data are reported in Goodsell, *The Education of Women* (*op. cit.*), pp. 34–61.

153. *Ibid.*, p. 37. The women's colleges were Barnard, Bryn Mawr, Mount Holyoke, Radcliffe, Smith, Vassar, Wellesley, Wells.

154. *Ibid.*, p. 43.

155. *Ibid.*, p. 39.

156. Woody, *op. cit.*, p. 333; Committee W Report, *op. cit.*

were strong incentives to remain single. For those women who did not hold jobs, marriage made it almost impossible to find one; for single women with jobs, marriage produced a high risk of being fired. Thus, the woman with aspirations to work had to consider the consequences of marriage for her career. Many well-educated women of the day were profoundly ambivalent about the alternatives of marriage or a career. One Vassar graduate of 1902, Ruth Benedict, made a 1917 entry in her journal on cross-pressures confronting educated women:

> So much of the trouble is because I am a woman. To me it seems a very terrible thing to be a woman. There is one crown which perhaps is worth it all—a great love, a quiet home, and children. We all know that is all that is worth while, and yet we must peg away, showing off our wares on the market if we have money, or manufacturing careers for ourselves if we haven't. We have not the motive to prepare ourselves for a "life-work" of teaching, of social work—we know that we would lay it down with hallelujah in the height of our success, to make a home for the right man.
>
> And all the time in the background of our consciousness rings the warning that perhaps the right man will never come along. A great love is given to very few. Perhaps this make-shift time filler of a job *is* our life work after all.[157]

Historical data on marriage rates provide context for these sentiments. The Hutchinson survey of 1924 indicates that three-quarters of the women Ph.D.'s were single; one-fifth, married; and the others either widowed or divorced.[158] Rossiter's analysis of women appearing in *American Men of Science* before 1920 found that 19 percent were married. Marital rates varied by academic discipline. Female zoologists (28 percent) and psychologists (26 percent) were more apt to be married than were physicists (0 percent) or mathematicians (13 percent).[159] These independent studies corroborate each other's findings: 75 to 85 percent of female Ph.D.'s prior to 1920 were not married.

Data on the marital status for Ph.D.'s in the five historical cohorts are presented in Table 6–13. Historical data for the entire United States population fourteen years of age and older suggest that the proportion of unmarried men remained at roughly 30 to 33 percent from 1890 to 1940. These figures include, of course, many men not yet of marrying age. Among men thirty-five to forty-four years of age, roughly 10 to 15 percent are single. The marital rate increases slightly from 1940 to 1957. Comparable figures for women are equally constant over time: About three-quarters of the female population fourteen years and older are married, as are 90 percent of those from thirty-five to forty-four. Marital patterns for male and female Ph.D.'s are vastly different from these national averages.

In this century the marriage rates of men and women Ph.D.'s have undergone remarkable changes. Fully 86 percent of the women in the 1911 cohort were unmarried. Although it is based upon a very small sample, the proportion is so

157. As quoted in Margaret Mead, *Ruth Benedict* (New York: Columbia University Press, 1974), p. 8.

158. Hutchinson, *op. cit.*, p. 90.

159. Rossiter, *op. cit.*, Table 3, p. 315.

TABLE 6–13.   Unmarried Rates for Men and Women Scientists and for U.S. Population: 1911–60 (Percent)[1]

COHORT	ACADEMICS			U.S. POPULATION[2]	
	Total	Males	Females	Males	Females
1911–13	61 (51)	43 (30)	86 (21)	27	17
1927	38 (34)	23 (22)	67 (12)	25	16
1932	34 (137)	7 (75)	68 (62)	22	14
1942	35 (212)	16 (126)	63 (86)	22	15
1952	26 (229)	8 (142)	54 (87)	14	10
1960			45[3]	12	7

NOTES:
  1. The U.S. data and the data from my historical samples are for slightly different years. The census data were taken for 1910, 1920, etc. My data are for Ph.D.'s who received their degrees two years into each decade.
  2. The marital-status data for the U.S. population is for males and females between the ages of 25 and 44. This range of ages seemed the best comparison for Ph.D.'s, who tend to marry somewhat later than the general population.
  3. The marital data for academic women listed under 1960 are drawn from the work of Helen Astin, *op. cit.* They represent data for the year 1965; no comparable data were available for men.

Source: U.S. Department of Commerce, Bureau of the Census, *Historical Statistics of the United States, Colonial Times to 1970,* Bicentennial Edition (Washington, D.C.: Government Printing Office), pp. 20–21, Series A 160–171. The 1960 data are based upon the 25 percent sample; the 1950 data, on the 20 percent sample.

consistent with those found by Hutchinson and by Rossiter that it is probably close to the mark. In the same cohort, 43 percent of the male Ph.D.'s were single. There is, then, a difference of 43 percentage points between men and women; the marital status of men and women academics in this earliest period is more apt to differ than be the same. But what of the historical trend for women? Data on marriage rates, presented in Table 6–13, show a sharp increase during the decade of the 1920s in the proportion of women Ph.D.'s who were married. The rate levels off for the cohorts of the 1920s, 1930s, and 1940s, and increases again significantly in the 1950s cohort. Although the proportion of single female Ph.D.'s remains far above the national averages, by the 1950s almost half of the women were married. These figures are remarkably close to those found for psychologists in 1946 by Alice I. Bryan and Edwin G. Boring, who report that 47 percent of the women and 93 percent of the men were married. Rita James Simon, Shirley Merritt Clark, and Kathleen Galway, in their study of over 1,700 women who received their doctorates between 1958 and 1963, also found that about 50 percent were married. They identify some field differences. Sixty-two percent of the female Ph.D.'s in the physical and natural sciences were married, as compared with 57 percent of those with social science degrees, 47 percent with degrees in the humanities, and only 35 percent of those with degrees in education. In contrast, 95 percent of the men with doctorates were married. Helen Astin's 1965 data indicate that about 55 percent of females with the doctorate were married.[160] Women Ph.D.'s in the different cohorts have changed their marital and family statuses such that they increasingly look like the status-sets of nonprofessional women. There have been marked increases in the marriage rate of male Ph.D.'s since the first decade of the century. By the 1930s they had already approached the national average.

Alternative interpretations of these data are of course possible. One might argue that little has changed inasmuch as the relative difference between men and women Ph.D.'s has not changed since 1910; after all, there was a 43 percent difference in 1911 and a 46 percent difference in 1951. Or one could point out that the percentage change among men exceeds the comparable figure for women. Such interpretations play games with the import of the data. The social meaning represented by these changes for the sexes is vastly different from the purely statistical.

The extraordinarily high level of unmarried women in the first part of the century is consistent with the marginality thesis. When they opted for academic careers women were cut off from traditional female roles. But the more powerful point is that this was not a free-choice situation, since married women were denied jobs and single women who married while working were often dismissed. Men never faced such prospects. The male-dominated academic establishment forced women to abandon their traditional domestic roles and simultaneously denied them access to new professional roles. Consequently, women remained marginal figures in the academic community.

This situation has changed. Rules about marital status no longer overtly impede access to jobs, although marriage often restricts the mobility of women academics more than that of their male counterparts. The increase in the proportion of married women in academic life has still further increased the similarity between male and female status-sets. Furthermore, the awkwardness associated with being single in a couple-dominated social system no longer affects many female academics. Thus, the woman academic, married or single, is less apt to be treated as an outsider.

The marital rates discussed in this section bear directly upon female marginality, and among other things deliver a severe blow to the explanation of productivity differentials in terms of female family obligations. If sex-related productivity differentials varied directly with the marital rates of female Ph.D.'s, there would be a prima facie case supporting this interpretation. But historically the correlation between sex status and published productivity has remained remarkably constant. If there is a modest association between gender and scholarly productivity when almost *no* women are married, productivity differences cannot be explained by marital status or number of children. And this, in fact, is the case for the first three cohorts.

---

160. Astin, *The Woman Doctorate in America* (*op. cit.*), p. 27; Alice I. Bryan and Edwin G. Boring, "Women in American Psychology: Factors Affecting Their Professional Careers," *American Psychologist* 11 (1947): 3–20; Rita James Simon et al., "The Woman Ph.D.: A Recent Profile," *Social Problems* 15 (1967): Table 1.

TABLE 6–14.   The Relationship between Research Productivity and Marital and Family Statuses for 1932, 1942, 1952 Cohorts (Zero-Order Correlations)*

		COHORT (YEAR OF PH.D.)		
		1932	*1942*	*1952*
Men	Marital Status	.18 (78)	.21 (126)	.09 (148)
	Number of Children	−.07 (78)	−.14 (123)	−.09 (147)
Women	Marital Status	.01 (63)	.18 (84)	.04 (92)
	Number of Children	.00 (62)	−.25 (83)	−.04 (91)

*The slightly varying numbers on which these correlations are based result from missing data.

To clinch the point that marriage does not adversely affect research productivity, I examined the trend for men and women separately. Again, the small number of cases in the first two cohorts precludes robust correlations. However, even with these small samples, the patterns are consistent with those found for the later cohorts. Nonetheless, let us review only the last three cohorts (see Table 6–14). Marriage influences research productivity somewhat differently for men and women. There is no indication that marriage hurts published productivity of either sex, but it helps the productivity of men somewhat more than it does that of women. Among women there is either no correlation or a low correlation between marital status and productivity: It is .01 for the 1930s cohort; .18 for the 1940s cohort, indicating a slight benefit of marriage; and .04 for the 1950s cohort. Among men the correlations for these groups are respectively .18, .21, and .09. The lower correlations for women in each cohort comparison result from several artifacts. First, the low average published productivity of women restricts the variance of the variable. Second, until recently there was not much variance in marital status. Consequently, we cannot expect to find a significant relationship between the two variables. If anything the data corroborate the inference drawn in Chapter 3 that marriage promotes social stability and thereby facilitates rather than impedes research. Because of the weight of both historical and contemporary evidence, it is time to abandon the notion that marriage and children seriously hinder the scientific and scholarly productivity of men *or* women in the academic community.

THE CHANGING CHARACTER OF THE PROFESSORIATE

While changes in the nonscientific statuses of women scientists reduced their marginality, other changes in institutional character also influenced the rate at which women were more fully assimilated into the mainstream of the academic community. Most of these changes took place after World War II, and can be seen largely in terms of changing social characteristics of the male professoriate. The reduction in the marginality of women in the late 1950s and 1960s was in part a function of marked changes in the status-set structure of academic men, most notably in terms of their religious backgrounds. These changes, which reflected the overcoming of discrimination by other social groups, produced an altered academic community—one notably different from the communities constituting great American universities and colleges in the first half of the century, and one which, I believe, has become more receptive to women faculty members. I will focus here on only one basic change—the religious composition of the professoriate—but other characteristics such as social class background could well have produced a set of new attitudes toward women in the academic community.

It has been well established by now that positions on university and college faculties were largely closed to Jewish-Americans until after World War II.[161] Anti-Semitism in academe reached its height in the 1920s and 1930s, when Jews were attending colleges in increasing numbers. Quotas for Jewish graduate students and faculty were openly defended by the then president of Harvard, A. Lawrence Lowell, and of Columbia, Nicholas Murray Butler.[162] As late as 1945 quotas were defended by Ernest M. Hopkins, president of Dartmouth College, who asserted: "Dartmouth is a Christian college founded for the Christianization of its students."[163] It was extremely difficult for Jews to find faculty positions. Thus, Ludwig Lewisohn was prevented from teaching English; Lionel Trilling was the first Jew appointed to the English department at Columbia, and not without direct intervention by Butler into the appointment process; Paul Freund and Milton Katz were the first Jews appointed to the Harvard Law School (1939) since Felix Frankfurter. Heywood Broun and George Britt, examining bias against Jewish faculty at the ultraliberal City College of New York in the 1920s, wrote: "Only five [Jews] have the rank of full professors. . . . All five are men of exceptional attainments. The percentage of Jews in the lower orders . . . is much higher than among the more desirable positions. Even in a friendly college, the openings for Jewish professors are distinctly limited."[164]

To observers of the fate of women in academe, it all sounds familiar. In the late 1940s Harvard's Albert Sprague Coolidge, speaking to a Massachusetts legislative committee, said: "We know perfectly well that names ending in 'berg' or 'stein' have to be skipped by the board of selection of students for scholarships in chemistry."[165] Many bright Jews with academic aspirations

161. Seymour Martin Lipset and Everett Carll Ladd, Jr., "Jewish Academics in the United States: Their Achievements, Culture and Politics," *American Jewish Year Book 1971-72* (New York: Jewish Publication Society of America, 1972), pp. 89–130.

162. *Ibid.*, p. 90.

163. As quoted in *ibid.*, p. 90.

164. Heywood Broun and George Britt, *Christians Only* (New York: Vanguard Press, 1931), pp. 72–124.

165. As quoted in Lipset and Ladd, *op. cit.*, p. 91.

changed their names, hid their pasts, and assumed manners that enabled them to successfully pass, adaptations not open to women with regard to sex. Admissions officers argued that it was a waste of valuable resources to train Jewish academics in the absence of employment opportunities for them. Another self-fulfilling prophecy was created.

The point is not, of course, to recount the history of anti-Jewish feelings and behaviors on American campuses, but to suggest how, in fact, the campuses were opened to Jews in the 1950s and 1960s to such an extent as to qualitatively change the social composition of the professoriate. Although good time-series data on the religious identification of American professors are lacking, synthetic cohort data based on 60,000 faculty members' questionnaire responses are available from the 1969 Carnegie Commission Study on Higher Education. By dividing professors into age groups, Seymour Martin Lipset and Everett Carll Ladd, Jr., examine the religious composition of faculty entering the community at different historical points. They report significant changes in the religious backgrounds of the professoriate. The generation of scientists and scholars who entered the community in the 1920s—that is, those around age sixty-five—were less than 4 percent Jewish, and were almost four-fifths Protestant.[166] By the late 1940s and early 1950s the proportion of Jews had more than doubled to slightly more than 9 percent; the proportion of Catholics rose from about 14 to 20 percent in the same period; and the number of Protestants dropped to about 63 percent. The cohorts entering academe between 1945 and 1960 were much more apt than previous ones to list their religious background as "Other" or "None." A significant portion of those claiming no religious background were probably Jews, who feared that disclosure of religion would limit their opportunities.

The changing proportions of Jewish and Catholic professors at "elite" colleges and universities were still more dramatic.[167] During the late twenties and early thirties, in the age group of sixty or over, roughly 10 percent of the faculty was Jewish, 9 percent Catholic, and 75 percent Protestant. Since the 1950s Jews have constituted about one-fifth of the faculty at the elite schools, Catholics a somewhat lower proportion, and Protestants about half of the faculties. At Ivy League schools the proportions are even higher. In the late 1960s roughly 18 percent of professors in their fifties were Jewish, as were about 25 percent of those under fifty.[168]

The data indicate that the religious composition of the professoriate has become far more heterogeneous than it was in the 1920s and 1930s, and that the changes have brought into the academic community large numbers of a minority group that has a long history of liberal attitudes on social and political questions. Jewish faculty perceive themselves as more liberal than other religious groups,

---

166. *Ibid.*, p. 92.

167. *Ibid.*, p. 93. Elite status is based on a three-item index, including SAT scores required for admission, revenue per student, and research expenditures per student.

168. *Ibid.*, p. 93.

and Catholics, who also have grown in faculty representation, perceive themselves as somewhat more liberal than Protestants.[169] Jews have been more liberal in their attitudes on social issues, ranging from views about Vietnam to the legalization of marijuana. They also have been more liberal than other religious groups about relaxing standards in appointing members of other minority groups to faculty positions, but the difference here was smaller than on most other issues.

Although the Carnegie study included no direct question on positions for women scholars and scientists, it seems a reasonable inference that Jews are more apt to support the hiring and promoting of qualified women than are members of other groups. Although this must remain speculation, it does not require an excessively great logical leap to conclude that the increasing proportion of religious minorities in colleges and universities—especially in the highly prestigious ones—has contributed indirectly to the increased academic rewards gained by women since World War II. Of course, basic cultural or ideological change regarding exclusivity may well have contributed to the general opening up of opportunites for all heretofore excluded groups.

CONCLUSIONS: OVERCOMING THE TRIPLE PENALTY

The history of American women scientists and scholars in this century is a history of struggle to overcome the triple penalty. These women traditionally paid treble damages for their gender. First, they had to overcome cultural barriers to their entering science and scholarly careers; second, after entering such careers, they had to overcome the widespread belief among scientists and scholars that they were either physiologically or psychologically incapable of creative work; and third, they had to overcome structural impediments to success—that is, actual discrimination in the allocation of opportunities and rewards within the academic community.[170]

The three components of the triple penalty interact to produce the necessary conditions for a self-fulfilling prophecy. Discrimination often reduces motivation to perform. Those subjected to it feel that nothing they do will make any real difference in the rewards they receive. A reduction in motivation, in turn, leads to lower performance levels and a withdrawal from the race. Temporary withdrawal, whether or not in the form of actually leaving the system, leads to obsolescence of skills, reduced opportunities to acquire resources needed for research performance, and diminished interest of the academic community in hiring someone with dated skills. For individual women, this leads to still further reductions in motivation, self-doubts about their competence, and perceptions of

---

169. *Ibid.*, p. 113.
170. This argument was outlined in Zuckerman and Cole, "Women in American Science" (*op. cit.*).

"failure." For the academic audience, the cumulative force of the triple penalty tends to confirm and reinforce beliefs that women are less able and less motivated than men. Under such conditions, the gap between women's and men's scientific achievements widens.

The consequences of the triple penalty materialized historically in the form of female marginality. Consider the various ways that the status-sets of men and women differed in the first part of this century. Work was virtually sex-segregated. Women worked predominantly at women's colleges and later in some state university systems; positions at the major private universities and in the better state university systems were largely reserved for men. In short, many men held prestigious departmental and university appointments; women rarely did. Similarly, positions in honorific and professional societies were dominated by men; women tended to be excluded. Men tended to move up the academic-rank hierarchy; women could rarely do so except at the women's colleges. Men's salaries were influenced significantly by their research performance; women's tended not to be. Consequently, the status-sets of academic men and women rarely overlapped, and men almost invariably occupied positions superordinate to those of women in terms of power, income, and prestige. It is no wonder that Pearl Buck declared as early as 1941 that "the truth is that women suffer all the effects of a minority."[171]

And yet today the signs and symbols of sex discrimination in the academic community have diminished to a significant degree. This is not simply a result of the development of increasingly sophisticated techniques for masking what had, heretofore, been blatant discrimination. Historical data clearly suggest a nontrivial reduction in the marginality of academic women. In fact, since the late 1950s and certainly in the 1970s, women have been just as apt as men to find jobs in distinguished departments; they have received roughly equal salaries when in the lower ranks, and have been honored with as many awards or prizes (although not the most prestigious ones) and postdoctoral fellowships. Women are working in increasing numbers at centers of scientific and scholarly activity. Their extra-academic statuses are more likely than not to look like those of their male colleagues: While they still are not as apt to marry and have families as men, they are opting increasingly for these statuses, and perhaps more important, they no longer face formal sanctions for selecting them.

This is not to say that all aspects of the triple penalty have vanished. The culture continues to sex-label certain occupations, creating obstacles to the choice by women of certain scientific pursuits. Within the academic marketplace, women still face formidable barriers to promotion to high rank. While their not being promoted is due in part to their producing less scholarly work, and work of lesser impact, there does remain strong evidence that independently of these other explanatory elements, bias in promotions continues to be a formidable problem for academic women.

---

171. Pearl S. Buck, *Of Men and Women* (New York: John Day Co., 1941), p. 170.

Nor has the triple penalty operated uniformly in all aspects of the scientific and scholarly career. The small number of women who have earned the Ph.D. are just as likely as men to have been trained at the most distinguished university departments. And this was as much so in the 1920s as it is today. But although the admissions procedures to graduate schools evidence no significant sex biases, this does not mean that the quality of the graduate experience, the working and training relationships between faculty and students, has been the same for men and women. There is some indication that this aspect of academic life for women has recently improved.

How do we account for the historical trend toward reduced marginality and increased equality between the sexes? Five interrelated factors have been identified as significant in changing the position of academic women. First, the generally hostile climate of opinion about women's place, about married women working, about women's capacity to be creative scientists and scholars, has changed over time. A survey of public attitudes today might still find an absolute majority of Americans believing that women should not place careers on a par with or ahead of marriage and a family, but the attitude toward women who make that choice has undoubtedly been fundamentally changed in the last fifty years.[172] Thus, one important feature of the triple penalty, the cultural pressure not to pursue a career, has been largely neutralized in recent years. And it surely is not an anomaly to find married women working. It is difficult, of course, to determine precisely what produced this attitudinal flexibility, but certainly it was in part a response to demand for more women workers during and after World War II. Whether changing attitudes about working women in general have been paralleled by increased attitudinal flexibility toward women entering scientific careers remains largely unstudied.

Second, economic necessity and increased demand were a strong influence on change not only in the country as a whole but in the academic community. This century witnessed two periods of significant expansion in the American higher education system: the 1920s, and the late fifties and early sixties. Growth approached exponential proportions in the number of undergraduates enrolling in college, the number of graduate students, the number of new colleges and universities opening their doors, and as a consequence, the number of faculty

---

172. *New York Times,* Sunday, Nov. 27, 1977, p. 75. A 1977 national survey of 1,063 adult Americans by the *New York Times* and CBS News focused on attitudes toward the participation of women in the labor force—particularly women who were married. Fifty-four percent of all respondents said that a woman should work "even if she has a husband capable of supporting her," while 40 percent disagreed. Subdividing the sample into men and women showed that 50 percent of the men and 58 percent of the women were in favor of women working under that condition. The responses were highly variable in terms of the age of respondents. Roughly three-quarters of those between the ages of eighteen and twenty-nine believed women should work, compared with 57 percent of those aged thirty to forty-four, 48 percent of those between forty-five and sixty-five, and 28 percent of those over sixty-five. If the age differential is taken as a synthetic cohort, we can see significant changes in attitudes toward women pursuing careers, even when they "don't have to." Perhaps more telling was the response to the question: "What kind of marriage do you think makes the more satisfying way of life?" Forty-eight percent favored the situation described as: "The husband and wife both have jobs, both do housework and both take care of the children," while 43 percent preferred the situation described as: "The husband provides for the family and the wife takes care of the house and children."

positions available. The tremendous increase in demand for new young faculty also opened opportunities for women who had earned the doctorate.

Third, there was significant structural change in the academic community. Whereas the large, eastern, privately endowed universities had dominated the scene of higher education in the first quarter-century, the increases in size of the whole university system functioned to dilute somewhat the importance of the elite private institutions and their sister schools. State universities, which increasingly were constructing excellent research facilities, not only expanded in size but advanced in quality. Increasingly able to attract bright youngsters and senior figures of stature, these universities began to account for an increasing proportion of high-quality research and training. And since their inception these state schools, largely coeducational, have had more permissive attitudes toward filling faculty positions with women. Ironically, most of the male Ivy League schools have had a superior record in the admission and training of female graduate students, but have been reluctant historically to hire their female graduates. For them, the separate elite women's colleges existed in part to absorb these female Ph.D.'s.

Fourth, the changing social composition of the professoriate seems to have helped reduce sex discrimination. This must remain conjectural because the direct link between the changing composition and the changed attitudes toward female scientists and scholars is missing. But the diversification of social backgrounds of the professoriate—especially the increased inclusion of members of religious minority groups, who had themselves experienced exclusion from faculty posts—changed the ideological tone on many campuses, and most especially on the campuses of leading universities. The generally more liberal social and political attitudes of these new academics constituted an enlarged base of support for admission of women into central academic positions.

Fifth, and finally, the pressure from women's interest groups and ultimately from the federal government over the past decade have strongly influenced shifts in behavior if not in attitudes on many campuses. Feminist activism had its effect on change in the early 1920s, and today political activism, lawsuits, and affirmative-action anti-bias legislation have all contributed somewhat to reducing the level of marginality further. As is now commonly known, colleges and universities, under the threat of losing federal grants and contracts, have been challenged to show that they do not pursue policies that discriminate against women and minorities in hiring, promotions, and salaries. This governmental effort, largely itself a response to pressure by the women's movement, has prodded many campuses into constructing affirmative-action plans for future faculty appointments. Of course, the exact effects of affirmative-action policy on the extent of sex-based discrimination have yet to be estimated precisely.

Having discussed where we have come from and where we are now in the matter of women in science, I turn now to the problematics inherent in affirmative-action policy—that is, the question of where we are going.

Chapter 7

AFFIRMATIVE ACTION POLICY: PROBLEMATICS

WITHIN A DECADE of its inception, affirmative action policy has become a public controversy. As with so many controversies that focus on solutions to social problems, there is a tendency for debate to move from emphasis on the content of ideas and their merits to social and ideological conflict.[1] This pattern has characterized a significant portion of the debate on the merits of affirmative action. That ideology predominates over empirically based ideas should not surprise us, since affirmative action calls into question a number of fundamental principles and values in American society.[2]

In fact, there remain many pointed questions to be raised about affirmative action as a general social policy and method of producing social change. Only recently have some of these questions begun to be systematically studied or debated. Many of the problematics of affirmative action have not been clearly articulated. Few among the working press, or among scholars, administrators, and others who are directly affected by it, are familiar with the range of problems associated with this innovative public policy. Some are content to let the courts decide on the legality and procedures of affirmative action.

Affirmative action policy actually goes far beyond legal briefs and court decisions. It is an attempt in part to restructure methods of locating individuals

---

1. Robert K. Merton, "Social Conflict over Styles of Sociological Work," in *The Sociology of Science: Theoretical and Empirical Investigations*, 1961 (reprinted, Chicago: University of Chicago Press, 1973), pp. 47–69.

2. Bernard Barber, "Function, Variability, and Change in Ideological Systems," in *Stability and Social Change*, ed. Bernard Barber and Alex Inkeles (Boston: Little, Brown & Co., 1971), pp. 244–264.

for, and assigning them to, social positions.[3] In short, it is a procedure designed to increase "fairness" in the occupational structure of American society. But "fairness" is plainly a difficult concept on which to reach social consensus and indeed much of the ideology and the controversy rests, either explicitly or implicitly, upon different ideas about fairness.

In broad theoretical terms, affirmative action policy deals with one of the most basic issues of social stratification: the bases on which the relatively scarce resources and rewards of the society will be distributed among its population. The passion generated by the *Bakke* case, recently decided by the Supreme Court, represents in a fundamental way the outlines of the areas of contention. There is an attack on the old rules in the distribution game. It is a profound attack, because it not only relies on arguments about overtly hostile or invidious acts of discrimination against racial minorities but it threatens a system that for years was viewed as prototypical of one based upon principles of meritocracy. Various kinds of standardized tests—such as civil service examinations, IQ tests, MCATs, Law School Admissions Tests, and Graduate Record Examinations, among others that rely on "blind" scoring—are now being questioned as improper methods for determining an order of merit and for social selection.

If there are advocates for changing the rules of the game, there are equally outspoken advocates who are resisting these rule changes. The conflict over means and ends has touched most social groups in the country and is surely directly related to what they perceive as the multiple consequences of the outcome for them as individuals, for members of their families, for the social groups to which they are tied, and for the general distribution of rewards throughout the larger society.

Affirmative action policy is only one means, of course, for achieving social change and the redistribution of resources, power, prestige, and influence. Antidiscrimination litigation not involving affirmative action plans is another. Here I shall concentrate only on problematics of affirmative action, since the careers of women in the scientific community are probably affected more by affirmative action policy than by alternative means of structuring opportunities.

Affirmative action policy is, of course, directly related to the problems of recruitment and retention of women in the scientific community. Efforts to increase the proportion of women in academic-science departments and among the higher-ranked members of those departments rely significantly both upon the actual affirmative action policies within universities and colleges and upon the implied threat that if women applicants are treated unjustly legal action will be taken to redress grievances.

This chapter raises sets of questions about the substance and methods of affirmative action policy. Many of these questions are tough ones; they are not easily answered. But they are questions that must be confronted and examined in

---

3. In proper context, affirmative action policy should be seen as linked to problematics in social stratification—in particular, the bases on which social rewards are distributed.

detail, whether or not the answers obtained carry implications that run counter to our personal predispositions. Many of the problematics discussed here have been touched on throughout this book. I now focus attention on those that are particularly significant in the formulation and evaluation of affirmative action policy.

PROBLEMATICS

The history of affirmative action, and in particular its legal history, has now been well documented.[4] Consequently there is no need to outline that history here. This chapter focuses on eight problematics of affirmative action. Specifically I discuss:

1. The problems of determining which of the factors that govern policy decisions are actually facts.
2. The problem of analyzing pools of available talent.
3. The problem of measuring job performance and discrimination in different institutional settings.
4. Affirmative action as part of larger social processes, particularly social patterns of accumulative advantage and disadvantage.
5. The problem of specifying conditions under which sex, ethnic, and racial discrimination obtain.
6. The effects of factors exogenous to universities, colleges, and other parts of the scientific community on fulfilling affirmative action goals.
7. Problems of implementing policy within different settings.[5]
8. The competition between affirmative action and older principles of equality of opportunity; the relationship between principles of individual and group justice.

I am interested in determining what we currently know, what we need to know, and what we can do to produce increments in our knowledge about affirmative action.

---

4. For discussions of the recent history of affirmative action, see among others, Richard Lester, *Antibias Regulations of Universities: Faculty Problems and their Solutions* (New York: McGraw-Hill Book Co., for the Carnegie Commission on Higher Education, 1974), pp. 61–69; Nathan Glazer, *Affirmative Discrimination, Ethnic Inequality and Public Policy* (New York: Basic Books, 1975); Marshall Cohen, Thomas Nagel, and Thomas Scanlon, eds., *Equality and Preferential Treatment* (Princeton, N.J.: Princeton University Press, 1977); John Kaplan, "Equal Justice in an Unequal World: Equality for the Negro—The Problem of Special Treatment," *Northwestern University Law Review* 61, No. 3 (July-Aug. 1966): 363 ff.; Lino A. Graglia, "Special Admission of the 'Culturally Deprived' to Law School," *University of Pennsylvania Law Review* 119 (1970): 351–363; Derrick A. Bell, Jr., "In Defense of Minority Admissions Programs: A Response to Professor Graglia," *University of Pennsylvania Law Review* 119 (1970): 364–370; Robert M. O'Neil, "Preferential Admissions: Equalizing the Access of Minority Groups to Higher Education," *Yale Law Journal* 80, no. 4 (March 1971): 699–767; *Disadvantaged Students and Legal Education—Programs for Affirmative Action, Toledo Law Review* (1970): 227 ff. (this 700-page volume is based on a symposium on the pros and cons of affirmative action, particularly in reference to special admissions programs for professional schools); Malcolm J. Sherman, "Affirmative Action and the AAUP," *AAUP* 61, no. 4 (Dec. 1975): 293–303; Thomas Sowell, "'Affirmative Action' Reconsidered," *The Public Interest* 42 (winter 1976): 47–65.

5. There are other problematics that I could address, two of which would be a discussion of the necessity of frequently differentiating the problem of affirmative action for women and for minorities, and a discussion of the multiple consequences of affirmative action for different social groups in this society.

Determining the Facts that Govern
Affirmative Action Policy Decisions

Policy decisions are often based upon alleged fact, yet it is often a difficult problem to establish what are the facts, and the level of consensus about these facts. Much social policy is based upon a high quantity of fiction and a low quantity of agreed-upon fact. Of course, the level of consensus on facts depends a great deal upon social values, attitudes, and beliefs. While few individuals would disagree with the general *fact* that there is widespread racial and sexual discrimination in American society, there is a great deal of well-reasoned disagreement over facts about the location and magnitude of this discrimination.

Some of the confusion over the factual basis for affirmative action results from the critical problem of identifying appropriate units of analysis. This is most striking, of course, when arguments about the existence of discrimination take the form: "Well, I know an individual at college X who clearly was not promoted because she was a woman." Explanations drawn from impressionistic evidence based on single cases rarely are compelling. But faulty inferences also can be based upon sound statistical comparisons. Take, for example, the argument that women are systematically excluded from graduate schools. Data were reported in 1975 by P. J. Bickel et al. on 1973 admissions to the University of California, Berkeley.[6] There were almost 13,000 applicants to graduate departments and programs; about two-thirds of the applicants were males. Of those who applied, 44 percent of the males and 35 percent of the females were admitted.[7] This is a fact. But are these data sufficient to make inferences about sex discrimination? Are they a basis for affirmative action? Some scholars have used comparative figures such as these to support contentions of discrimination. Indeed, most affirmative action plans define "deficiencies" by just such simple calculations of proportions.

Even if we were to assume that there were no measurable differences between men and women in terms of their relative intelligence qualifications, skills, or promise, an inference that discrimination is a *fact* in Berkeley's admissions procedures would be premature. If discrimination was a fact in admissions at Berkeley, then we would have to investigate individual departments, since that is where admissions decisions are made. When this was done by Bickel and his collaborators, they found two new *facts* that were directly related to the overall bivariate relationship between sex status and admissions.

First, at the department level they found an extremely skewed distribution of female applicants: "Two-thirds of the applicants to English but only 2% of the applicants to mechanical engineering are women."[8] A chi-square test of signifi-

---

6. The discussion in this example draws heavily upon a recent article published by P. J. Bickel, E. A. Hammel, and J. W. O'Connell, "Sex Bias in Graduate Admissions: Data from Berkeley," *Science* 187, no. 7 (Feb. 1975): 398–404.

7. *Ibid.*

8. *Ibid.*, p. 399.

cance for this relationship indicated that the possibility of this distribution existing by chance was near zero. Second, it turned out that the rejection rates in the departments varied considerably. "Now, the odds of getting into a graduate program are in fact strongly associated with the tendency of men and women to apply to different departments in different degree. The proportion of women applicants tends to be high in departments that are hard to get into and low in those that are easy to get into.'"[9] Consider the effect of aggregating data for departments that have different proportions of female applicants and different rejection rates. We start with two departments, physics and English at a university. A total of 400 men have applied to the physics department but only 200 women. One half of each sex are admitted. In the English department, which has more applicants and is harder to gain admission to, 150 men and 450 women have applied. Two-thirds of both men and women are rejected—that is, 100 men and 300 women. If we combine the two departments we find a total of 1,200 applicants: 550 men and 650 women. But now roughly 45 percent of male and 38 percent of female applicants are admitted—is this a factual demonstration of sex discrimination at the university? One might wish to ask, of course, why the rejection rates are higher in general among departments with high ratios of female applicants. Several speculative answers are plausible, but we need not go into them. The basic point here is that even if we do not take performance variables into account, there are still other pitfalls in drawing inferences from simple proportions.

Query: For affirmative action purposes how do we decide what is the proper analytic unit and level? Should the level of analysis be the level at which decisions are made? What are the strengths and weaknesses of using various levels of analysis for defining discrimination and the effectiveness of affirmative action plans? To what extent do the problems of units of analysis and consensus on what are facts differ in different types of organizations? Are there any identifiable patterns to these differences?

I have been discussing how inferences about social facts relating to discrimination are often derived from modes of analysis that are either imprecise or incomplete. I shift now to a different type of fact problem and consider how essentially agreed-upon empirical facts, which can act as a basis for important policy decisions, are often supported weakly, if at all, by existing data.

For purposes of illustration, let us consider some of the underlying assumptions of social facts that can be found by an examination of materials produced in *The Regents of the University of California* v. *Bakke*,[10] which was decided in July 1978 by the U. S. Supreme Court. Although this case involves questions of race-based discrimination under the Fourteenth Amendment and is not a gender

---

9. *Ibid.* This pattern might well be a product of differential types of self-selection in applications to departments in engineering and in the humanities.

10. *The Regents of the University of California v. Allan Bakke,* 18 Cal. 3d 34, *cert. granted,* 45 U.S.L.W. 3570 (1977). This discussion of the *Bakke* case is part of a larger monograph that I am working on that deals with uses and abuses of social science evidence in legal decision-making.

case, the *Bakke* decision is viewed as one of the first dramatic tests of affirmative action policy. Most proponents of affirmative action policy see the *Bakke* case as setting significant precedent for future affirmative action litigation for both race and gender discrimination. It is important to emphasize that I am not concerned here with the normative questions involved in the decision by either the California Supreme Court or the U. S. Supreme Court.

The case involved a claim by Allan Bakke that as a Caucasian he had been invidiously discriminated against on the basis of race by the admissions committee of the University of California at Davis Medical School.[11] Without going into detail, much of which is widely known, Davis had two separate admissions procedures: one for majority applicants and one for minority applicants. Some 2,644 people applied for 100 places in the medical school class in 1973; 3,727 applied in 1974, the two years in which Bakke applied. He was denied admission in both years. Sixteen places out of the 100 were set aside for minority applicants. Race was used as a classification in the admissions procedure. Bakke claimed that this classification was unconstitutional under the Equal Protection Clause of the Fourteenth Amendment. His personal claim of discrimination rests upon a comparative analysis of several criteria for admissions, including the items making for a composite index of quality, which is based upon the Medical School Aptitude Tests (MCATs), grade-point average, a score given to personal interviews, and so on. Bakke had an undergraduate grade-point average of 3.51 out of 4.0; his MCAT scores were 96th percentile on the verbal test, 94th on the quantitative, 97th on the science, and 72nd on the general information test. Bakke's index score in 1973 was 468 out of a possible 500, and in 1974 was 549 out of 600.[12] "The mean percentage scores on the [MCAT] test of the minority students admitted to the 1973 and 1974 entering classes under the special program were below the 50th percentile in all four areas tested. In addition, the combined numerical ratings of some students admitted under the special program were 20 to 30 points below Bakke's rating."[13]

The trial court concluded that the Davis admissions procedure constituted invidious discrimination against Bakke on the basis of race and violated his rights under the Fourteenth Amendment. The Supreme Court of the State of California upheld the lower court decision, but importantly the majority opinion did not argue against there being a compelling state interest that would allow for a classification by race; it argued that there were alternative means to reaching the goals desired by the state.[14]

The question of whether or not there is a compelling state interest that justifies a racial classification and whether there exist alternative means to reach-

---

11. *Allan Bakke* v. *the Regents of the University of California*, S.F. 23311, (Super. Ct. No. 31287), p. 2.

12. *Ibid.*, p. 11.

13. *Ibid.*, p. 12.

14. An informative discussion about the means-and-ends fit under the Equal Protection Clause can be found in Owen M. Fiss, "Groups and the Equal Protection Clause," in Cohen et. al., *op. cit.*, pp. 84–155.

ing the desired goals has been framed to an extraordinary degree in terms of empirical fact. The parties to this affirmative action case, in arguing for or against a compelling state interest and about alternative means to achieving the desired ends, largely agreed that several basic aspects of the case rested on a firm fact basis. Let us consider several of the assumptions of fact and the evidence that exists to support these so-called facts. A number of such assumptions are illustrated in a single paragraph from an essay by Ronald Dworkin:

> According to the 1970 Census, only 2.1 percent of U.S. doctors were black. Affirmative action programs aim to provide more black doctors to serve black patients. This is not because it is desirable that blacks treat blacks and whites treat whites, but because blacks, for no fault of their own, are now unlikely to be well served by whites, and because a failure to provide the doctors they trust will exacerbate rather than reduce the resentment that now leads them to trust only their own. Affirmative action tries to provide more blacks as classmates for white doctors, not because it is desirable that a medical school class reflect the racial makeup of the community as a whole, but because professional association between blacks and whites will decrease the degree to which whites think of blacks as a race rather than as people, and thus the degree to which blacks think of themselves that way. It tries to provide "role models" for future black doctors, not because it is desirable for a black boy or girl to find adult models only among blacks, but because our history has made them so conscious of their race that the success of whites, for now, is likely to mean little or nothing for them.[15]

Within this one statement there exist some basic empirical assumptions of fact that I want to review because of their relevance to much of the debate over affirmative action for both minority group members and for women. I have cataloged roughly a dozen theoretical and empirical assumptions made by the participants in the *Bakke* case. I will discuss seven of these briefly.

The Assumption About Role Models   It is assumed by Dworkin and by other parties to the dispute that a critical element in effective socialization into the professions and the selection of specific occupations is the presence of persons of the same race or sex who will act as models for potential students. It is further assumed that not only is it necessary to have role models, but that women depend on other women for role models, minorities depend on other minority group members for role models, and so on.[16] The argument proceeds that without a

---

15. Ronald Dworkin, "Why Bakke Has No Case," *New York Review of Books,* 10 Nov. 1977, p. 11.

16. This is an interesting variation of the "insider" versus "outsider" argument. The "insider" argument holds that only persons with statuses in common have access to the special knowledge required for specified tasks. Thus, it has been argued that only blacks can study blacks; only women can study women; only Poles can study Poles, and so on. Here the variant holds that only people of like kind can act as models for others. For detailed discussion of this phenomenon, see Robert K. Merton, "The Perspectives of Insiders and Outsiders," in *The Sociology of Science (op. cit.),* pp. 96–136; and for similar views consider the following excerpt from Justice William O. Douglas' dissent in the DeFunis case:

> The reservation of a proportion of the Law School class for members of selected minority groups is fraught with similar dangers, for one must immediately determine which groups are to receive such favored treatment and which are to be excluded, proportions of the class that are to be allocated to each and even the criteria by which to determine whether an individual is a member of a favored group. There is no assurance that a common agreement can be reached, and first the schools, and then the courts, will be buffeted with

sufficient critical number of minority doctors who can act as role-models for minority youth, it will be very difficult, if not impossible, to increase both the absolute and relative number of minority doctors in American society. Consequently, affirmative action that admits minorities to medical school programs under a special admissions policy serves the important, indeed compelling, state interest of producing role models who will in due course significantly influence the numbers of young minority group members who want to become physicians. The argument is similar for women in science and other professions in which women represent a small fraction of the total population.

There are two empirical social-science assumptions here. First, it is assumed that role models are important factors in the recruitment and learning process for minority (and female) students. Second, it is assumed that the role-model effect on students involves race or sex matching—that is, black students are more influenced in career decisions by black role models than by white role models; female students are more influenced in career decisions by female instructors than by male instructors. Consider several illustrations of this position:

> Minority doctors will, moreover, provide role models for younger persons in the minority community, demonstrating to them that they can overcome the residual handicaps inherent in past discrimination.[17]

> Diversity in background, including race, within faculties is important, enriching the interchange of ideas and offering role models to minority students.[18]

The California Supreme Court's majority opinion, which in the end holds against affirmative action in the Davis form, discusses and accepts the argument that the production of role models is necessary and will have secondary benefits:

> The University seeks to justify the program on the ground that the admission of minority students is necessary in order to integrate the medical school and the profession. The presence of a substantial number of minority students will not only provide diversity in the student body, [we will come to this argument] it is said, but will influence the students and the remainder of the profession so that they will become aware of the medical needs of the minority community and be encouraged to assist in meeting those demands.[19]

---

competing claims. The University of Washington included Filipinos, but excluded Chinese and Japanese; another school may limit its program to Blacks, or to Blacks and Chicanos. Once the Court sanctions racial preferences such as these it cannot then wash its hands of the matter, leaving it entirely in the discretion of the school, for then we would have effectively overruled Sweatt v. Painter, 339 U.S. 629, and allowed imposition of a "zero allocation" . . . but what standard is the Court to apply when a rejected applicant of Japanese ancestry brings suit to require the University of Washington to extend the same privileges to his group? The committee might conclude that the proportion of Washington is now 2% Japanese, and the Japanese also constitute 2% of the bar, but that had they not been handicapped by a history of discrimination, Japanese would now constitute 5% of the bar, or 20%. Or alternatively the Court could attempt to assess how grievously each group has suffered from discrimination, and allocate proportions accordingly; if that were the standard the current University of Washington policy would almost surely fall, for there is no western State which can claim that it has always treated Japanese and Chinese in a fair and even-handed manner [DeFunis v. Odegaard, 94 S. Ct. 1704 (1974)].

17. 132 Cal. Rptr. 680, 692 (1976).

18. Brief of Columbia University, Harvard University, Stanford University, and the University of Pennsylvania as amici curiae in the case of The Regents of the University of California v. Allan Bakke, June 7, 1977.

19. 132 Cal. Rptr. 680, 692 (1976).

The role-model argument is used to support the idea of a compelling state interest, and it is offered by the defendants as justification for a benign racial classification. On this point Justice Tobriner, in his dissent to the California Supreme Court opinion, cites an argument by Robert O'Neill:

> Finally, over and above the benefits accorded to the medical school and to the medical profession, the special admission program was implemented to serve the larger national interest of promoting an integrated society in which persons of all races are represented in all walks of life and at all income levels. As Professor O'Neill has explained: "For minority youth... professionals from and within their community offer essential role models. Success and expanding opportunities suggest there are ways of 'making it' without resort to violence. Conversely, the denial or closing off of opportunities for education cannot help but breed frustration, resentment and anger at predominantly white Anglo society."[20]

There is a fundamental assumption here that the presence of role models will significantly affect the rate at which minorities (and women) seek entrance into the powerful professions. This is an empirical question. Is it a fact? What evidence is there in the extant literature that role models have an effect in the socialization process leading to occupational choice?

Actually, the literature on the influence of role models is rather thin. There has been some work on modeling effects on children, but for the older population the literature is particularly sparse. There is very little empirical material that addresses role models for minority group members. The best review of the literature on gender role-models was recently produced by Eleanor Maccoby and C. Nagy Jacklin. First they pose the issue:

> To recapitulate: the fact that observational learning occurs is not in doubt. It is also clear enough that children learn many items in their behavioral repertoires through imitation of their parents. The problem is why children of the two sexes should learn *different things*—sex-typed things. Two explanations have been offered: (1) that the same-sex model is more available, and (2) that children select same-sex models among those that are available on the basis of perceived similarity between themselves and the model.[21]

Maccoby and Jacklin conclude:

> On the whole it simply cannot be said that young children spontaneously imitate people of their own sex more than people of the opposite sex. This is true of imitations of parents as well as of models who are unfamiliar to the child.[22]

And they state further:

> Our analysis of the arguments concerning the role of modeling in sex typing and our review of the research on selective imitation have led us to a conclusion that is very difficult to accept, namely that modeling plays a minor role in the development of sex-typed behavior.[23]

---

20. *Ibid.*
21. Eleanor Emmons Maccoby and Carol Nagy Jacklin, *The Psychology of Sex Differences* (Stanford: Stanford University Press, 1975), Chapter 8, "Sex Typing and the Role of Modeling," pp. 287–288.
22. *Ibid.*, p. 295.
23. *Ibid.*, p. 300.

Most discussions of the effects of role modeling are anecdotal accounts of the supposed influence of one or more individuals on the careers of women or minorities. But as Elizabeth Douvan has pointed out:

> We take it on faith that knowing a woman academician and being close enough to see something of the reality of her life and action will help the young intellectual woman to concretize her own role conception, invest her aspirations with greater reality, and perhaps offer her some useful clues about ordering her spheres of action. It seems plausible. But we must take it on faith because identification and modeling have [not] been studied systematically except in preschool children.[24]

And at that, there are considerably more data on the effects of role models for women than there are for minorities. One recent study by Reitzes and Elkhanialy compared sources of influence on career choice among current medical students and practicing physicians, all of whom were black. They conclude: "The data on the influence of others on the career choice suggest that the students are more inclined to make this decision for themselves than were the physicians: 34 percent of the physicians compared with 44 percent of the students stated that 'no particular person' was influential."[25]

But the problems with the empirical data on role model effects goes beyond whether or not the findings reinforce the presumptions. Even if the findings were consistent with the assumption that there is a homogeneous role model influence, they would have to be questioned in terms of their applicability to the questions raised in *Bakke*.

First, they are the products of an experiment involving single exposure, often outside of natural settings, which are not followed up to test for longer term influences of the models.[26] Second, they often involve pre-school children—a long time from the point of career choice.[27] Third, frequently they make use of small samples (most often under 100), which are frequently unrepresentative, drawing predominantly on middle and upper-middle class children, such as sons and daughters of Stanford faculty members.[28] Fourth, there is no distinction made in the legal arguments about what form of role model effects the Court

24. Elizabeth Douvan, "The Role of Models in Women's Professional Development," *Psychology of Women Quarterly* 1, no. 1 (fall 1976): 5.

25. Dietrich C. Reitzes and Herkmat Elkhanialy, "Black Students in Medical Schools," *Journal of Medical Education* 51 (Dec. 1976): 1001–1005.

26. See, among others, A. Bandura, D. Ross, and S. A. Ross, "Imitation of Film-Mediated Aggressive Models," *Journal of Abnormal and Social Psychology,* 66 (1963): 3–11; A. Bandura, D. Ross, and S. A. Ross, "A Comparative Test of the Status Envy, Social Power, and Secondary Reinforcement Theories of Identificatory Teaching," *Journal of Abnormal and Social Psychology* 67 (1963): 527–534; L. Kohlberg and E. Zigler, "The Impact of Cognitive Maturity on the Development of Sex-Role Attitudes in the Years 4–8," *Genetic Psychology Monographs,* 75 (1967): 84–165; David J. Hicks, "Imitation and Retention of Film-Mediated Aggressive Peer and Adult Models," *Journal of Personality and Social Psychology* 2 (1965): 97–100; E. M. Hetherington, "A Developmental Study of the Effects of Sex of the Dominant Parent on Sex-Role Preference, Identification, and Imitation in Children," *Journal of Personality and Social Psychology* 2 (1965): 188–194; A. D. Leffer, "The Relationship between Cognitive Awareness in Selected Areas and Differential Imitation of a Same-Sex Model," Unpublished M.A. Thesis, Stanford University, 1966.

27. Bandura, *et al., op. cit.;* J. F. Rosenblith, "Learning by Imitation in Kindergarten Children," *Child Development* 30 (1959): 211–213.

28. See, among others, W. D. Ward, "Processes of Sex-Role Development," *Developmental Psychology,* 1 (1969).

expects to obtain. Are they looking for "imitative" effects, "learned behavior" effects, or "expectational" effects? The Court opinions allude to alterations in "expectations" about occupational possibilities.

> ... It has been argued that ethnic preferences "compensate" the group by providing examples of success whom other members of the group will emulate, thereby advancing the group's interest and society's interest in encouraging new generations to overcome the barriers and frustrations of the past ... [29]

A reasonable hypothesis, but one that has no basis in empirical evidence. There are in fact virtually no studies that even attempt to measure changing "expectations of the possible" that result from the existence of "like-kind" models. Again, it is not that the data are in. There are no data to speak to the issue. Fifth, are results obtained for gender models generalizable for questions of race models? The answers do not exist.

What these studies tell us about the influence of "like-kind" individuals on the selection of medical careers is thoroughly problematic. Thus the literature on the actual effects of role models is uninformative in multiple ways: the data that do exist are inconsistent with the hypothesized outcome and the studies that produced these results are so far removed from the context in question that the results, regardless of their direction, would require an enormous leap of inference to assume similar results within the altered context.

Of course, it is arguable therefore that the results which run counter to the assumption made by participants in the *Bakke* case do not refute the claim. And this is so. The real point is that for all practical purposes the role model argument like so many of the other empirical issues in *Bakke* remains entirely at the conjectural level and should have been treated as such.

In sum, there simply is no strong evidence that gender matching or racial matching between young people and their elders has a significant influence on career choice. We might like to think that there is such an effect; it may have intuitive appeal; but the facts that are assumed to exist simply don't.

This is an interesting example of how intuitively appealing theories and concepts often get elevated to "facts" in the social sciences, without going through the rigorous processes of what Karl Popper has called "criticism," or Robert K. Merton has called "organized skepticism." There is something intuitively appealing about the concept of role models; it seems plausible that they should have some influence on career selection; and furthermore, it may even by somewhat logical to expect people of "like" kind to have more influence on each other (especially given probabilities of proximity) than people who are "unlike." But we must be honest enough to recognize that all of this lies on a hypothetical, or conjectural, plane. The danger lies in elevating the hypotheses into facts before their time, because it is often the case that once elevated, appealing conceptions are resilient against counterfactual evidence.

It is essential to note, however, that the concept of role models can be transformed into an empirical inquiry that could yield far more useful answers

---

29. United States Law Week 46 LW 4906 at footnote 43.

than those obtained to date. The instruments for studying such a problem with quantifiable empirical data are available, but thus far the work has not been done.

As with the other matters that I shall discuss, we may not have much more useful or sound data today on some of the issues in *Bakke* than were available to the social psychologists and sociologists who presented material to the Court in *Brown I,* and as in *Brown I* it may be argued that a result one way or the other in Bakke need not have rested on any empirical data, but today we do have the tools to do a lot more to test reliably these assumptions of fact than were available a generation ago to Kenneth Clark and his colleagues.

In way of illustration, returning to Chief Justice Vinson's opinion in *Sweatt v. Painter,* 339 U.S. 629 (1950), where the Court based its decision not only on the tangible differences between the University of Texas Law School and the hastily constructed alternative for blacks, but also on "those qualities which are incapable of objective measurement but which make for greatness in a law school," you find that almost all of the items that could not be measured in 1950, could be measured fairly easily today, and with substantial reliability. Today we can easily measure the reputational standing of law schools, we can easily measure the relative "quality" of the faculties at the school, etc. While the measurements may be based upon appraisal of peers rather than how the schools stand in the sight of God, they will be "objective" in terms of measurement, with high levels of validity and reliability.[30]

The Heterogeneity of Intellectual Environments   A second assumption found in Dworkin, in Justice Powell's controlling opinion, and in many other discussions of the *Bakke* case, is the presumed positive effects on the intellectual development of students of a heterogeneous intellectual environment. One illustration of this argument can be seen in Justice Tobriner's dissent in the California Supreme Court decision: "The special admission process at issue here, of course, was in fact implemented for just such an educational purpose, to provide a diverse, integrated student body in which all medical students might learn to interact with and appreciate the problems of all races so as to adequately prepare them for medical practice in a pluralistic society."[31]

The position of the University of California and of other supporters of the dual-admissions program contends that a diverse intellectual student body is part of the socialization or maturation process in medical school, and that heterogeneity better prepares the student to participate responsibly in a pluralistic society. Heterogeneity may be a value in itself, but what exactly the supposed effects of heterogeneity are and what can be demonstrated in the way of its consequences remains wholly ambiguous both in the briefs and opinions in this case and in scholarly papers on the subject. Will heterogeneity make the majority students better physicians, and if so, in what specific ways? Will it influence

---

30. Jonathan R. Cole and James A. Lipton, "The Reputations of American Medical Schools," *Social Forces* 55 (March, 1977): 62–84.

31. 132 Cal. Rptr. 680, 715 (1976).

their interaction with patients of all races, of minority groups? Will it influence their ethical standards, their competence as surgeons, as internists, as gynecologists? The specific consequences heterogeneity is supposed to have remain totally unarticulated; there are only vague hypotheses, and almost no empirical evidence of effects. The arguments are made *a priori*. It all remains conjecture, and the articulation of values and ideology.

It is true that schools, particularly the prestigious private colleges in the United States, have argued that the heterogeneity of a student body, along geographic and other lines, enhances the "student experience," but even here there are no actual empirical data to suggest what types of positive, and possibly negative, effects such selection criteria have on a set of designated outcomes.[32] The simple fact that universities have in the past used criteria such as athletic ability, genealogy, or cash contributions in admissions decisions does not mean that *a fortiori* their use was fair and should act as precedent for the setting of current standards.

Surely, the argument for heterogeneity is probably not well grounded in empirical fact if it rests on the belief that heterogeneity among medical school students will result in lower levels of prejudice and discrimination toward minority patients. What little evidence exists suggests that such prejudicial attitudes are formed while people are fairly young, and that once prejudices are formed people are selectively attentive to those facts and events that support them. In any event, there is little evidence that directly tests the counter-hypothesis. The question remains: Is there any evidence that the heterogeneity or homogeneity of the educational environment has any effect on the intellectual or *moral* development of students? Furthermore, can we expect heterogeneity at the professional school level to have effects similar to those that it might have at the elementary and secondary school level? It is more plausible, a priori, that heterogeneity will have significant effects on attitudes, beliefs, perceptions, etc. if experienced early on in life than after the race is essentially run—at the graduate school or professional level. In fact, the few studies that have been done on the effects of college environments on changes in aptitudes and attitudes, suggest little change from the freshman to the senior years. To a large extent, inputs equal outputs.

There is a paradox here, which is not lost on the participants in the *Bakke* case, and which is epitomized by the literature on the effects of desegregation on the intellectual performance of minorities. Is it possible to facilitate integration by selecting students on the basis of their race, or to put it another way, will short-run color consciousness produce long-term colorblindness? If students are selected through special admissions procedures as representatives of minority communities, will their participation in the educational process be affected by the method of selection itself?[33]

The basic point is: There may be beneficial effects of diversity (and on the

32. Columbia University et al., *amici curiae* brief, *op. cit.*

33. Dworkin, *op. cit.*; I want to thank Mark Johnson, a Columbia University graduate student in sociology, for suggesting this line. He also was extremely helpful in locating sources for this section and in providing me with brief summaries of various articles.

basis of personal values we may strongly hope that there are), but if there are, there has been little work to specify the dimensions of those effects, and there has been little empirical work that demonstrates either positive or negative consequences of diversity for the various subpopulations that are affected.

The Assumption about Test Scores and Role Performance   Two questions that continually appear in discussion of the *Bakke* case and in the opinions and briefs themselves are: Is there any relationship between scores on standardized admissions tests and performance in professional schools? Is there any relationship between such tests (combined with performance in professional schools) and "success" as a practitioner? Almost invariably the answers are either a flat no, or that there is at best a very slight positive association between test scores and performance. As examples, first note Justice Tobriner's dissent in *Bakke*:

> Indeed, the medical school's decision to deemphasize MCAT scores and grade point average for minorities is especially reasonable and invulnerable to constitutional challenge in light of *numerous empirical studies which reveal that, among qualified applicants, such academic credentials bear no significant correlation to an individual's eventual achievement in the medical profession....* As medical school admissions officials themselves acknowledge, these studies raise questions of the most serious order as to the propriety of the continuing use of traditional admission criteria.[34]

Or, to cite one more illustration:

> The Court's opinion ... is based on the unfounded assumption that Bakke and many of the other white applications who were rejected are "better qualified" than the minority students who were admitted. But there appears to be little correlation between applicants' test scores and their subsequent performance as medical students.[35]

Indeed, there does exist some empirical evidence, cited by the California court, that suggests that test scores are not good predictors of subsequent performance both in medical school and in professional life. For example, Peterson et al. conclude: "There was no demonstrable relationship between the test score and subsequent grades in medical school, nor was there a significant relationship between the test score and physicians' subsequent performance in practice."[36] Turner et al. suggest that cognitive tests and medical school grades are unreliable predictors of physicians' performance.[37] Similarly, Price et al. found that "97 percent of [the] intercorrelations [between 3 measures of academic performance and 77 measures of on-the-job performance] were of a zero-order magnitude

---

34. 132 Cal. Rptr. 680, 714 (1976); emphasis added.

35. Charles Lawrence III, "The Bakke Case: Are Racial Quotas Defensible?" *Saturday Review*, 15 Oct. 1977, pp. 14–16.

36. O. L. Peterson, L. P. Andrews, R. S. Space, and B. B. Greenberg, "An Analytical Study of North Carolina General Practice, 1953–1954," *Journal of Medical Education* 31, pt. 2 (Dec. 1956).

37. Edward V. Turner, M.D., Malcolm M. Helper, Ph.D., and S. David Kriska, "Predictors of Clinical Performance," *Journal of Medical Education* 49 (April 1974): 338–342.

[sic]. Conversely, only three percent were of sufficient magnitude to indicate a significant relationship between undergraduate grades and physician performance, and more of those were negative than positive.[38]

Wingard and Williamson recently reviewed the literature on this subject and concluded: "The major finding of the search was that little concrete information exists on attempts to correlate medical training and postacademic performance. . . . Although studies in this area are sparse, available research findings have demonstrated that little or no correlation exists between academic and professional performance."[39]

In many ways this test-score argument is the most powerful argument that has been made for affirmative action. If test scores are weighted heavily in the overall evaluation process and they bear little or no relation to actual performance either in medical school or on the job, then their use to "rank-order" the qualifications of applicants is plainly suspect. Furthermore, there is good precedent in fair-employment cases for questioning the use of testing procedures of selection when they cannot be demonstrated to have a direct relation either to training programs or to on-the-job performance.[40] If the social science evidence to support this position were robust, a very strong case could be made against the use of test scores as selection criteria. But can we accept the extant data as "fact"? Let us consider the matter in greater detail.

Before addressing the central weakness of those studies which Justice Tobriner has accepted as convincing, there are several technical weaknesses worth noting. For example, the Peterson study was based on a total of thirty physicians studied during the 1950s, some twenty years before this litigation was initiated. Furthermore, all of the studies that have examined role performance of physicians have significant measurement problems with the dependent variable. When they deal with the performance of physicians who are faculty members at medical schools, the performance measures tend to be better than when they deal with general practitioners. But since most medical school students go into private practice, the measurement problems, which have not been worked out, are formidable. Indicators of performance include, among others, proportion of time spent seeing patients, number of patients referred to other physicians, number of referrals received from other physicians, cost of office equipment (physician's estimate), and number of medical journals reviewed regularly.[41] While these are reasonable indicators, it is not altogether clear what one might hypothesize to be the significance of the correlations between them and "high quality" or "low quality" role performance. We are dealing, then, with basic findings that are

38. Philip B. Price, M.D., Calvin W. Taylor, Ph.D., James M. Richards, Jr., Ph.D., and Tony L. Jacobsen, "Measurement of Physician Performance," *Journal of Medical Education* 39 (Feb. 1964): 203–210.

39. John R. Wingard and John W. Williamson, M.D., "Grades as Predictors of Physicians' Career Performance: An Evaluative Literature Review," *Journal of Medical Education* 48 (April 1973): 311–312.

40. Among other cases, see *Griggs* v. *Duke Power Co.*, 401 U.S., at 432 (1971); *Albemarle Paper Co.* v. *Moody* 422 U.S. 405 (1975); *Washington* v. *Davis* 422 U.S. 405 (1976); *Dothard* v. *Rawlinson* 97 S. Ct. 2720 (1977).

41. Price, et al., *op. cit.*

questionable because of the conceptualization and measurement of what is meant by high-quality performance as a physician.[42]

But the much more fundamental problem with these findings—and one which has been overlooked or avoided, to my knowledge, by all of the discussants of the test-score issue—is the problem of attenuated variance in the independent variable, test scores. What do I mean by "attenuated variance"? To begin with, the audience for these arguments seems not to realize that the analysis of the relationship between test scores and role performance in professional school and on the job is not based upon a sample of all applicants to the professional schools, or on any kind of random sample from the overall distribution of scores on the test. And of course this is not possible, since only a fraction of the applicant pool is actually admitted to the schools. But consider the scores that we are actually dealing with. Since there is such great competition to get into medical school, law school, and other professional schools, and since in the past, test scores on the MCATs or the LSATs were a principal criterion in selection, those students who were actually admitted had, on average, very high scores. For the very top schools, those students who are actually admitted most often rank in the top 5 percent of the overall distribution.

Let me cite as one example material that appears in the 1976 *Columbia University Bulletin for the School of Law,* which gives prospective applicants an idea of their chances of admission on the basis of data drawn from the "current" student population.

> Recent applicants who were within a college GPA range of 3.5 to 4.0 *and* LSAT range of 675 to 800 had about a seven-out-of-ten chance for admission; those with a GPA range of 3.2 to 3.5 *and* LSAT of 675 or higher, or a GPA range of 3.7 to 4.0 *and* test scores of 650 to 675 had about a four-out-of-ten chance for admission; those with the GPA below 3.3 *and* a test score below 650 were not likely to be admitted.[43]

Although Columbia's admissions standards are far higher than those of the average law school, the studies of the relationship between GPA and LSATs and performance are based upon data that use individual schools as the unit of analysis. The range of scores that would make an applicant a likely candidate for admission would be no different for a major medical school, as is clear from the data generated by the *Bakke* case.

Since we are dealing with such a restricted range on the test-score variable, it would simply be foolish to expect small differences—between, say, 720 and 740 on the LSATs—to be fine-grained predictors of either Law school performance or later performance as lawyers (even assuming that we had a good measure of the latter two variables). Aside from the error factors that may be involved in grade-point averages within medical school, it is simply a substantive and probably a statistical error to expect that such a small subsample of the total distribution of scores, and a sample that is almost exclusively at the upper tail of the

---

42. *Ibid.*
43. *Columbia University Bulletin, School of Law,* 26 July, 1976, p. 97; emphasis in original.

distribution, will be good predictors of almost anything, much less such complex phenomena as professional competence and performance.

A much better test of the test-score hypothesis may shortly be possible and could be done. If we were to take all students who were admitted to a professional school under a special admissions program—that is, students with significantly lower MCATs or LSATs than those obtained by students who gain admission through the standard procedure—and we matched those students with a random sample from those admitted in the standard way, we could observe the correlation between test score and role performance among this special sample. Here we would at least be extending the variance on the test-score variable. Assuming that there is no preferential grading policy within schools, this could produce a more adequate assessment of the relationship between test scores and performance. This would be an even stronger test of the hypothesis than we might obtain were we simply to select a random sample of lower scores, because supposedly those students selected in the special admissions process are "poor testers" and have other indicators in their past educational records that they are particularly capable of handling work in the professional school.

If adequate empirical studies were carried out and we were still to find no association between measures of performance and test scores, the use of such scores as admissions criteria would have to be seriously questioned. Plainly, with such findings any suggestion that relatively small differences in scores, when accepted applicants have lower scores than nonaccepted applicants, represent invidious discrimination would hold little force. But without adequate empirical inquiries such inferences simply cannot be made and we cannot claim a factual basis for them.

At the risk of becoming excessively repetitive, let me reiterate that the essential point is not so much that we know what the outcome of such an experiment would be, but that as of now we simply do not have the appropriate data to draw correct inferences about the relationship in question. Finally, it is surely incorrect to reify minor differences in test scores and to assume that they distinguish differential capacity to carry out professional work, but if we examine differences between 550 and 700 on the MCATs rather than differences between 690 and 700, we might well find that the correlations that were supposed to be insignificant turn out to be considerably higher than those found with the currently used, inadequate data-sets.

The Assumption That Minorities Will Serve Their Communities   Another fundamental argument for special medical-school admissions procedures is that only through the greater production of minority doctors can essential medical needs of minority communities be met. Several examples of this position are:

> The special admission program will assertedly increase the number of doctors willing to serve the minority community, which is desperately short of physicians.[44]

---

44. 132 Cal. Rptr. 693, 716 (1976).

By increasing the number of minority students, many of whom could be expected to practice in ghetto areas, it would improve health care to the poor and minorities.[45]

There can be no doubt—and there is good statistical evidence to support this—that minority communities are desperately short of physicians.[46] Furthermore, minority student applicants state that they intend to serve their communities after they complete their studies. Intent and actual practice are sometimes two different matters, especially when rewards can be had for an expression of intent that need not be followed up by action and that entails no contractual obligation. The California Supreme Court's majority rejected the argument that the special admissions procedure could be justified on the probability that minorities would return to serve their communities: "The record contains no evidence to justify the parochialism implicit in the . . . assertion. . . . There is no empirical data to demonstrate that any one race is more selflessly socially oriented or by contrast that another is more selfishly acquisitive."[47]

What is the evidence? The evidence suggests—and there is not a great deal to rely on—that black physicians do predominantly have black clientele, but not necessarily for altruistic reasons, and not necessarily within those communities of blacks that are in the greatest need of improved medical care. Historically there have been, and continue to be, barriers to blacks who want to practice in predominantly white communities. But perhaps the most important finding on this issue is that black doctors tend to serve the black middle class and that they are no more apt to serve lower-class populations than are other doctors. Reitzes and Elkhanaily studied doctors who had received National Medical Fellowships, awarded in a program designed to improve the quality of training of black and other minority group physicians. Responses to a questionnaire about the social class of patients of fellowship recipients and "other" physicians suggest that there is little difference in the clienteles.

> Information on the class composition of patients was provided by 163 NMF [National Medical Fellowship recipients] and by 247 "other" physicians. These returns indicate that, in the NMF group, 38 percent served primarily lower economic groups and that among the "other" group 36 percent did so.
> Our data show that the black physicians serve an overwhelmingly black patient group. However, they also indicate that within the black group they serve primarily middle and upper class groups rather than the lowest economic strata. This is true for physicians who are Board certified as well as for those who are not. . . .
> The impact that an increase in the number of minority group physicians has on the medical care of minority group communities is not clear. Our data suggest that black physicians serve primarily middle-class blacks. The major problem in the

45. *Newsweek*, 26 Sept. 1977, pp. 53–54.

46. Theodis Thompson, "Curbing the Black Physician Manpower Shortage," *Journal of Medical Education* 49 (1974): 944–950.

47. 132 Cal. Rptr. 680, 693, 695 (1976).

health care delivery system today, however, is in the services provided to lower-class inner-city communities.[48]

Although we could raise questions of whether the courts or medical schools should expect minority doctors to return to inner-city communities, the point again is that there is no factual basis for this argument placed before the Supreme Court and consistently found in articles about the *Bakke* case.

The Assumption About the Quality of Health Care   This assumption, which is directly related to the location-of-service assumption, is another variant on the "insider" argument. It holds that doctors and patients with similar *social* characteristics have a better relationship, and consequently better health care will be provided. The California Supreme Court summarized the position: "It is urged, black physicians would have a greater rapport with patients of their own race and a greater interest in treating diseases which are especially prevalent among blacks, such as sickle cell anemia, hypertension, and certain skin ailments." [49]

Does this hold for all ethnic, religious, gender, and racial groups? Is there any sound evidence to support this assertion?

All of this raises the possible total displacement of fundamental goals. Is the goal to improve medical treatment for minority communities, or to produce a match between doctors and patients in terms of racial status? If we demonstrate that racial, ethnic, or religious characteristics have no effect on the quality of medical care offered, then the argument for a status match is weakened. Conversely, if status matches actually are related to the quality of health care, and they are not "spurious" relationships, then there may be merit to the argument for homogeneous racial statuses of doctor and patient.

But there simply is no evidence that homogeneity between doctor and patient is related to the quality of medical care.

> The assertion that health professionals experience great difficulty in communicating with and involving the culturally different client provides a rationale for the hypothesis that the psychosocial accessibility problems of blacks in obtaining health care would be alleviated through the existence of an appropriate number of black physicians to meet the black demands for health care services. However, this hypothesis assumes that the black physician operates within the "life space" of the black client—poor and non-poor alike—that is, that he is more likely to share the values of the black health care consumer with respect to the consumer's orientation toward health and medical care than are non-black physicians. Unfortunately, there is no quantitative data to confirm or deny this assertion.[50]

Assumptions About Locus of Discrimination   The problems of statistical discrimination, discussed in Chapter 2, are directly related to affirmative action

---

48. Dietrich C. Reitzes and Herkmat Elkhanialy, "Black Physicians and Minority Group Care—The Impact of NMF," *Medical Care* 14, no. 12 (Dec. 1976): 1052.

49. 132 Cal. Rptr. 680, 693 (1976).

50. Thompson. *op. cit.,* pp. 944–945.

cases such as *Bakke*. Although blacks (or women) as a group have unquestionably been subjected to extensive discrimination in the labor market and in access to training required for a wide variety of occupations, are the same individuals who are benefiting from affirmative action programs those who have actually been subjected to invidious discrimination—or are they simply "representative" of a group that has experienced widespread discrimination? There is little empirical evidence that addresses this question. While some scholars, such as Owen M. Fiss, forcefully argue that the social ends justify the tolerance of individual "mistakes" that result from the application of affirmative action policy, the fact question precedes agreement or disagreement with this value position.[51] Have the members of minority groups who enter professional schools under special admissions procedures actually been objects of racial discrimination, except in the most global institutional sense? Where did these individuals go to college? What types of socioeconomic backgrounds do they have? Have they been educationally or socially deprived?

More specifically, were the minority students admitted to the Davis medical school program under the special procedures for the sons and daughters of impoverished or lower-class families that had been subjected to direct racial discrimination? Were these students deprived of educational opportunities from grade school through college, or did they actually have educational experiences at superior colleges, such as Berkeley, Stanford, Harvard, and Swarthmore? Who, in short, are the beneficiaries of this new social policy?

When affirmative action discrimination cases involve claims by individuals, whether women or minority group members, then the merits of the case can be weighed for these individuals. But when the focus is on the implementation of a general policy that involves race or gender classifications, the actual recipients of benefits from affirmative action programs may not be the members of the group that have actually experienced deprivation. And whether or not this is so remains largely unknown.

A feature of this assumption that is particularly ironic is that generally statistical discrimination has been used against minority groups and women. That is, in a series of Title VII and Fourteenth Amendment cases leading to *Dothard* v. *Rawlinson* (97 S. Ct. 2720 [1977])—in which statistical parameters about height and weight of women were used to exclude them from jobs as prison guards—women and minorities have been deprived of opportunities because of the application of group statistics that may not apply in individual cases or that may not be related to job performance.[52] In some affirmative action cases, group correlations are being used now to exclude some individuals who might qualify on an individual basis. Furthermore, the decision rules are being applied in such a way that even within minority groups individuals from the middle class without clear

---

51. Fiss, "Groups and the Equal Protection Clause," in Cohen et al., *op. cit.*

52. Kenneth M. Davidson, Ruth B. Ginsburg, and Herma H. Kay, *Sex-Based Discrimination* (St. Paul, Minn.: West Publishing Co., 1974).

evidence of past discrimination against them may be limiting the future opportunities of minority group members who really have had personal histories of limited opportunities because of discrimination.

The Assumption That There Is "No Other Way"   In the California decision on *Bakke*, the majority held that there were compelling state interests that would justify the use of racial classifications in the admissions procedure, but that there were alternative means of reaching the stated goals. Proponents of the special admissions procedures hold that there is in fact "no other way" to reach the socially desirable goals. The Court suggested several possible alternative means; the opponents of that decision argue that no alternative means exist that would produce a large enough absolute number of minority doctors. This position is stated straightforwardly in the *amici curiae* brief submitted to the Supreme Court by Columbia, Harvard, Stanford, and the University of Pennsylvania:

> The Supreme Court of California appears to acknowledge the constitutional propriety of selecting a racially diverse student body. But the court has held that this permissible end must be sought without taking race into account—an anomalous circuity insisted upon in the belief, unsupported by the record, that racially random processes would somehow produce a student body of sufficient racial diversity . . . [W]e disagree with the California court's conjecture—and it is only conjecture, flatly contradicted by the only testimony of record—that universities can achieve racially diverse student bodies without taking into account the race of those applying for admission. Our institutions' experience confirms that the substitute devices suggested by the California court are incapable of fulfilling this constitutionally legitimate objective.[53] . . . This case was decided by the Supreme Court of California upon a record almost devoid of relevant evidence. Apart from the pleadings, the record consisted principally of a declaration under oath of Associate Dean Lowery, and Dr. Lowery's deposition taken by plaintiff's attorney. In particular, on those issues crucial to the decision of the court below—the feasibility of other means, not race-oriented, for accomplishing the University's goals—the only evidence was that of Dr. Lowery, and its substance was that there existed no such means. Although uncontradicted, it was disregarded by the Supreme Court of California, which reached its own conclusions presumably on the basis of its own assumed expertise.[54]

While it does appear that there is no solid empirical evidence that would suggest clear-cut alternative means at arriving at the socially desirable goals of having more minority doctors, curiously there is no solid evidence presented by the *amici* that there are *no* alternative means. In short, there simply is no empirical evidence of any quality that could be used to argue one position or the other. There is no factual basis for a position on the question of alternative means; there is only room for an expression of opinions reinforced by questionable logic on both sides—opinions that do not have a demonstrated empirical foundation.

---

53. Columbia University et al., *amici curiae* brief, *op. cit.*, p. 9.

54. *Ibid.*, pp. 36–37.

It is extremely important to underscore again the point that a position pro or contra Bakke can be argued without the presence of empirical data to support one or another position. Traditional modes of legal argument could compete before the Court. But given that the Court and the participants in the most fundamental affirmative action case to yet reach the Court (with the possible exception of the mooted *DeFunis* and the pending *Weber* cases) have defined the case largely in empirical terms, we need to assess the empirical basis for factual assumptions and assertions. When we get down to the basis on which such "facts" have been put before the Court, we see that there is very little solid ground for inferences of fact. The data are simply nonexistent, scanty, or methodologically flawed. To the extent that affirmative action decisions are going to be influenced significantly by empirical data, new methods for evaluating existing data and collecting and analyzing new data need to be more fully developed.

Pools of Available Talent

A central problem in developing adequate affirmative action plans remains the difficulty in defining pools of individuals who might be considered for the same job.[55] Are pools an important problem? Since the working definition of the presence or absence of discrimination is based upon a comparison of the proportion of women and minorities in specific job classifications with the proportion in the pool of applicants or the pool who may be available for a job, the pool definition underlies the analysis of compliance with affirmative action. A "deficiency" or a prima facie case for discrimination is defined in terms of the difference between the pool statistic and the job-category statistic. For illustrative purposes, let me focus on universities.

Universities must develop two types of pool statistics: one for the faculty, the other for all the supporting staff. Consider problems in defining faculty pools. There are no standardized methods for establishing the pool of available women and minorities. Although the proportion of Ph.D.'s produced is often used as the definition of the pool, this is not always used, nor must the college or university use it. To the extent that it is used, the indicator produces a number of problems. It assumes that a Ph.D. is required for a faculty position. While this is undoubtedly so in large universities, it has not been the case for many smaller colleges. Furthermore, the figures may over- or underestimate the actual proportion of women and minorities in the pool. In some fields, such as chemistry, male Ph.D.'s are more apt than female Ph.D.'s to seek jobs in industry. Women self-select themselves out of the pool. Consequently, using the pool figure of 8 percent female Ph.D.'s as a standard is somewhat misleading.

Query: How can we begin to get a standard indicator of pools of available

---

55. Almost all discussions of affirmative action have dealt with the problematics of pool construction; most of these focus on academic settings. Again, see Lester, *op. cit.*, Chapters 3, 4.

women and minorities that can be applied to faculty hiring at all universities and colleges?

Using Ph.D. figures as comparative reference points for hiring and promotion assumes that the proportion of Ph.D.'s remains constant over time. This was so for an extended time period, but as a result of increasing opportunities for women, Ph.D. statistics have changed considerably since 1970. These changes will require more complex comparative pool data, which take into account age-specific ratios of women.

Also, the use of national Ph.D. data does not accurately reflect the hiring pool for universities of varying quality or in different geographical locations. Major universities, such as Harvard or Stanford, do not draw upon a pool of all Ph.D.'s.

Query: How should different types of universities define their pool of available talent? If the pool statistic used for estimates of deficiencies is the production of female Ph.D.'s, then there may be built-in pressure to reduce the proportion of female Ph.D.'s in order to conform to a lower national average.

Just as major universities have specialized pools, so must small colleges and minor universities. No one has adequately considered the structure of the educational community in terms of the extensive variation and definitions of talent pools.

The national pool data are incomplete in still another way. They are typically available only for entire disciplines and not for specialized subfields (the biological sciences is an exception). Thus, the number of women in the national pool of sociologists is known, but the number of women in that pool that is pertinent for professorships in mathematical sociology or sex roles is not known.

Measuring the Quality of Role Performance

There is a fundamental flaw in the formulation of almost all affirmative action plans in both academic and industrial settings: In attempting to determine the level of gender- and race-based discrimination they do not take into account the quality of role performance. Throughout this book I have repeatedly stressed the importance of measuring performance in assessing levels of discrimination. This is particularly important when considering affirmative action policy. I will not repeat my arguments for the need for measuring the quality of role performance, except to underscore the finding that there do seem to be patterned differences in the quantity and quality of scientific research produced by men and women. Since these differences are not accounted for by any obvious combination of factors, differential rewards in science may result from performance differences rather than from invidious discrimination. As soon as we consider the formulation of affirmative action goals and we discuss deficiencies or the proportion of newly hired or promoted faculty members who must be women or minorities, we move directly to the question of performance.

Why are not indicators of performance built into affirmative action plans? The answer varies somewhat for different contexts. Universities and colleges could generate performance measures. Three factors make this possible: a virtually uniform reward system in all academic fields, an organizational apparatus to assess quality, and an unusually high level of agreement about bases of evaluation.[56] Although there is agreement on the bases, however, it is extremely difficult to generate a consensus on the weights assigned to professorial roles. The basic problem lies in applying good aggregate-level indicators to specific cases. Citation counts, which are fairly highly correlated with independent peer evaluations of the quality of research performance and receipt of honorific awards, can be used for large samples as an indicator of the quality of research performance, but it is hazardous to evaluate individuals on the basis of citations.[57] We would be reifying the indicator if we drew inferences about two individuals from knowledge that one received ten citations to his or her work and the other fifteen. Although an anomaly, one can find a Nobel laureate whose work has received only five or ten citations in a given year, when in the aggregate Nobel laureates average well over two hundred citations per year, and among scientists cited at all the average is six. Some scientists who have not made significant discoveries receive more citations to their work than a given Nobelist. In a number of recent affirmative action cases citations have been misused to make comparative judgments at the individual level about the relative competence of several chemists.[58] However, if we want to consider patterns of recognition and gender discrimination in a number of academic settings, the use of statistical indicators such as citations becomes an extremely valuable tool.

The idea of using citation counts as a measure of performance often elicits enormous resistance within the academic community. There is, of course, a significant drawback to using such a measure, even at the aggregate level: its total failure to measure teaching or administrative performance. Unless we accept the primacy of the research function, key aspects of performance within academe must remain unmeasured, even at the aggregate level.

Problems of measuring quality of role performance are compounded in industrial and public-sector employment. For some activities it is possible to develop quantifiable indicators of performance, particularly for jobs involving piecework or sales records. But most jobs are not of this type. Performance of athletes can be measured in a wide variety of ways, although we would be hard-pressed, for

---

56. Stephen Cole, "Scientific Reward Systems: A Comparative Analysis." Paper presented to the American Sociological Association, Aug. 1973.

57. There are a number of good sources for discussions of the validity of citations as a rough measure of the quality of performance. See among others, Kenneth E. Clark, *America's Psychologists: A Survey of a Growing Profession* (Washington, D.C.: American Psychological Association), Chapter 3; Alan E. Bayer and John Folger, "Some Correlates of a Citation Measure of Productivity in Science," *Sociology of Education* 39 (fall 1966): 381–90; Jonathan R. Cole and Stephen Cole, *Social Stratification in Science* (Chicago: University of Chicago Press, 1973).

58. For a discussion of the recent uses of citation analysis, and in particular the use in affirmative-action legal suits, see Nicholas Wade, "Citation Analysis: A New Tool for Science Administrators," *Science* 188 (2 May 1975): 429–432.

example, to decide whether batting averages or slugging percentages are the better measure of performance for baseball players. Lawyers' performance might be measured by the amount of business they bring into their firms or by their case records; prosecutors might be rated by percentage of convictions. But for the most part, there is no reward system in these sectors of the social system that is as uniform as the one found in academic science. As Katz and Kahn among many other social psychologists have pointed out, a salient feature of most organizations is the high level of structured ambiguity about the basis on which individual performance is being evaluated.[59] The question arises whether there exist clear bases of rewards in these organizations other than seniority. Are there any formal and well-articulated bases for rewards and promotions in most organizations? Long ago, Bales and others suggested that the prerequisite for the effective functioning of any organization was balance between instrumental leaders (i.e., individuals who suggest new ways of solving problems) and expressive leaders (i.e., individuals who serve integrative functions—who smooth out social relations between contending parties).[60] It is far easier to develop indicators of instrumental than of expressive contributions.

Where there is ambiguity over definitions of high-quality role performance, there is also apt to be the wholesale importation of functionally irrelevant criteria in the reward process. It is precisely in these situations that affirmative action can influence organizations to make explicit the parameters of their evaluation systems. Of course, not every aspect of the reward process can be articulated, but without some conception of these criteria, an estimate of the actual state of affairs within an organization is virtually impossible.

Affirmative Action as Part of Larger Social Processes

Affirmative action policy and its outcomes are often viewed statically, resulting in a simplified conception of the social problem to which the policy is addressed. To understand the consequences of affirmative action, it must be related to social processes, particularly that of accumulative advantage and disadvantage.

Accumulative Advantage and Disadvantage   One problem of using affirmative action as a means to reduce gender-based discrimination is that the policy is often applied when the race has already been run. Typical cases in academia come at the time of tenure decisions; in industry, at the time of promotion to supervisory positions—at T_2, if you will. That is, the qualifications of male and female candidates are considered at T_2. Frequently, "fair" decisions are made, which

---

59. For a discussion of the effects of ambiguity of evaluation on job satisfaction, see Daniel Katz and Robert L. Kahn. *The Social Psychology of Organizations* (New York: John Wiley & Sons, 1966).

60. Robert F. Bales, *Interaction Process Analysis, A Method for the Study of Small Groups* (Cambridge, Mass.: Addison-Wesley Press, 1950).

are defensible in court or elsewhere, that favor a disproportionate number of "majority males."

On the basis of universalistic performance criteria, solid evidence may support these decisions. What is not considered is that T_2 qualifications represent the end-point in a lengthy process that may start to take shape early on in life.

Analysis of the underrepresentation of women in science must focus eventually on the cultural forces that lead women to select themselves out of science careers. If women are treated differentially from men at time T_1 then the probability of their performing as well as men between T_1 and T_2 shifts dramatically. We have to consider social barriers to female achievement that make their virtual absence in certain occupations and their low standing within others all but inevitable. It should be remembered that the obstacles that confront women may have an additive or possibly a multiplicative effect that cumulates over time.

Until very recently, women have been told throughout much of their schooling that they should not compete with men and that some occupations are not appropriate for them. All these forces have led to a self-selective process of women out of certain occupations, and further contributed to the reinforcement of sex-labeling of occupations.[61] Thus, from the earliest point in their lives women have faced a set of social conditions that differ from those confronting men. They have started with greater disadvantages in pursuing a career than do males, and these disadvantages have probably accumulated throughout their childhood and adolescence.

Although it seems intuitively obvious that women suffer from accumulating disadvantage throughout their careers, the particularities of this process are largely uninvestigated. We know very little about specific points in the life histories of women that make more or less of a difference in distinguishing their aspirations from those of men; we know next to nothing about key pressure points in these histories; we know almost nothing about the ways that cultural pressures interact with individual traits, such as talent, to influence specific occupational choices. We have hypotheses. We need adequate data and analysis.

The possible effects of accumulative disadvantage on women's careers in science are directly related to affirmative action policy. If women are disadvantaged in terms of personal mobility, in access to facilities and resources, in diminished motivation resulting from cultural expectations, then the bases on which rational decisions are made at subsequent points in time are bound to be affected. By ten years after the doctorate a comparison of the careers of two similarly able individuals may not involve a real comparison at all: One is clearly "superior" to the other. While advantages accrue for some, disadvantages accumulate for others—dampening motivation, and opportunity to perform well and to gain concomitant rewards. This process can work to the advantage of the

---

61. Valerie K. Oppenheimer, *The Female Labor Force in the United States: Demographic and Economic Factors Governing Its Growth and Changing Composition*, Population Monograph Series no. 5 (Berkeley: University of California, 1970).

larger social system only if one precondition is met—that those individuals who are initially labeled as potential star performers, or simply as being able, are in fact more talented. If this is so, the efficacy of the resource and reward system is optimal. However, if the basis for initial social and self-selection is based on criteria that are irrelevant to quality of work, then the reward process operates to maximize differences between individuals with similar initial ability. It is important to see that this process need not involve conscious discriminatory behavior at any point in time. Individuals need not be capricious for accumulating disadvantages to obtain.

Failure to see affirmative action as part of larger social processes plainly increases the chances of making empirical errors. Take only one example of the pitfalls of a static conception. Until recently empirical studies that attempted to explain academic salary differences between men and women have controlled for academic rank, among several other independent variables. This seemed logical, since it would hardly be appropriate to compare the salaries of male assistant professors with female full‐professors. It should not be surprising that only limited differences are found for such rank-specific comparisons. The practice of holding academic rank constant actually underestimates the effect of sex status on salaries, because it fails to allow for the possibility that women are discriminated against in terms of rank. Two forms of gender discrimination can simultaneously affect the salaries of women: the direct effect of discrimination within a given academic rank and the indirect effect of gender on academic rank that in turn affects salaries. To correctly estimate salary discrimination, both the direct and indirect cost of being a woman must be calculated. This two-stage process is illustrated below:

FIGURE 7-1

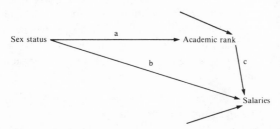

Since we do have solid evidence that there is some measure of gender discrimination in promotion to high academic rank, this angle of vision on salary discrimination becomes essential. Actually salary differences, even without the necessary controls for performance, should include the value of the direct effect, b, plus the product of the indirect effect, a and c. Without the idea of social process and of accumulative disadvantage, the model for gender-based salary differences is incomplete.

The social standing for women of science is probably determined within the first ten years after their entry into the labor market. Due to widely standardized up-or-out promotion rules in academia, differential treatment within these years is probably more apt to influence the overall pattern of careers than would the same span for women in other occupations, although there is a paucity of data on this point. Are there critical segments in the careers of women within other occupations? What are the accumulating effects of self-selection processes? If, for example, a woman chooses to enter a second-rate graduate department when she could have gone to a superior school, how do such early decisions influence later opportunities for mobility? Are women of varying ages as motivated as men to advance in their careers? How extensive is the "will to fail"? Empirical answers to these queries are needed if we are to estimate the impact of affirmative action policy.

Specifying Conditions Under Which Discrimination Obtains

I have noted the problem of identifying the conditions under which discrimination can be demonstrated. Knowledge of the conditions under which women are likely to experience salary, appointment, and promotion discrimination should inform affirmative action policy. For academic settings, the evidence presented in this book suggests that there are different effects of gender on different forms of recognition: There is little evidence of discrimination in hiring, but considerably more supporting evidence for discrimination in promotion and salaries. The clear signal in terms of affirmative action policy is that maximum attention and efforts at redressing grievances should be focused on the issue of promotions rather than on initial appointments.

In Chapter 3, I argued that when a college science department is faced with promoting one or another of two thoroughly unproductive individuals, one a man, the other a woman, there is a tendency to import sex status into the situation and use it in the promotion decision. Whether affirmative action policy has changed this pattern is unknown. Affirmative action should be based on models that identify the conditional probabilities of gender-based discrimination. At this point they rarely are.

The Influence of Exogenous Variables on Affirmative Action

The basic problem in measuring the effectiveness of affirmative action programs is that exogenous variables, particularly macroeconomic forces, operate to constrain the effects of interventions in particular organizations or institutions. Furthermore, since we are dealing with social interventions that cannot by their nature involve controls such as those found as part of clinical trials, it is exceedingly difficult to isolate the effects of these exogenous factors in relation to the interventions.

If pressure for affirmative action had first been applied in the expansionary period of the early 1960s, larger short-term gains for women in science probably could have been realized. The size of academic-science departments in colleges and universities was growing, and the level of federal funding of research and development rose sharply. Expansion of the job market was equally notable in industry. The 1970s has, thus far, been a period of general economic retrenchment, clearly indicated by simple supply-and-demand functions within American universities and colleges. Faced with large budget deficits, expansion has stopped in most cases and the size of faculties and administrative staffs has been reduced. For affirmative action these conditions represent serious impediments to change, even where good faith efforts are being made. Since most affirmative action plans are formulated in terms of newly hired personnel and do not require current overrepresentation to adjust for past "deficiencies," progress toward an occupational structure devoid of sex bias is apt to proceed slowly.[62] If there are virtually no jobs to be had in academic-science disciplines for either men or women, there clearly is no room for producing significant change in the overall sex composition of these fields. Furthermore, since there are even fewer tenure-level positions available, the movement toward an equitable distribution of men and women where it is more required moves at an even slower pace.

The economy not only affects affirmative action in terms of new appointments and promotions but also in terms of terminations. Since women and minorities comprise a high proportion of recently hired personnel, these two groups tend to be victimized to a greater extent than earlier appointments by economic retrenchment. Wherever a formal or informal seniority system determines employment terminations, those groups who have made the most significant recent gains will suffer the greatest immediate losses. Thus, when the proportion of junior faculty is cut in half, women and minorities are the most likely to suffer, and movement toward affirmative action goals is impeded. The recent AT&T case, in which the consent decree circumvents union seniority rules when they conflict with the implementation of affirmative action goals or quotas, represents one of the most significant developments for the future of affirmative action programs, although this decision does not apply to the academic-science community.

Court decisions in specific affirmative action cases, in conjunction with changing cultural values, should affect the formulation of affirmative action plans and compliance with decisions. It remains an open question whether or not there has been fundamental change as a result of affirmative action in attitudes toward admission of women into professional schools, and to the hiring and promotion of women once they are in the professions. Some observers claim that without court decisions on basic affirmative-action procedures there will be no compliance with the spirit of affirmative action within universities, professional schools, and industry. However, it is possible that within some environments

---

62. See, as examples of affirmative action goal formation, the plans of Columbia University and the University of California, Berkeley.

there has been basic attitudinal and value changes that will not be significantly influenced by the Supreme Court's decisions. For example, even with the Court's decision in *Bakke* there may not actually be a reduction in the proportion of minority admissions to medical schools. If the value of having a certain proportion of a medical school class composed of minority group members is strongly felt by both faculty and administration, the admissions committees will find methods for achieving these goals—in precisely the same ways that colleges have found methods for obtaining "geographical balance," a certain number of student athletes, and places for sons and daughters of alumni.

It might be argued, however, that without a direct and clear set of standards many employers and institutions of higher education, which have not participated fully in the change in values, will not comply with affirmative action directives. The extent to which we are witnessing fundamental changes in social values that will have consequences similar to those intended by affirmative action policy remains unmeasured but worth examination.

In sum, these exogenous forces, and many others that could be discussed, will interact and influence the development of affirmative action policy.

Implementing Affirmative Action Plans in Different Types of Organizations

Although public attention has been focused principally on affirmative action plans within universities, the development and implementation of these plans for industrial organizations probably will have a greater effect on the sex composition and rewards structure of industry than the universities' plans have on them. Yet almost nothing is known about the range of problems associated with implementation in industrial organizations. Compliance with affirmative action directives has yet to be studied systematically.[63]

Studies of compliance behavior should compare patterns within several sectors of society: universities, industry, labor unions. Particular attention should be paid to forms of institutional resistance to affirmative action.[64] Consider a list of potential sources of resistance.

Time-budget problems: When scientists in university departments or in industry spend time administering affirmative action plans, they are not teaching, doing research, or fulfilling other functions which they define subjectively as fundamental to their social status. Time-budget role conflict emerges; resistance to compliance may develop.[65] Resistance is often difficult to detect, because mechanisms are developed to produce the appearance of an exhaustive search. Institutionalized methods of evasion often become accepted practice.

---

63. Of course, developing an adequate research design to study various modes of compliance is full of its own methodological problems.

64. Although his work on resistance focuses on the history of science, see Bernard Barber, "Resistance by Scientists to Scientific Discovery," *Science* 134 (Sept. 1961): 596–602.

65. Robert K. Merton, *Social Theory and Social Structure* (New York: Free Press, 1968); see especially the chapters on role conflict and its resolution.

Conflicting organizational priorities: Organizations have competing objectives. Increasing the proportions of women within an organization is only one of these. Within universities, specific departments may have legitimate objectives that conflict with affirmative action goals. In biology, for example, roughly 16 percent of all Ph.D.'s are women, but a significantly smaller proportion specialize in molecular biology. It is often difficult to find highly qualified women in these specialties. To what extent should departments alter their substantive needs to meet affirmative action goals?

Resistance to compliance resulting from conflicting values: Some individuals perceive affirmative action principles as conflicting with longstanding attitudes about meritocracy and academic freedom.[66] These people contend that affirmative action is a camouflaged quota system and a system that may result in reverse discrimination. The consequence may be resistance.

Methodology: Individuals who in principle believe in affirmative action may disagree with the methods of defining pools, of defining underutilization and apparent discrimination, of setting specific timetables to achieve proportionate increases of women and minorities. These people may resist implementation because they believe the methodology of the policy defeats its ideals.

Personal interests: Some social groups see affirmative action as a direct threat to their own interests. There is concern among recent male Ph.D.'s about job opportunities and access to training for certain occupations. However, the extent to which able men are self-selecting themselves out of certain scientific occupations because of the perceptions that jobs for men are not available remains unknown.

Costs: Resistance can also obtain because of the costs of compliance. The direct cost of building an effective plan that is acceptable to the Department of Labor or to HEW may run into hundreds of thousands of dollars for large organizations. The creation of organizational structures to handle affirmative action problems entails substantial costs. Furthermore, affirmative action plans usually create another bureaucratic layer in organizations that are already burdened by bureaucracy. Once in place, the new structure of review reduces the speed with which decisions can be made.

Individual vs. Group Justice: Equality of Opportunity and Equality of Outcomes

Finally, I want to consider briefly two competing principles of distributive justice that are related to affirmative action and are the foci of controversy in contemporary American society: the principles of equality of opportunity and equality of outcomes. It has been argued that affirmative action in higher education and industry involves the practical application of the principles of equality of outcomes for social groups.[67] It is unnecessary to discuss the definitions or distinc-

---

66. Amitai Etzioni, *A Comparative Analysis of Complex Organizations* (New York: Free Press, 1961).

67. For one discussion that attacks the equality of outcomes argument, see Glazer, *op. cit.*

tions between the two ideas, since there are by now other extensive and excellent discussions of these differences.[68] I will note here only how the problems of affirmative action fit into this more global question of what principle should govern the distribution of social and economic rewards.

Before doing this, a distinction must be sharpened between the content of the norms of affirmative action and the regulatory feature of the norms of affirmative action. Some of the confusion about this new policy results from the conflation of these two fundamentally different dimensions of affirmative action policy. As a set of regulatory normative procedures affirmative action represents a powerful control on gate-keepers who are responsible for hiring and promotion decisions, but it has little to do with prescribed social outcomes. It simply normatively prescribes a methodology that an employer must follow before making personnel appointments or promotions. And importantly, it insists upon a written record of compliance with these rules. If proper procedures are demonstrable, the outcome, whatever its content, is judged equitable. If affirmative action were limited to this regulatory function, there undoubtedly would be fewer complaints about its ideological implications. More importantly, if it were only a set of procedures to follow, it would not be considered in some quarters a radically new method for changing the social composition of the occupational structure of American society. As a set of procedures affirmative action is wholly consistent with the older idea of equality of opportunity, which highlights selection among competitors without regard to their status characteristics. For the policy to offer a radical alternative, a normative content had to evolve. The policy was given a normative content, whether or not by conscious design, by those responsible for overseeing its implementation.

Ironically, the content grew out of the methodology itself. Affirmative action, by assuming that social outcomes will tell whether the regulatory norms and procedures have been followed, constructs its own content. And the position that is implicit in the policy is closely akin to the value position represented in the principle of the equality of social outcomes. Social outcomes in hiring, promotions, or salaries that show a disproportion of one group to another relative to their simple availability are taken as prima facie evidence of discrimination, and a fortiori imply that the proper methods or procedures were not followed. Thus, in terms of its content, affirmative action tends to create a grand tautology, based upon a possibly false assumption.

Affirmative action as we know it has its origins in the 1964 Civil Rights Act, in Title VII of that statute, in the Executive Orders that elaborated its sphere of influence, and in the EEOC guidelines for carrying out the policy. Affirmative action litigation that involves constitutional issues constitutes debates over the criteria that courts, particularly the Supreme Court, should use in interpreting the Equal Protection Clause of the Fourteenth Amendment.

Until recently, the Equal Protection Clause has been treated as if it contained

---

68. Charles Frankel, "Equality of Opportunity," *Ethics* 81 (April 1971): 191–211; Glazer, *op. cit.*; Fiss, *op. cit.*

an explicit "antidiscrimination principle"—which in fact it does not. As Paul
Brest, of the Stanford Law School, has noted:

> The roots of the antidiscrimination principle lie in the observation that race, ethnic
> origin, sex, and like characteristics are morally neutral, that is, they are not
> grounds for moral praise or blame. For this reason, it is unjust to impose burdens
> or confer benefit on people for the ultimate reason of their race, or to allow the
> supposed moral superiority or inferiority of a social group to be the basis for
> decision. . . . Under the antidiscrimination principle, it is presumptively improper
> to use race as a ground for treating people differentially even when it is used for a
> morally neutral purpose, e.g., as a proxy for some legitimate decision-making
> criterion that may correlate with race.[69]

Recent judicial and administrative reinterpretations of the antidiscrimination
principle lie at the heart of the national affirmative action debate. The debates
over goals versus quotas, over individual versus group justice, are specific conflicts in this larger debate, and the outcomes have multiple practical consequences for the distribution of individual rewards and for the pattern of rewards
among social groups.

Of course, the Fourteenth Amendment was intended to protect blacks against
invidious discrimination. In terms of intent, it had nothing to do with protection
of women's rights. But in the last decade the Equal Protection Clause has been
used by many disadvantaged groups as a constitutional pillar of protection
against discrimination. Women's groups have joined forces with minorities in
common purpose to ensure change in the interpretation of the clause when
affirmative action is at issue.

The essence of the earlier interpretation of the Equal Protection Clause was
its reliance on old principles of individual rights and freedoms that were independent of a social accounting scheme of historical disadvantages. Case-by-case
examinations of the presence or absence of discrimination was the rule; status
characteristics as a selection device or organizing principle could be used only
when a compelling state interest was involved. Examination of the history of
interpretations of the Fourteenth Amendment reveals only one instance when the
Supreme Court acceded to the compelling state interest argument—in the
Korematsu case concerning Japanese relocation camps.[70]

Why is the controversy surrounding the principles of equal opportunity and
equal outcome so important to identifiable social groups in the United States?
The answer lies in the impact that acceptance of one or the other of these
principles of distributive justice would have on the social composition of the

---

69. Paul Brest, "Justice and the Cultural Division of Labor in the United States." Preprint of a paper delivered at
Stanford University Works in Progress Seminar, 1975. I have benefited a great deal from several discussions of this
general area with Professor Brest and from his course in constitutional law offered at the Stanford University Law
School, which I attended in 1975 while at the Center for Advanced Study in the Behavioral Sciences.

70. *Korematsu* v. *United States*, 323 U.S. 214, 216 (1944). Some legal scholars might claim that an exception to
this can be found in *Swain* v. *Alabama*, 380 U.S. 202 (1965). In this case the Supreme Court seemed to grant
permission to the prosecution to exclude black jurors for discriminatory purposes through the use of preemptory
challenge. For a discussion of this, see Robert M. O'Neil, "Preferential Admissions: Equalizing the Access of
Minority Groups to Higher Education," *Yale Law Journal* 80, no. 4 (March 1971): 699–767, esp. 708.

labor force. The older principle, of so-called equality of opportunity, was believed for years to be founded on the transcendent idea that what a person does is more significant than who he is; that achievement was more important than ascription; that universalistic, meritocratic criteria of rewards should predominate over particularistic criteria. While this principle of equal opportunity to succeed or fail seemed to be embodied in constitutionally guaranteed rights, it was abridged, of course, continuously in the actual treatment of minorities and women. But the equality of opportunity principle never had any built-in assumption that there existed equal talent in the population. Quite the contrary, it seemed to supply strong evidence that there was not an equal distribution of talent, motivation, or skills among all social groups in the population. The important point was not that there is inequality of talent among individuals—no one would deny that—but that the inequality is not supposed to be socially patterned.

The equal-opportunity, antidiscrimination principle was made operational by use of hiring and promotion methods that were believed to be as meritocratic or universalistic as any that could be devised—the use of standardized examinations. This method was not applicable to many sectors of the American economy, but it became widely used in several important domains, particularly those involving public service employment and those involving access to higher learning and professional training. In fact, the institutionalization of the equality of opportunity principle can be seen perhaps most clearly in the construction and application of so-called standardized tests of aptitude, achievement, potential, knowledge, etc. Civil service examinations personify in concrete form the application of the abstract principle in the marketplace. These exams are designed supposedly to maximize universalism and minimize the utilization of irrelevant social characteristics in appointment and promotion to jobs. Jobs are allocated largely on the basis of test scores (with the use occasionally of personal interviews). Those individuals who score highest on the test are first to be appointed. Those lower on the list are appointed later if jobs are still available. Such apparently meritocratic procedures have led, however, to the unequal social distribution of economic, educational, and occupational rewards.

Over the past-quarter century we have witnessed an extraordinary expansion in the reliance on these tests. Institutions of higher learning and professional training, as examples, use a variety of standardized testing instruments that weigh heavily in the competition among applicants. The standardized tests were supposed to minimize invidious discrimination because of their impersonal character. The problem was that these gate-keeping mechanisms produced very different outcomes for different social groups. American Jews and Chinese-Americans, two minorities with long histories of suffering from ethnic discrimination in the United States, have benefited remarkably by the use of the impersonal testing criteria because they have as groups achieved comparatively high scores on the examinations. The use of the testing criteria has been one important factor, among many others, for the disproportionately high number of Jewish and Chinese-Americans in high-prestige and high-income occupations—positions in many cases that were previously closed to them because of overt or invidious discrimination. But not all minorities have benefited by the older rules, and this has been particularly true for black Americans.

Standardized tests are now under attack as reinforcers of inequality of opportunity, both inside and outside the courts.[71] Arguments pro and con are familiar to us all. Almost all the arguments typically involve concept and indicator problems. Advocates of testing do not claim that tests are perfect indicators of potential or current skills, but claim that they remain better, less expensive, and less open to bias than most other evaluation mechanisms. Those in opposition find cultural biases built into the indicators, and claim that in many cases these tests are unrelated to on-the-job performance. They point to the modest or weak correlation between test scores and later performance as evidence for restricting the widespread use of tests in decision-making processes. The test validation question is of great importance in current Title VII cases.[72] To some who make the cultural bias argument, the perfect test would be one that by design eliminated all variance between social groups, and allowed for variations only within the entire population which were uncorrelated with ethnic, racial, religious, or sex characteristics. What such a set of indicators of performance or potential would be like is difficult to imagine. In short, the application of the principle of equality of opportunity in the form of examinations produces skewed social distributions within occupations and by definition "deficiencies," "underutilization," or "discrimination."

While the testing criterion is the most visible one under attack, the rules governing the selection game more generally are also under attack, because, it is claimed, they are ineffective in producing desirable social change. Some legal scholars now argue that if the rules of the mobility game remain the same—that is, continue to be based largely on the use of the results of standardized tests—there is virtually no chance to make significant gains in redistributing American resources to members of heretofore discriminated-against minorities and women.[73] The argument is now made that the Equal Protection Clause, which, as we noted, has traditionally focused on individual rights, must be somewhat reinterpreted to take into account the effects of decisions on specific disadvantaged groups. These are the group justice advocates. Their basic position, articulated very clearly by individuals such as Owen M. Fiss or Ronald Dworkin, is that we should be concerned with the impact that affirmative-action or fair-employment legal decisions have on social groups, above and beyond their impact on individuals.[74] And the argument goes on to say that if racial or other classifications are necessary to achieve greater distributive justice between social

---

71. For more extended discussion and references, see the section of this chapter on the problems of testing and admissions to medical schools and the cases discussed in Chapter 2. Test validation is a major feature of Title VII cases today, and standards are still evolving.

72. Among others, see George Cooper, Harriet Rabb, and Howard J. Rubin, *Fair Employment Litigation* (St. Paul, Minn.: West Publishing Co., 1975); Barbara Allen Babcock, Anne E. Freedman, Eleanor Holmes Norton, and Susan C. Ross, *Sex Discrimination and the Law* (Boston: Little, Brown & Co., 1975); Davidson, Ginsburg, and Kay, *op. cit.*.

73. This is the position argued in the Columbia University et al. *amici curiae* brief filed in the *Bakke* case.

74. Fiss. *op. cit.*; Dworkin, *op. cit.*

groups, then the interpretation of the relevant statuses or clauses of the constitution should be revised.

The indicator of group justice is frequently the proportional representation of a particular group in an occupational or wage category. When the proportions are consistent with those in the pools or in the population, group justice is indicated; when they are at variance (unless there is overrepresentation of the disadvantaged group), there is an indication of the absence of justice.

The proponents of the group justice principle argue that the Equal Protection Clause must be reinterpreted to protect individuals who are members of socially disadvantaged groups. Owen M. Fiss expresses this succinctly:

> I would therefore argue that blacks should be viewed as having three characteristics that are relevant in the formulation of equal protection theory: (a) they are a social group; (b) the group has been in a position of perpetual subordination; and (c) the political power of the group is severely circumscribed. Blacks are what might be called a specially disadvantaged group, and I would view the Equal Protection Clause as a protection for such groups. Blacks are the prototype of the protected group, but they are not the only group entitled to protection.[75]

The argument is made that distributive justice can only be achieved through programs of preferential hiring, promotions, admissions, and so on. Advocates for this theory of equal protection doctrine argue that even if some form of preferential treatment for disadvantaged groups involves what is called noninvidious reverse discrimination, without the preferential treatment the larger societal goals of greater social equality cannot be achieved.

Concretely, advocacy of this position frequently involves the development of new standards for hiring and promotion, which are not only color- or gender-conscious but also involve separate classification on the basis of these statuses. It is easy to see why some social groups would be vehemently opposed to these new rules.

Preferential systems of hiring and advancement within occupations involve a complex set of relationships between past injury (in the form of prejudice and discrimination) and current remedies. As Brest has noted:

> Preferential hiring may be designated: (a) to compensate an applicant who was personally discriminated against by the employer who now must prefer him; (b) to compensate the applicant who was personally discriminated against by other employers and institutions; (c) to compensate the applicant who is the indirect maleficiary of discrimination against other members of his social group; (d) indirectly to compensate other members of the applicant's social group who are the direct or indirect maleficiaries of discrimination.[76]

Only the first of these alternatives is covered by the traditional interpretation of the antidiscrimination principle read into the Equal Protection Clause. The three others represent an explicit use of a status category not directly related to an

---

75. Fiss, *op. cit.*, p. 96.
76. Brest, *op. cit.*, pp. 19–20.

individual case of discrimination but related to remedies for past behavior toward the individual or toward the group of which he or she is a member. What we do not know is the extent to which the beneficiaries of "group justice" are individuals who have experienced direct discrimination, past or present.

This takes us to a particularly thorny aspect of this new line of analysis. It focuses more on social results for different groups and less on what might be termed "errors" in individual decisions. It is paramountly concerned with errors of underinclusion rather than overinclusion; that is, it will allow some members of historically disadvantaged groups to be given special consideration through the use of special classifications even if their personal histories do not warrant it, rather than possibly exclude members of these groups who do warrant special consideration. Furthermore, the longer-term social benefits are claimed to outweigh the limited cost of noninvidious discrimination against some individuals who are excluded from opportunities as a result of the special classifications. The limiting case would go something like this: Minority group members who have attended middle-class elementary and secondary schools, who have attended elite colleges, still should be specially classified by entrance committees to professional schools and be given special preference over majority candidates who come from lower-class backgrounds with poorer educational training. Ironically, truly economically and educationally disadvantaged minority group individuals within the special classification may be at a competitive disadvantage in the mobility process because the special positions are taken by the more advantaged minority group individuals who are also placed in a special applicant category.

One of the difficulties with the group justice position lies in the implicit assumption of greater homogeneity of experience *within* social groups than actually exists and the corresponding assumption of greater heterogeneity of experience *between* social groups than really exists. Until quite recently, the social and economic life chances of minority groups and women were so different from their racial or gender counterparts as to make their opportunities virtually mutually exclusive. This is no longer the case. The question is whether or not the assumptions of the group justice principle are too costly to those men and women of the nonpreferred classes who have experienced deprivation or who are extraordinary candidates for positions; to those members of the preferred groups who have experienced significant deprivations and lose out in admissions, hiring, and promotion decisions to other members of the same group who have not experienced substantial deprivation.

CONCLUSION

From this large array of problematics you might infer that I do not support the concept of affirmative action. On the contrary, I do believe that affirmative action, if properly defined and applied, can be of distinct, if limited, use in improving the opportunities of women in the scientific community. The problem for me lies in the fact that "affirmative action" is a blanket concept and in its application is easily distorted and abused. In summing up, let me suggest both the ways that affirmative action can be an important instrument for broadening opportunities for women in science and its limitations.

First of all, my position on the benefits of affirmative action is similar to the one held by Ruth Bader Ginsburg of the Columbia University Law School. For lack of better formulation I will quote at length from her recent articulation of these views.

As in the case of discrimination against racial and ethnic minorities, the ultimate goal with respect to sex-based discrimination should be a system of genuine neutrality. Movement in that direction, however, requires remedies necessary and proper to alter deeply entrenched discriminatory patterns. But changing those patterns entails recognition that the generators of race and sex discrimination are often different. Neither ghettoized minorities nor women are well served by lumping their problems in the economic sector together for all purposes.

With respect to race, the effects of officially sanctioned segregation are still very much with us. Doctrine directed to the continuing impact of race segregation is not necessarily applicable to gender discrimination. . . . The difference is perhaps best illustrated by reference to educational experience. Females have not been impeded to the extent ghettoized minorities have been by lingering effects of officially sanctioned segregation in education, housing and community life. Females have been significantly disadvantaged by some aspects of educational programs, vocational and athletic training most conspiciously. . . . But on the whole, girls tend to do at least as well as boys in elementary and high school academic programs. For many females, the record of achievement continues in higher education. For example, since 1971, women have outscored men on aptitude tests for the study of law. . . . In short, most non-minority females do not encounter a formidable risk of "death at an early age." . . .

The problem growing up female is that from the nursery on, an attitude is instilled insidiously. . . . To cure the problem . . . the overriding objective must be an end to role delineation by gender, and in its place, conduct at every school level, later in the job market, signalling that in all fields of endeavor females are welcomed as enthusiastically as males are.

To achieve that objective "affirmative action" is needed. . . . Pursued with intelligence and good faith, affirmative action should ultimately yield neither a pattern of "reverse discrimination" nor abandonment of the merit principle. On the contrary, it should operate to assure more rational utilization of human resources. . . .

In some settings, redressing gender discrimination can be accomplished effectively simply by altering recruitment patterns and eliminating institutional practices that limit or discourage female participation. This seems to be the case with respect to educational opportunity. Indeed, numerical approaches may operate to the disadvantage of females in academic programs. . . . Non-numerical affirmative action seems the course appropriate to the challenge of female enrollments in law school, for example. . . .

In the labor market, abandonment of standards and tests not reliably related to job performance will sometimes prove sufficient remedial action. For example, courts have invalidated arbitrary height-weight specifications that operated to screen out women and members of certain ethnic groups fully capable of quality performance on the job. . . . Catching-up training programs and abbreviated waiting periods before promotions are remedies particularly appropriate when artificial barriers have blocked the way to equal opportunity....

Finally, there are situations in which numerical relief is "the most feasible mechanism for defining with clarity the obligation to move employment practices in the direction of true neutrality." [*United States* v. *Lee Way Motor Freight*, 7 EPD 9066, at 6500 (W.D. Okla. 1973)]....

In academia too, goals and timetables seem an essential part of the initial program necessary to accomplish equal opportunity....

In sum, transition period affirmative action, tailored to the particular setting, far from compromising the equality principle, is an essential part of a program designed to realize that principle.[77]

Ginsburg touches on many of the possibilities of effective use of affirmative action policy. For women in science, more specific conclusions can be reached about the potential benefits of affirmative action.

Affirmative action acts as a very necessary control on the basic gate-keeping structure in the scientific community. It provides needed constraints on the conduct of job searches and promotions within that community, despite the attempts that are often made to evade its procedural requirements. It increases the accountability of departments and industrial-science organizations who must comply with the legislation. Since noncompliance can result in expensive litigation and the withholding of government contracts, there is an incentive for organizations to make a good-faith effort to comply with the procedural norms of affirmative action. This certainly will not entirely solve the problem of qualified women being overlooked or not approached for high-ranking jobs, but it should have a nontrivial influence and it may over the longer haul even contribute to changes in values toward women in science held by many of the gate-keepers. Sharp discrepancies in salaries will be more difficult to justify when they are open to review through an organized complaint system. Equal employment benefits are apt to be achieved under threat of litigation. I think that we are already seeing the positive effects of affirmative action policy in some of these areas.

Affirmative action can be of value if the muscle behind it is applied selectively to those pressure points that really require pressure because of strong evidence of injustice. In academic science, this seems to be at the time of promotion to tenure and high rank. The application of pressure at the first job stage, beyond requiring adherence to the procedural norms, simply does not require a quota system. There is little evidence of inequity in the initial job location of men and women Ph.D.'s. Properly applied pressure at the critical tenure stage of the scientific career should improve the chances of women with publications records that are just as distinguished as those of the men who are also "up" for promotion. In short, if adequate controls for scientific role-performance can be part of affirmative action models, they can become very good devices for reducing gender bias. In such situations, the policy can stimu-

77. Ruth Bader Ginsburg, "Gender and the Constitution." *University of Cincinnati Law Review* 44, no. 1 (1975): 1–42; quoted from pp. 29–34.

late further the movement of women of science from the margins to the center of the community.

Although this must remain conjecture, affirmative action may influence indirectly the perceptions of young women about career opportunities in science. Perhaps the greatest potential indirect effect of this policy would be for it to help alter the sex stereotyping of occupations, such that girls in elementary and secondary schools begin to view science as a viable career. In the final analysis, the real battle for improving the situation for women in science will be fought out before women ever reach the borders of the scientific community. Programs that will change the perceptions of young girls about the barriers to scientific careers will be valuable. Affirmative action as a method of assuring opportunities for girls who are gifted in science may have some diffuse influence on the selection of science among career-oriented girls.

Plainly, I am less sanguine about other aspects of affirmative action policy as it frequently operates. Quotas or numerical goals with strict timetables for compliance seem unwarranted for the scientific community for several reasons. First, it is quite possible that the basic objectives of affirmative action can be reached for women in science without the application of quotas. Second, the setting of quotas or numerical goals is often done simply because they will "fly in Washington," not because they are estimates based upon reasonably sophisticated models of gender discrimination in the scientific community. Most affirmative action plans are based upon models that so truncate reality that their value becomes suspect. Plans without any assessment of performance, either subjective or objective, other than length of service can be of limited value only. The strict application of the proportional representation rule not only does an injustice to qualified women but also produces conditions that can result in increased rather than reduced amounts of gender discrimination. In short, when many affirmative action plans are constructed they often are less concerned with the facts, or the actual standing of women in university or industrial organizations, than with whether the plan will "wash" with government civil rights officials. Thus, the plans frequently include assumptions of fact that are anything but clearly established.

The assumption that there can be no patterned relationship between role performance and status characteristics that is not a product of prior discrimination seems to me totally untenable. Moreover, the explanation for differing patterns of scientific productivity and quality of work is rarely conceived of as a product of social forces lying just as often outside the scientific community as inside it. The processes that lead to differences in scientific role-performance are not viewed longitudinally—as evolving over time as a result of the interplay between social and cultural forces that influence the performance of men and women.

Furthermore, there tends to be a lack of attention to the "choreography of scientific careers." The absence of an extended temporal perspective frequently does not allow observers to correctly understand how steps in the career at one point in time become conditions for possible steps in a later period. Careers are choreographed by both the performers and the audience. Affirmative action plans often underestimate levels of gender bias by not taking a larger perspective. Thus, as has been pointed out repeatedly, if there is gender discrimination in promotion, then controls on rank when looking at salary differences will under-estimate the differences.

There are several other concluding observations. I believe that the historical and contemporary conditions that face women and racial minorities, particularly black Americans, are so different that viewing the two aspects of affirmative action as one is only counterproductive. As Ruth Ginsburg suggests, the two groups simply do not face similar cultural types of deprivation, nor are the opportunity structures open to them in education really similar. Thus, while the *Bakke* case may suggest general problems of establishing "facts" in affir-mative action cases, it seems to me a mistake for women and racial minorities to hitch themselves to the same wagon. In fact, if gender discrimination had been at issue in the *Bakke* case, the arguments unquestionably would have had to be vastly different from those presented to the Supreme Court. Although both groups have a stake in the constitutionality of affirmative action, the problematics are so different that a great deal is lost by conflating the affirmative action problems faced by minorities and women.

In this connection, it seems to me that gender discrimination deserves "strict scrutiny" by the courts, and it is unfortunate that it has not yet been given that status by the Supreme Court. But strict scrutiny does not necessarily require accepting the principle of gender or racial classifications or the principle of group justice. The absence of greater homogeneity of conditions faced by women and greater heterogeneity of conditions faced by both men and women seems to suggest that acceptance of gender classification in order to achieve national goals must not be done without the strongest possible evidence that a compelling state interest exists and that no alternative means exist to achieve the desired social goals.

Finally, perhaps the greatest constraint on the positive influences of affirma-tive action lies outside of the scientific community altogether. General societal conditions and more specifically the economics of the scientific community can limit the potential gains made by women, regardless of the presence or absence of affirmative action. In recent years, constriction of the labor market for scien-tists, male or female, has been placing a limit on the amount of progress that can be made in the short run. Ironically, in the academic-science community the proportion of females entering the job market after the Ph.D. has increased dramatically since 1970. Is this a result of a decline in bias or of affirmative action? Perhaps in part, but the increased proportion of females to the total number of Ph.D.'s might just as easily mean that the perception of a diminished job market has had greater impact on the career choices of men than on those of women. The increased proportion of female Ph.D.'s could just as easily result from a disproportionate number of men selecting themselves out of this particular market as from a reduction in bias about admitting female graduate students or an increase in the number of women who want to pursue scientific careers.

In the end, I have to emphasize again that although there do appear to be pockets of patterned sex-based discrimination in the academic-science community (most notably in the effects of gender historically and in the present-day on promotion of women to high rank), to an extraordinary degree the scientific community distributes its resources and rewards in an equitable fashion. At the beginning of this inquiry I thought that an anomalous situation might exist in the community's treatment of women. But there is only limited evidence to support this view for most aspects of the scientific reward system. Perhaps matters are quite different in industrial science, and surely we must emphasize rather than gloss over those conditions in academic science that do lead to differential treatment of women, but the evidence presented in this book suggests that remedial action in the form of affirmative action quotas is not needed for achieving still greater equality between the sexes in science. Strict adherence to the fundamental procedural norms of affirmative action and a willingness among the maleficiaries of discrimination to litigate their cases when these norms are abridged should prove sufficient remedial action for assuring that existing opportunities for women in science are maintained and that greater opportunities for full participation among the elite of science are achieved.

APPENDIX

TABLE A-1. Mean Perceived-Quality Scores for Men and Women Scientists

FIELD	MEN	s.d.	N	WOMEN	s.d.	N	BOTH SEXES COMBINED	N
Entire Sample of 582*	4.39	.63	446	4.15	.48	136	4.33	582
Sociology*	4.30	.53	129	4.05	.47	44	4.23	173
Psychology*	4.42	.55	129	4.15	.48	44	4.35	173
Biological Sciences**	4.42	.73	188	4.24	.50	48	4.38	236
Botany	4.36		47	4.20		12	4.33	59
Molecular and Developmental	4.28		47	4.22		12	4.26	59
Biochemistry	4.56		46	4.35		12	4.52	59
Physiology	4.49		47	4.17		12	4.43	59
			446			136		582

*Differences between male and female perceived-quality scores statistically significant at .01 level.
**Differences between male and female perceived-quality scores statistically significant at .05 level.

TABLE A-2.   Visibility Scores of Male and Female Scientists (Percent)

VISIBILITY SCORE	%	NUMBER OF SCIENTISTS	%	NUMBER OF MALE SCIENTISTS	%	NUMBER OF FEMALE SCIENTISTS
80–100	9	52	9	42	7	10
60–79	9	52	11	48	3	4
50–59	8	47	8	34	10	13
40–49	8	49	10	44	4	5
30–39	12	67	12	54	10	13
20–29	20	115	21	94	16	21
10–19	22	131	21	92	29	39
0–09	12	69	9	39	22	30
	100	582	101	447	101	135

Statistics:
Total Sample: Mean = 35.5
Standard Deviation = 25.6
Coefficient of Variation = .721
Males: Mean = 37.6
Standard Deviation = 25.7
Coefficient of Variation = .684
Females: Mean = 28.3
Standard Deviation = 24.0
Coefficient of Variation = .848
Difference between male and female mean visibility scores are statistically significant at the .01 level.

TABLE A-3.   The Relationship between Sex Status and Two Forms of Reputational Standing (Zero-Order Correlation Coefficients)

FIELD	REPUTATIONAL STANDING		N
	Visibility	*Perceived Quality*	
Entire Sample†	−.19*	−.18*	582
Sociology	−.25*	−.21**	173
Psychology	−.22**	−.22**	173
Biological Sciences†	−.12***	−.12***	236
Botany	.00	−.11	59
Molecular and Developmental	−.08	−.03	59
Biochemistry	−.07	−.18***	59
Physiology	−.33**	−.16	59

NOTES:
  * = Statistically significant at .001 level.
  ** = Statistically significant at .01 level.
  *** = Statistically significant at .05 level.
  † = Visibility and perceived-quality scores were standardized for the entire sample and for the biological sciences before the correlations were computed.
Sex status was coded: 1 = male; 2 = female.

TABLE A-4. Predicting Reputational Standing of Scientists by Pairs of Statuses in the Status-Set: OLS Regression Analysis

PAIRS OF STATUSES	DEPENDENT VARIABLE: REPUTATIONAL STANDING							
	Visibility				Perceived Quality of Work			
	r	B	Std. Error of B	Beta Coefficient	r	B	Std. Error of B	Beta Coefficient
1. Sex Status	−.19	−8.25	2.41	−.17	−.18	−.21	.06	−.16
Rank of Current Dept.	.30			.29	.29			.28
2. Sex Status	−.19	−9.31	2.44	−.19	−.18	−.24	.06	−.19
Rank of Ph.D. Dept.	.21			.21	.13			.13
3. Sex Status	−.19	−8.40	2.51	−.17	−.18	−.21	.06	−.16
No. of Honorific Awards	.28			.27	.27			.26
4. Sex Status	−.19	−6.24	2.34	−.15	−.18	−.19	.06	−.14
Academic Rank	.35			.33	.30			.28
5. Sex Status	−.19	−3.67	2.46	−.07	−.18	−.09	.06	−.07
Research Performance	.56			.54	.56			.54

TABLE A–5. Matrix of Correlations Including Indicators of Reputational Standing and Seven Scientific Statuses

	X_1	X_2	X_3	X_4	X_5	X_6	X_7	X_8	X_9
X_1 Perceived Quality	—								
X_2 Visibility	.677	—							
X_3 Sex Status	-.176	-.191	—						
X_4 Rank of Dept.	.290	.300	-.064	—					
X_5 Rank of Ph.D. Dept.	.128	.212	.002	.203	—				
X_6 No. of Awards	.267	.282	-.066	.141	.081	—			
X_7 Academic Rank	.302	.350	-.138	.016	.032	.084	—		
X_8 Age	.096	.174	.050	.055	-.016	.531	.575	—	
X_9 Scientific Research Performance	.559	.557	-.226	.193	.090	.277	.274	.047	—
Mean†	4.332	35.479	1.232	25.658	18.548	.920	3.515	48.749	.007
Standard Deviation	.607	25.581	.422	18.811	18.371	1.239	.681	9.237	1.744

NOTE: †The means and standard deviations presented are for the variables before standardization. The correlations presented in this matrix involving perceived quality, visiblity, rank of current department, rank of Ph.D. department, and scientific research performance are all based upon standardized variables. The other variables did not need to be standardized because of their common metric. For most pairs of correlations, there is little difference between the standardized and unstandardized correlations.

TABLE A–6. Correlation Matrix of Reputational Standing and Statuses in Scientists' Status-Sets (Men above the Diagonal; Women below the Diagonal)

		X_1	X_2	X_3	X_4	X_5	X_6	X_7	X_8
					MALE SCIENTISTS				
	X_1 = Visibility	—	.320	.307	.687	.571	.285	.197	.386
	X_2 = No. of Awards	.079	—	.175	.298	.291	.102	.046	.128
	X_3 = Rank of Current Dept.	.259	.007	—	.300	.190	-.038	.218	.006
	X_4 = Perceived Quality	.581	.068	.237	—	.550	.192	.125	.328
FEMALE	X_5 = Scientific Research Performance	.390	.083	.232	.529	—	.104	.108	.276
SCIENTISTS	X_6 = Age	-.063	-.116	-.096	-.161	-.188	—	.046	.575
	X_7 = Age of Ph.D. Dept.	.278	.096	.169	.154	.022	-.027	—	.048
	X_8 = Academic Rank	.170	-.122	-.106	.141	.202	.459	.003	—

BIBLIOGRAPHY

ALLISON, P. D., AND STEWART, J. A.
1974   "Productivity Differences Among Scientists: Evidence for Accumulative Advantage." *American Sociological Review* 39: 596–606.

ARISTIDES
1975   "Sex and the Professors." *American Scholar* 3:357–63.

ARMOR, D.
1972   "The Evidence of Busing." *Public Interest* 28.

ARROW, K. J.
1972   "Models of Job Discrimination." In *Racial Discrimination in Economic Life*, ed. A. H. Pascal, Chapter 2. Lexington, Mass.: D. C. Heath & Co., Lexington Books.

ASTIN, H. S.
1969   *The Woman Doctorate in America*. New York: Russell Sage Foundation.

ATWOOD, C.
1972   "Women in Fellowship and Training Programs." In Association of American Colleges, "Women in Graduate Education: Clues and Puzzles Regarding Institutional Discrimination," unpublished mineo.

BABCOCK, B. A., A. E. FREEDMAN, E. NORTON, AND S. C. ROSS
1975   *Sex Discrimination and the Law: Causes and Remedies*. Boston: Little, Brown & Co.

BALES, R. F.
1950   *Interaction Process Analysis: A Method for the Study of Small Groups*. Cambridge, Mass.: Addison-Wesley Press.

BANDURA, A., D. ROSS, AND S. A. ROSS
1963   "Imitation of Film-Mediated Aggressive Models." *Journal of Abnormal and Social Psychology*, 66, 3–11.
1963   "A Comparative Test of Status Envy, Social Power, and Secondary Reinforcement Theories of Identificatory Teaching." *Journal of Abnormal and Social Psychology*, 67, 527–534.

BARBER, B.
1952   *Science and the Social Order*. New York: Free Press.
1961   "Resistance by Scientists to Scientific Discovery." *Science* 134: 596–602.
1971   "Function, Variability, and Change in Ideological Systems." In *Stability and Social Change*, ed. B. Barber and A. Inkeles, pp. 244–64. Boston: Little, Brown & Co.

———, J. J. LALLY, J. MAKARUSHKA, AND D. SULLIVAN
1973   *Research on Human Subjects: Problems of Social Control in Medical Experimentation*. New York: Russell Sage Foundation.

BAVELAS, A.
1951   "Communication Patterns in Task-Oriented Groups." in *The Policy Sciences*, ed. D. Lerner and H. Passwell, pp. 193–202. Stanford: Stanford University Press.

BAYER, A. E.
1973   *Teaching Faculty in Academe: 1972–73*. Research Report no. 8. Washington, D. C.: American Council on Education.

———, AND H. S. ASTIN
1968   "Sex Differences in Academic Rank and Salary Among Science Doctorates in Teaching." *Journal of Human Resources* 3, no. 2:191–201.
1975   "Sex Differentials in the Academic Reward System." *Science* 188: 796–802.

BAYER, A. E., AND J. FOLGER
1966   "Some Correlates of a Citation Measure of Productivity in Science." *Sociology of Education* 39:381–90.

BECKER, G. S.
  1971   *The Economics of Discrimination,* 2nd ed. Chicago: University of Chicago Press.

BELL, D.
  1973   *The Coming of Post-Industrial Society.* New York: Basic Books.

BELL, D. A., JR.
  1970   "In Defense of Minority Admissions Programs: A Response to Professor Graglia."
          University of Pennsylvania Law Review 119:364-70.

BELLER, A. H.
  1977   "The Impact of Equal Employment Opportunity Laws on the Male/Female Earning
          Differential." Paper presented at the Department of Labor and Bernard College
          Conference on Women in the Labor Market. New York.

BERELSON, B.
  1960   *Graduate Education in the United States,* p. 32. New York: McGraw-Hill Book Co.

BERGMANN, B.
  1971   "The Effect of White Incomes on Discrimination in Employment." *Journal of
          Political Economy* 79:294-313.

BERNARD, J.
  1964   *Academic Women.* University Park, Pa.: Pennsylvania State University Press.

BICKEL, P. J., E. A. HAMMEL, AND J. W. O'CONNELL
  1975   "Sex Bias in Graduate Admissions: Data from Berkeley." *Science* 7:398-404.

BIRD, C., AND S. W. BRILLER
  1968   *Born Female: The High Cost of Keeping Women Down.* New York: David McKay
          Co.

BLALOCK, H. M.
  1967   *Toward a Theory of Minority Group Relations.* New York: John Wiley & Sons.

BLAU, P. M.
  1975   "Structural Constraints of Status Complements." In *The Idea of Social Structure,* ed.
          L. A. Coser, pp. 117-38. New York: Harcourt Brace Jovanovich.

———, AND O. D. DUNCAN
  1967   *The American Occupational Structure.* New York: John Wiley & Sons.

BLOCK, N. J., AND G. DWORKIN, EDS.
  1976   *The IQ Controversy.* New York: Pantheon Books.

BOWEN, W., AND T. A. FINEGAN
  1969   *The Economics of Labor Force Participation.* Princeton, N.J.: Princeton University
          Press.

BREST, P.
  1975   "Justice and the Cultural Division of Labor in the United States." Reprint of a paper
          delivered at Stanford University Work in Progress Seminar.
  1975   *Processes of Constitutional Decision Making: Cases and Materials.* Boston: Little,
          Brown & Co.

BRIM, O. G., D. GLASS, J. NEULINGER, AND I. J. FIRESTONE
  1969   *American Beliefs and Attitudes about Intelligence.* New York: Russell Sage
          Foundation.

BRITTAIN, J.
  1976   *The Inheritance of Socioeconomic Status.* Washington, D.C.: Brookings Institution.

BROUN H., AND G. BRITT
  1931   *Christians Only,* pp. 72-124. New York: Vanguard Press.

BROWNLEE, W. E., AND M. M. BROWNLEE
    1976   *Women in the American Economy: A Documentary History, 1675 to 1929.* New Haven, Conn.: Yale University Press.

BRYAN, A. I., AND E. G. BORING
    1947   "Women in American Psychology: Factors Affecting Their Professional Careers." *American Psychologist* 11:3–20.

BUDNER, S., AND J. MEYER
    1964   "Women Professors." In *Academic Women,* ed. J. Bernard, University Park, Pa.: Pennsylvania State University Press.

BUCK, P.
    1941   *Of Men and Women.* New York: John Day Co.

CAHN, E.
    1955   "Jurisprudence." *New York University Law Review* 30.

CAIN, G. C.
    1966   *Married Women in the Labor Force.* Chicago: Chicago University Press.

CALABRESI, G.
    1975   "Concerning Cause in the Law of Torts: An Essay for Harry Kalven, Jr." *University of Chicago Law Review* 1.

CALKINS, M. W.
    1930   "Autobiography." In *The History of Psychology in Autobiography,* ed. Carl Murchison, vol. 1. Worcester, Mass.: Clark University Press.

CAMPBELL, D., AND J. STANLEY
    1963   "Experimental and Quasi-Experimental Designs for Research on Teaching." In *Handbook of Research on Teaching,* ed. N. L. Gage, pp. 171–246. Chicago: Rand McNally & Co.

CANDOLLE, M. A. DE
    1913   *Histoire des Sciences et des Savants depuis Deux Siècles.* Quoted in H. J. Mozans, *Women in Science,* p. 392. Cambridge, Mass.: MIT Press, 1974.

CANTRIL, H.
    1951   *Public Opinion 1935–46.* Princeton, N.J.: Princeton University Press.

CAPLOVITZ, D.
    1960   "Student-Faculty Relations in Medical School." Ph.D. diss., Columbia University.

CAPLOW, T., AND R. J. McGEE
    1958   *The Academic Marketplace.* New York: Basic Books.

CARTTER, A. M.
    1966   *An Assessment of Quality in Graduate Education.* Washington, D.C.: American Council on Education.

CENTRA, J. A.
    1974   *Women, Men, and the Doctorate.* Princeton, N.J.: Educational Testing Service.

CHAFE, W.
    1972   *The American Woman: Her Changing Social, Economic, and Political Role.* London: Oxford University Press.

CHOWDHRY, K., AND T. M. NEWCOMB
    1952   "The Relative Abilities of Leaders and Non-Leaders to Estimate Opinions of Their Own Groups." *Journal of Abnormal and Social Psychology* 47:51–57.

CLARK, K.
    1960   "The Desegregation Cases: Criticism of the Social Scientists' Role." *Villanova Law Review* 5.

CLARK, K. E.
　　　　　America's Psychologists: A Survey of a Growing Profession. Washington, D.C.:
　　　　　American Psychological Association.

COHEN, M.; T. NAGEL, AND T. SCANLON, EDS.
　　1977　*Equality and Preferential Treatment.* Princeton, N. J.: Princeton University Press.

COLE, J. R., AND S. COLE
　　1973　*Social Stratification in Science.* Chicago: University of Chicago Press.
　　1976　"The Reward System of the Social Sciences." In *Controversies and Decisions: The
　　　　　Social Sciences and Public Policy.* ed. C. Frankel, pp. 55–88. New York: Russell
　　　　　Sage Foundation.

COLE, J. R., AND J. A. LIPTON
　　1977　"The Reputations of American Medical Schools." *Social Forces* 55: 662–84.

COLE, J. R., AND H. ZUCKERMAN
　　1975　"The Emergence of a Scientific Specialty: The Self-Exemplifying Case of the
　　　　　Sociology of Science." In *The Idea of Social Structure: Papers in Honor of Robert K.
　　　　　Merton,* ed. L. A. Coser, pp. 139–74. New York: Harcourt Brace Jovanovich.

COLE, S.
　　1975　"The Growth of Scientific Knowledge: Theories of Deviance as a Case Study." In
　　　　　The Idea of Social Structure: Papers in Honor of Robert K. Merton, ed. L. A. Coser,
　　　　　pp. 175–220. New York: Harcourt Brace Jovanovich.

———, J. R. COLE, AND L. DIETRICH
　　1977　"Measuring the Cognitive State of Scientific Disciplines." In *Toward a Metric of
　　　　　Science,* ed. Y. Elkana, J. Lederberg, R. K. Merton, A. Thackray, and H. Zuckerman.
　　　　　New York: John Wiley & Sons, Wiley-Interscience.

　　1978　"Scientific Reward Systems: A Comparative Analysis." In *Research in Sociology of
　　　　　Knowledge, Sciences and Art,* ed. R. A. Jones, pp. 167–190. Greenwich, Conn.: JAI
　　　　　Press.

　　1979　"Age and Scientific Performance." *American Journal of Sociology* 84: 958–977.

COLE, S., L. RUBIN, AND J. R. COLE
　　1978　*Peer Review in the National Science Foundation.* Washington, D.C.: National
　　　　　Academy of Sciences.

COLEMAN, J. S.
　　1960　"Adolescent Subculture and Academic Achievement." *American Journal of
　　　　　Sociology* 65:337–47.
　　1973　Review symposium, *Inequality: A Reassessment of the Effect of Family and Schooling
　　　　　in America,* by D. Jencks et al. *American Journal of Sociology* 78: 1523–1527.

Columbia University Bulletin
　　1976　*School of Law:* 7/26. p. 97.

COMMITTEE W, PRELIMINARY REPORT
　　1921　"Status of Women in College and University Faculties." *Bulletin of the American
　　　　　Association of University Professors* 7, no. 6:21–32.

COOPER, G.; H. RABB, AND H. J. RUBIN
　　1975　*Fair Employment Litigation.* St. Paul, Minn.: West Publishing Co.

COSER, L. A., ed.
　　1975　*The Idea of Social Structure: Papers in Honor of Robert K. Merton.* New York:
　　　　　Harcourt Brace Jovanovich.

COSER, R. L.
　　1975　"The Complexity of Roles as a Seedbed of Individual Autonomy." In *The Idea of
　　　　　Social Structure,* ed. L. A. Coser, pp. 237–64. New York: Harcourt Brace
　　　　　Jovanovich.

COUNCIL OF THE AMERICAN PHYSICAL SOCIETY
1972   Report of the Committee in Physics. "Women in Physics," p. 26, unpublished mimeo.

COURNAND, A., AND H. ZUCKERMAN
1970   "The Code of Science: Analysis and Some Reflections on Its Future." *Studium Generale* 23:941–62.

CRANE, D.
1965   "Scientists at Major and Minor Universities: A Study of Productivity and Recognition." *American Sociological Review* 30:699–714.

CROSSLAND, F. E.
1971   *Minority Access to College.* New York: Schocken Books.

D'ANDRADE, R. G.
1966   "Sex Differences and Cultural Institutions." In *The Development of Sex Differences,* ed. E. E. Maccoby. Stanford: Stanford University Press.

DAVIDSON, K. M., R. B. GINSBURG, AND H. H. KAY
1974   *Sex-Based Discrimination.* St. Paul, Minn.: West Publishing Co.

DOERINGER, P. B., AND M. J. PIORE
1971   *Internal Labor Markets and Manpower Analysis.* Lexington, Mass.: D. C. Heath & Co., Lexington Books.

DOUVAN, E.
1976   "The Role of Models in Women's Professional Development." *Psychology of Women Quarterly* 1:5.

DUNCAN, O. D.
1957   "The Measurement of Population Distribution." *Population Studies* 11, no. 1:40.
1967   "Discrimination against Negroes." *Annals of the American Academy of Political and Social Science* 371:87.
1968   "Ability and Achievement." *Eugenics Quarterly* 15:1–11.
1968   "Inheritance of Poverty or Inheritance of Race?" In *On Understanding Poverty,* ed. D. P. Moynihan, p. 108. New York: Basic Books.

———, D. FEATHERMAN, AND B. DUNCAN
1973   *Socioeconomic Background and Achievement.* New York: Seminar Press.

DUNCAN, O. D., AND R. W. HODGE
1963   "Education and Occupational Mobility." *American Journal of Sociology* 68:629–44. As reported in P. M. Blau and O. D. Duncan, *The American Occupational Structure.* New York: John Wiley & Sons, 1967.

DURKHEIM, E.
1897   *Suicide.* Translated from the French. New York: Free Press, 1951.

DWORKIN, R.
1977   "Why Bakke Has No Case." *New York Review of Books,* 10 Nov., p. 11.

EDGE, D. O., AND M. J. MULKAY
1976   *Astronomy Transformed.* New York: John Wiley & Sons.

EDGEWORTH, F. Y.
1922   "Equal Pay to Men and Women for Equal Work." *Economic Journal* 32:431–57.

EHRLICH, I.
1977   "Capital Punishment and Deterrence: Some Further Thoughts and Additional Evidence." *Journal of Political Economy* 85, no. 4:741–788.

ELIOT, G.
1871–72 *Middlemarch.* Reprinted, Middlesex, England: Penguin Books, 1965.

ELLIS, H.
1900   *Man and Woman.* New York: Charles Scribner's Sons.

EPSTEIN, C. F.
1970   *Woman's Place: Options and Limits in Professional Careers.* Berkeley: University of California Press.
1973   "Positive Effects of the Multiple Negative: Explaining the Success of Black Professional Women." *American Journal of Sociology* 78:912.

ETZIONI, A.
1961   *A Comparative Analysis of Complex Organizations.* New York: Free Press.

FAVA, S. F.
1960   "The Status of Women in Professional Sociology." *American Sociological Review* 25:271.

FEATHERMAN, D.
1972   "Achievement Orientations and Socioeconomic Career Attainments." *American Sociological Review* 37:131-43.

FIDELL, L. S.
1970   "Empirical Verification of Sex Discrimination in Hiring Practices in Psychology." *American Psychologist* 12:1094-1098.

FISS, O. M.
1977   "Groups and the Equal Protection Clause." In *Equality and Preferential Treatment,* ed. M. Cohen, T. Nagel, and T. Scanlon. Princeton, N.J.: Princeton University Press.

FOLGER, J. K., H. S. ASTIN, AND A. E. BAYER
1970   *Human Resources and Higher Education.* New York: Russell Sage Foundation.

FORMBY, J. P.
1968   "The Extent of Wage and Salary Discrimination Against Negro-White Labor." *Southern Economic Journal* 35:140-150.

FRANKEL, C.
1971   "Equality of Opportunity." *Ethics* 81:181-211.

FRANKFURTER, F.
1916   "Hours of Labor and Realism in Constitutional Law." *Harvard Law Review* Vol. 29, no. 4 (February):333-373.

FRANKLIN, M. A.
1971   *Tort Law and Alternatives.* New York: Foundation Press.

FREUD, S.
1959   " 'Civilized' Sexual Morality and Modern Nervous Illness." In *Collected Works,* vol. 9, pp. 198-99. London: Hogarth Press.

FRIEDAN, B.
1963   *The Feminine Mystique.* New York: Dell Publishing Co.

FRIEDMAN, M.
1953   *Essays in Positive Economics.* Chicago: University of Chicago Press.

FUCHS, V.
1971   "Differences in Hourly Earnings Between Men and Women." *Monthly Labor Review* vol. 44, no. 5, 9-15.

GASTON, J.
1973   *Originality and Competition in Science.* Chicago: University of Chicago Press.

GATES, M. J.
1976   "Occupational Segregation and the Law." *Signs* 1, no. 3, pt. 2:61-74

Gerstel, N.
    1978   "Commuter Marriage." Ph.D. diss., Columbia University.

Ghiselli, E. E.
    1955   "The Measurement of Occupational Attitude," *University of California Publication in Psychology*, vol. 8, Berkeley: University of California.
—————, and C. W. Brown
    1948   "The Effectiveness of Intelligence Test in the Selection of Workers." *Journal of Applied Psychology* 6:575–80.

Gildersleeve, V. C.
    1954   *Many a Good Crusade*. New York: Macmillan Co.

Gilman, C. P.
    1898   *Women and Economics*. Boston: Small, Maynard & Co.

Ginsburg, R. B.
           "Women, Men and the Constitution: Key Supreme Court Rulings." Preprint, *Columbia University Law School*, p. 14.
    1975   "Gender and the Constitution." *University of Cincinnati Law Review* 1:1–42.

Glazer, N.
    1975   *Affirmative Discrimination, Ethnic Inequality and Public Policy*. New York: Basic Books.

Goffman, E.
    1959   *The Presentation of Self in Everyday Life*. New York: Doubleday & Co., Anchor Books.

Goodenough, F. P.
    1949   *Mental Testing: Its History, Principles, and Application*. Reprints, New York: Holt, Rinehart & Winston, 1969.

Goodman, F.
    1972   "De Facto School Segregation: A Constitutional and Empirical Analysis." *California Law Review* 60.

Goodsell, W.
    1924   *The Education of Women: Its Social Background and Its Problems*. New York: Macmillan Co.
    1929   "The Educational Opportunities of American Women—Theoretical and Actual." *The Annals* 143:1–13.

Gordon, N., T. Morton, and I. Braden
    1974   "Faculty Salaries: Is There Discrimination by Sex, Race, and Discipline?" *American Economic Review* 64:419–427.

Graglia, L. A.
    1970   "Special Admission of the 'Culturally Deprived' to Law School." *University of Pennsylvania Law School Review* 119:351–63.

Graham, P. A.
    1969   "Women in Academe." *Science* 395:1284–1289.

Hagstrom, W. O.
    1965   *The Scientific Community*. New York: Basic Books.
    1971   "Inputs, Outputs, and Prestige of American University Science Departments." *Sociology of Education* 44:375–97.
    1974   "Competition in Science." *American Sociological Review* 39:1–18.

Hahn, O.
    1966   *Otto Hahn: A Scientific Autobiography*. New York: Charles Scribner's Sons.

HARGENS, L.
    1971   "The Social Context of Scientific Research." Ph.D. diss., University of Wisconsin.
——, AND W. O. HAGSTROM
    1967   "Sponsored and Contest Mobility of American Academic Scientists." *Sociology of Education* 40:24–38.
    1969   "Patterns of Mobility of New Ph.D.'s Among American Academic Institutions." *Sociology of Education* 42:18–37.

HARMON, L. R.
    1963   "High School Backgrounds of Science Doctorates." *Science* 133:679–88.
    1963   "The Development of a Criterion of Scientific Competence." In *Scientific Creativity: Its Recognition and Development,* ed. C. W. Taylor, and F. Barron, pp. 44–53. New York: John Wiley & Sons.
    1965   *High School Ability Patterns: A Background Look from the Doctorate.* Scientific Manpower Report no. 6. Washington, D.C.: National Academy of Sciences–National Research Council.
——, AND H. SOLDZ
    1963   *Doctorate Production in United States Universities.* Washington, D.C.: National Academy of Sciences–National Research Council. Publication no. 1142. Table 26.

Harvard Law Review
    1975   "Beyond the Prima Facie Case in Employment Discrimination Law: Statistical Proof and Rebuttal." Notes, *Harvard Law Review* 2:387–422.

HAUSER, R.
    1970   "Educational Stratification in the United States." In *Social Stratification; Research and Theory for the 1970's,* ed. E. O. Laumann, p. 112. Indianapolis: Bobbs-Merrill Co.

HAWTHORNE, M. O.
    1929   "Women as College Teachers." *The Annals* 143:146–153.

HERRNSTEIN, R.
    1971   *IQ in the Meritocracy.* Boston: Little, Brown & Co.

HETHERINGTON, E. M.
    1965   "A Developmental Study of the Effects of Sex of the Dominant Parent on Sex-Role Preference, Identification, and Imitation in Children." *Journal of Personality and Social Psychology* 2:188–194.

HICKS, DAVID J.
    1965   "Imitation and Retention of Film-Mediated Aggressive Peer and Adult Models." *Journal of Personality and Social Psychology* 2:97–100.

HIESTAND, D. L.
    1970   *Discrimination in Employment: An Appraisal of the Research.* Policy Papers in Human Resources and Industrial Relations, no. 16. Washington, D.C.: Institute of Labor and Industrial Relations, University of Michigan–Wayne State University, and National Manpower Policy Task Force.

HO, C.-J.
    1928   "Personnel Studies of Scientists in the United States," pp. 23, 24. Master's thesis, Teachers College, Columbia University.

HOCHBAUM, G., et al.
    1955   "Socioeconomic Variables in a Large City." *American Journal of Sociology* 61:31–38. As reported in P. M. Blau and O. D. Duncan, *The American Occupational Structure.* New York: John Wiley & Sons, 1967.

HODGE, R. W., P. SIEGAL, AND P. ROSSI
    1964   "Occupational Prestige in the United States, 1925-63." *American Journal of Sociology* 70:286-302.

HOLTON, G.
    1973   *Thematic Origins of Scientific Thought: Kepler to Einstein.* Cambridge: Harvard University Press.

HORNER, M. S.
    1970   "Femininity and Successful Achievement: A Basic Inconsistency?" In *Feminine Personality and Conflict,* ed. J. Bardwick et al., pp. 45-77. Belmont, Calif.: Brooks/Cole.

HUDSON, L.
    1970   "Intellectual Maturity." In *The Ecology of Human Intelligence,* ed. P. Hudson. Middlesex, England: Penguin Books.

HUGHES, E. C.
    1945   "Dilemmas and Contradictions in Status." *American Journal of Sociology* 50:353-59.

HUGHES, R. M.
    1925   *A Study of the Graduate School of America.* Oxford, Ohio: Miami University Press.

HUTCHINSON, E. J.
    1929   *Women and the Ph.D.* Bulletin no. 2. Greensboro, N.C.: North Carolina College for Women.

HYMAN, H. H.
    1966   "The Value Systems of Different Classes." In *Class, Status, and Power,* 2nd edition, ed. R. Bendix and S. M. Lipset, pp. 488-99. New York: Free Press.

JACKSON, M.
    1973   "Affirmative Action—Affirmative Results." *American Sociological Association Footnotes* 9.

JAMES, W.
    1885   *The Meaning of Truth.* Reprint, New York: Longmans, Green & Co., 1932.

JAY, J. M.
    1971   *Negroes in Science: Natural Science Doctorates 1876-1969.* Detroit: Balump Publishing Co.

JENCKS, C., et al.
    1972   *Inequality: A Reassessment of the Effect of Family and Schooling in America.* New York: Basic Books.

JENSEN, A. R.
    1969   "How Much Can We Boost IQ and Scholastic Achievement?" *Harvard Educational Review* 30:1-123.

JOHNSON, D. M.
    1948   "Applications of the Standard Score IQ to Social Statistics." *Journal of Social Psychology* 27:217-27. As reported in *Intelligence in the United States,* p. 73. New York: Springer Publishing Co., 1957.

JOHNSON, G. E., AND F. P. STAFFORD
    1975   "Women and the Academic Labor Market." In *Sex, Discrimination, and the Division of Labor,* ed. C. Lloyd, New York: Columbia University Press.

JOHNSTON, J. D., JR., AND C. L. KNAPP
    1971   "Sex Discrimination by Law: A Study in Judicial Perspective." *New York University Law Review* 46:674-747.

KANOWITZ, L.
1973   *Sex Roles in Law and Society: Cases and Materials.* Albuquerque, N.M.: University of New Mexico Press.

KAPLAN, J.
1966   "Equal Justice in the Unequal World: Equality for the Negro—the Problem of Special Treatment." *Northwestern University Law Review* 3.

KATZ, D., AND R. L. KAHN
1966   *The Social Psychology of Organizations.* New York: John Wiley & Sons.

KELLER, E. F.
1977   "The Anomaly of a Woman in Physics." In *Working It Out,* ed. S. Ruddick and P. Daniels, pp. 77–91. New York: Pantheon Books.

KELLEY, J.
1973   "Causal Chain Models for the Socioeconomic Career." *American Sociological Review* 38:481–93.

KERN, S.
1975   *Anatomy and Destiny.* Indianapolis: Bobbs-Merrill Co.

KITSON, H. D.
1926   "Relation Between Age and Promotion of University Professors." *School and Society* 24:400–404.

KLEVORICK, A.
1975   "Jury Size and Composition: An Economic Approach." Preprint obtained from author.

KOHLBERG, L. AND E. ZIGLER
1967   "The Impact of Cognitive Maturity on the Development of Sex-Role Attitudes in the Years 4–8." *Genetic Psychology Monographs,* 75:84–165.

KREUGER, A. O.
1963   "The Economics of Discrimination." *Journal of Political Economy* 71:481–86.

KUHN, T.
1962   *The Structure of Scientific Revolutions.* Chicago: University of Chicago Press.

LAKATOS, I.
1970   "Falsification and the Methodology of Scientific Research Programmes." In *Criticism and the Growth of Knowledge,* ed I. Lakatos and A. Musgrave, pp. 91–195. Cambridge, England: Cambridge University Press.

1970   "History of Science and Its Rational Reconstructions." In *Boston Studies in the Philosophy of Science,* vol. 8, pp. 92–182. Dordrecht, Holland: D. Reidel Publishing Co.

LASORTE, M. A.
1971   "Sex Differences in Salary Among Academic Sociology Teachers." *American Sociologist* 6:304.

LAWRENCE, C., III
1977   "The Bakke Case: Are Racial Quotas Defensible?" *Saturday Review,* 15 Oct., pp. 14–16.

LAZARSFELD, P. F., AND W. THIELENS, JR.
1958   *The Academic Mind.* Glencoe, Ill.: Free Press.

LEIFER, A. D.
1966   "The Relationship between Cognitive Awareness in Selected Areas and Differential Imitation of a Same-Sex Model." Unpublished M.A. Thesis, Stanford University.

LENZER, G., ed.
1975   *Auguste Comte and Positivism: The Essential Writings.* New York: Harper & Row, Torchbooks.

LESTER, R.
1974   *Antibias Regulations of Universities: Faculty Problems and Their Solutions.* New York: McGraw-Hill Book Co., for the Carnegie Commission on Higher Education.

LEVIN, H. M.
1975   "Education, Life Chances, and the Courts: The Role of Social Science Evidence." *Law and Contemporary Problems* 39:217-40.

LEWIN, A. Y., AND L. DUCHAN
1971   "Women in Academia." *Science* 173:892-95.

LIPSET, S. M., AND E. C. LADD, JR.
1971   "Jewish Academics in the United States: Their Achievements, Culture and Politics." *American Jewish Year Book 1971-72*, pp. 89-130. New York: Jewish Publication Society of America.

LLOYD, C., ed.
1975   *Sex, Discrimination, and the Division of Labor.* New York: Columbia University Press.

LOEHLIN, J. C., G. LINDZEY, AND J. N. SPUHLER
1975   *Race Differences in Intelligence.* San Francisco: W. H. Freeman.

LONG, C. D.
1960   *Wages and Earnings in the United States, 1860-1890.* Princeton, N.J.: Princeton University Press.

LUBKIN, D. G.
1971   "Women in Physics." *Physics Today* 24.

MACCOBY, E. E., AND C. N. JACKLIN
1975   *The Psychology of Sex Differences.* Stanford: Stanford University Press.

MACKINNON, D. W.
1962   "The Nature and Nurture of Creative Talent." *American Psychologist* 17:487-88.

MADDEN, J. F.
1973   *The Economics of Sex Discrimination.* Lexington, Mass.: D. C. Heath & Co.

MAGOUN, H. W.
1966   "The Cartter Report on Quality in Graduate Education." *Journal of Higher Education* 37, no. 9:481-492.

MARSHALL, R.
1974   "The Economics of Racial Discrimination: A Survey." *Economic Literature* 12:849-71.

MARWELL, G., R. ROSENFELD, AND S. SPILERMAN
1976   "Residence Location, Geographic Mobility, and the Attainment of Women in Academia." Unpublished report.

MCCARTHY, J. L., AND D. WOLFLE
1975   "Doctorates Granted to Women and Minority Group Members." *Science* 189:856-59.

MEAD, M.
1974   *Ruth Benedict,* p. 8. New York: Columbia University Press.

MEREI, F.
1949   "Group Leadership and Institutionalization." *Human Relations* 2:23-29.

MERTON, R. K.
1949   "Discrimination and the American Creed." In *Discrimination and National Welfare*. New York: Harper & Brothers.
1949   Part 1; "Manifest and Latent Functions"; "Science and the Democratic Social Structure"; "The Self-Fulfilling Prophecy" (first published, *Antioch Summer Review* 9 [1949]:192–210). All in *Social Theory and Social Structure*. Reprint, New York: Free Press, 1968.
1957   "Priorities in Scientific Discovery: A Chapter in the Sociology of Science." *American Sociological Review* 22:635–59.
1963   "The Ambivalence of Scientists." *Bulletin of the John Hopkins Hospital* 112:77–97.
1968   "The Matthew Effect in Science." *Science* 199:55–63.
1973   "Social Conflict over Styles of Sociological Work" (originally published in 1961); "The Perspectives of Insiders and Outsiders" (originally published in 1972). Both in *The Sociology of Science: Theoretical and Empirical Investigations*. Reprint, Chicago: University of Chicago Press.

MEYER, M.
1961   Introduction to "Hedda Gabler and Three Other Plays. New York: Doubleday & Co., Anchor Books.

MILL, J. S.
1970   "The Subjection of Women." In *John Stuart Mill and Harriet Taylor Mill: Essays on Sex Equality*, ed. A. S. Rossi, p. 197. Chicago: University of Chicago Press.

MINCER, J.
1962   "Labor Force Participation of Married Women: A Study of Labor Supply." In *Aspects of Labor Economics*, ed. National Bureau of Economic Research. Princeton, N.J.: Princeton University Press.
1976   "Progress in Human Capital Analyses of the Distribution of Earnings." In *Personal Income Distribution*, ed. A. Atkinson. London: Royal Economic Society.

————, AND S. W. POLACHEK
1974   "Family Investments in Human Capital: Earnings of Women." *Journal of Political Economy* 2:76–110.

MITCHELL, J. M., AND R. R. STARR
1969–1971   *Aspirations, Achievement and Professional Advancement in Political Science: The Prospect of Women in the West*. Women in Political Science: Studies and Reports of the APSA Committee on the Status of Women in the Profession, 1969–1971. Washington, D.C.: American Political Science Association.

MITROFF, I.
1974   *The Subjective Side of Science*. New York: American Elsevier Publishing Co.

MOYNIHAN, D. P.
1968   *On Understanding Poverty*. New York: Basic Books.

MURCHISON, C., ed.
1930   *History of Psychology in Autobiography*, vol. I. Worcester, Mass.: Clark University Press.

MYRDAL, G.
1944   *An American Dilemma*, vol. 2. New York: Harper & Row.

National Academy of Sciences-National Research Council
1965   *Career Patterns*, Report no. 1. Washington, D.C.

National Education Association
1971   *Salaries Paid and Salary-Related Practices in Higher Education*. Research Report 1972-R5, p. 11.

Newsweek
1977   26 Sept., pp. 53–54.

O'NEIL, R. M.
1971   "Preferential Admissions: Equalizing the Access of Minority Groups to Higher Education." *Yale Law Journal* 4:699–767.

O'NEILL, R. M.
1970   "Disadvantaged Students and Equal Education—Programs for Affirmative Action." *Toledo Law Review* 227.

OPPENHEIMER, V. K.
1970   "The Female Labor Force Participation in the United States: Demographic and Economic Factors Governing Its Growth and Changing Composition." *Population Monograph Series No. 5*. University of California, Berkeley.

PARSONS, T.
1970   "Equality and Inequality in Modern Society, or Social Stratification Revisited." In *Social Stratification: Research and Theory for the 1970's*, ed. E. O. Laumann, pp. 3–72. Indianapolis: Bobbs-Merrill Co.

PEARCE, R. H., et al.
1972   *Women in the Graduate Academic Sector of the University of California: Report on an Ad Hoc Committee of the Coordinating Committee on Graduate Affairs.* Los Angeles: University of California.

PERRUCCI, C. C.
1970   "Minority Status and the Pursuit of Professional Careers: Women in Science and Engineering." *Social Forces* 49:245.

PETERSON, O. L.; L. P. ANDREWS, R. S. SPACE, AND B. B. GREENBERG
1956   "An Analytical Study of North Carolina General Practice, 1953–1954." *Journal of Medical Education* 31, pt. 2: (December):1–165.

PETTIGREW, T. F., E. L. USEEM, C. NORMAND, AND M. S. SMITH
1973   "Busing: A Review of 'the Evidence.'" *Public Interest* 30:88–118.

POLACHEK, S. W.
1975   "Discontinuous Labor Force Participation and Its Effects on Women's Market Earnings." In *Sex, Discrimination, and the Division of Labor*, ed. C. Lloyd, pp. 90–122. New York: Columbia University Press.

POLANYI, M.
1958   *Personal Knowledge.* London: Routledge & Kegan Paul.

POPPER, K.
1934   *The Logic of Scientific Discovery.* Reprinted, New York: Harper & Row, Torchbooks, 1965.

POUND, R.
1971   "Causation." *Yale Law Journal* 671:1–18.

PRICE, D. DE SOLLA
1963   *Little Science, Big Science.* New York: Columbia University Press.
1971   "Citation Measures of Hard Science and Soft Science, Technology and Non-Science." In *Communication Among Scientists and Engineers*, ed C. E. Nelson and D. K. Pollack, pp. 3–22. Lexington, Mass.: D. C. Heath & Co., Lexington Books.

PRICE, P. B.; C. W. TAYLOR, J. M. RICHARDS, JR., AND T. J. JACOBSEN
1964   "Measurement of Physician Performance." *Journal of Medical Education* 39:203–210.

REAGAN B., AND B. MAYNARD
1974   "Sex Discrimination in Universities: An Approach Through Internal Labor Market Analysis." *AAUP Bulletin* 60:11–21.

REITZES, D., AND H. ELKHANIALY
1976   "Black Physicians and Minority Group Care—the Impact of NMF." *Medical Care* 12:1052.
1976   "Black Students in Medical Schools." *Journal of Medical Education* 51: 1001–1005.

RESKIN, B. F.
1973   "Sex Differences in the Professional Life Chances of Chemists." Ph.D. diss., University of Washington.
1976   "Sex Differences in Status Attainment in Science: The Case of the Post-Doctoral Fellowship." *American Sociological Review* 41:597–612.
1978   "Scientific Productivity, Sex, and Location in the Institution of Science." *American Journal of Sociology* 5:1235–43.

RILEY, M. W., M. JOHNSON, AND A. FONER
1968   *Aging and Society,* vol. 1. New York: Russell Sage Foundation.
1969   *Aging and Society,* vol. 2. New York: Russell Sage Foundation.
1972   *Aging and Society,* vol. 3. New York: Russell Sage Foundation.

ROE, A.
1952   *The Making of a Scientist.* New York: Dodd, Mead & Co. .

ROENTHAL, D. A. H.
1970   "Starting Salaries—1970." *Clinical and Engineering News* 23 Nov.

ROOSE, K. D., AND C. J. ANDERSEN
1970   *A Rating of Graduate Programs.* Washington, D.C.: American Council on Education.

RORTY, A. O.
1977   "Dependency, Individuality, and Work." In *Working It Out,* ed. S. Ruddick and P. Daniels, pp. 38–54. New York: Pantheon Books.

ROSENBERG, R.
1975   "In Search of Woman's Nature, 1850–1920." *Feminist Studies* 1, no. 2:141–154.

ROSENBLITH, J. F.
1959   "Learning by Imitation in Kindergarten Children." *Child Development,* 30:211–223.

ROSSI, A. S.
1964   "Equality between the Sexes: An Immodest Proposal." *Daedalus:* 98–143.
1965   "Women in Science: Why So Few?" *Science* 3674:1196–1202.
1965   "Barriers to the Career Choice of Engineering, Medicine, or Science among American Women." In *Women and the Scientific Professions,* ed. J. A. Mattfeld and G. C. Van Aken. Cambridge, Mass.: MIT Press.
1970   "Status of Women in Graduate Departments of Sociology 1968–1969." *American Sociologist* 1:1–12.

——, AND A. CALDERWOOD, eds.
1973   *Academic Women on the Move.* New York: Russell Sage Foundation.

ROSSITER, M. W.
1974   "Women Scientists in America Before 1920." *American Scientist* 62, no. 3:312–323.

ST. JOHN, N.
1970   "Desegregation and Minority Group Performance." *Review of Educational Research* 40:111 ff.

SCHILPP, P. A., ed.
1974   *The Philosophy of Karl Popper.* 2 vols. La Salle, Ill.: Open Court.

SEGRÉ, E.
1970   *Enrico Fermi: Physicist.* Chicago: University of Chicago Press.

SELVIN, H., AND T. HIRSCHI
1967   *Delinquency Research.* New York: Free Press.

SHALLCROSS, R.
1940   *Should Married Women Work?* Public Affairs Pamphlet no. 49.

SHERMAN, M. J.
1975   "Affirmative Action and the American Association of University Professors." *American Association of University Professors Bulletin* 4:293-303.

SIEGAL, J. P.
1965   "On the Cost of Being a Negro." *Sociological Inquiry* 35:41-57.

SIEGAL, J. P., AND E. E. GHISELLI
1971   "Managerial Talent, Pay, and Age." *Journal of Vocational Behavior* 1:129-35.

SIMON, R. J.; S. M. CLARK, AND K. GALWAY
1967   "The Woman Ph.D.: A Recent Profile." *Social Problems* 221:228-29.

SMUTS, R. W.
1959   *Women and Work in America.* New York: Columbia University Press.

SOMIT, A., AND J. TANENHAUS
1964   *American Political Science: A Profile of the Discipline.* New York: Atherton Press.

SOWELL, T.
1976   " 'Affirmative Action' Reconsidered." *Public Interest* 42:47-65.

SPENCER, H.
1896   *Principles of Sociology,* vol. 1, 3rd ed. New York: D. Appleton.

SPREHE, J. T.
1967   "The Climate of Opinion in Sociology: A Study of the Professional Value and Belief Systems of Sociologists." Ph.D. diss., Washington University.

STANFORD UNIVERSITY
        The Study of Graduate Education at Stanford: Report of the Task Force on Women. Stanford: Stanford University.

STIGLITZ, J. E.
1973   "Approaches to the Economics of Discrimination." *American Economic Review* 63:287-95.

STINCHCOMBE, A. L.
1968   *Constructing Social Theories.* New York: Harcourt Brace Jovanovich.

STONEQUIST, E. V.
1937   *The Marginal Man.* New York: Charles Scribner's Sons.

STORER, N. W.
1966   *The Social System of Science.* New York: Holt, Rinehart & Winston.

SULLIVAN, D.
1975   "Competition in Bio-Medical Science: Extent, Structure, and Consequences." *Sociology of Education* 45:223-41.

SUTER, L. E., AND H. P. MILLER
1973   "Income Differences Between Men and Career Women." *American Journal of Sociology* 78, no. 4:962-974.

TANUR, J. M., AND R. L. COSER
1978   "Pockets of 'Poverty' in the Salaries of Academic Women." *American Association of University Professors Bulletin* 27.

TAVRIS, C., AND C. OFFIR
1977   *The Longest War: Sex Differences in Perspective.* New York: Harcourt Brace Jovanovich.

TERMAN, L. M.
1926   Genetic Studies of Genius. Vol. 1, *Mental and Physical Traits of a Thousand Gifted Children,* 2nd ed. Stanford: Stanford University Press.

———, AND M. H. ODEN
 1959 *The Gifted Child Grows Up: Twenty-five Years' Follow-up of a Superior Group.*
   Stanford: Stanford University Press.

THOMAS, W. I.
 1897 "On a Difference in the Metabolism of the Sexes." *American Journal of Sociology* 3,
   no. 11:31-63.

THOMPSON, T.
 1974 "Curbing the Black Physician Manpower Shortage." *Journal of Medical Education*
   49:944-50.

THORNDIKE, R. L., AND E. P. HAGEN
 1963 *10,000 Careers,* as quoted in L. E. Tyler, *Tests and Measurements.* Englewood
   Cliffs, N.J.: Prentice-Hall.

THUROW, L. C.
 1975 *Generating Inequality.* New York: Basic Books.

TOCQUEVILLE, A. DE
 1840 *Democracy in America,* vol. 2. Translated from the French. New York: Alfred A.
   Knopf, 1945.

TREIMAN, D. J., AND K. TERRELL
 1975 "Sex and the Process of Status Attainment: A Comparison of Working Men and
   Women." *American Sociological Review* 40, no. 2:174-200.
 1975 "Women, Work and Wages." In *Social Indicator Models,* ed. K. Land and S.
   Spilerman. New York: Russell Sage Foundation.

TUCKER, A.; D. GOTTLIEF, AND J. PEASE
 1964 *Attrition of Graduate Students at the Ph.D. Level in the Traditional Arts and Sciences.*
   Publication no. 8. East Lansing: Michigan State University.

TURNER, E. M.; M. HELPER, AND S. E. KRISKA
 1974 "Predictors of Clinical Performance." *Journal of Medical Education* 49:338-42.

TURNER, R.
 1960 "Sponsored and Contest Mobility and the School System." *American Sociological*
   Review 25:163-72.
 1966 "Some Aspects of Women's Ambition." *American Journal of Sociology* 72: 163.

TYLER, L. E.
 1963 *Tests and Measurements.* Englewood Cliffs, N.J.: Prentice-Hall.

U. S. DEPARTMENT OF COMMERCE
 1975 *Historical Statistics of the United States, Colonial Times to 1957.* Bureau of the
   Census, Series D. 654-688, p. 94. Washington, D.C.

WADE, N.
 1975 "Citation Analysis: A New Tool for Science Administrators." *Science* 2:429-32.

WAITE, L. J.
 1976 "Working Wives, 1940-1960." *American Sociological Review* 41, no. 1:65-80.

WALSTER, E., T. A. CLEARY, AND M. M. CLIFFORD
 1971 "Research Note: The Effect of Race and Sex on College Admissions." *Sociology of*
   Education 44:237-44.

WARD, W. D.
 1969 "Processes of Sex-Role Development." *Developmental Psychology* 1, no. 2:163-68.

WATSON, J. D.
 1966 "Growing Up in the Phage Group." In *Phage and the Origins of Molecular Biology,*
   ed J. Cairns, G. S. Stent, and J. D. Watson, p. 240. Cold Spring, N.Y.: Cold Spring
   Harbor Laboratory of Molecular Biology.

WATTS, P.
    1965   Introduction to *Ibsen: Plays*. Middlesex, England: Penguin Books.
WEITZMAN, L. J., D. EIFLER, E. HOKADA, AND C. ROSS
    1972   "Sex-role Socialization in Picture Books for Pre-School Children." *American Journal of Sociology* 77:1125–50.
WELCH, F.
    1967   "Labor Market Discrimination: An Interpretation of Income Differences in the Rural South." *Journal of Political Economy* 75:225–40.
WINGARD, J. R., AND J. W. WILLIAMSON
    1973   "Grades as Predictors of Physicians' Career Performance: An Evaluative Literature Review." *Journal of Medical Education* 48:311–12.
WOLFLE, D.
    1954   *America's Resources of Specialized Talent: A Current Appraisal and a Look Ahead.* New York: Harper & Brothers.
WOOD, A. D.
    1973   "The Fashionable Diseases: Women's Complaints and Their Treatment in Nineteenth-Century America." *Journal of Interdisciplinary History* 4, no. 1:25–52.
WOODY, T.
    1929   *A History of Women's Education in the United States,* vol. 4, bk. 2. New York: Science Press.
YOUNG, M.
    1958   *The Rise of Meritocracy.* Baltimore: Penguin Books.
ZEISEL, H., AND S. S. DIAMOND
    1974   " 'Convincing Empirical Evidence' on the Six-Member Jury." *University of Chicago Law Review* 2:281–95.
ZIMAN, J.
    1968   *Public Knowledge.* Cambridge, England: Cambridge University Press.
ZUCKERMAN, H.
    1967   "Nobel Laureates in Science: Patterns of Productivity, Collaboration and Authorship." *American Sociological Review* 32:391–403.
    1970   "Stratification in American Science." *Sociological Inquiry* 1:235–57.
    1971   "Women and Blacks in American Science: The Principle of the Double Penalty," pp. 34–35. Paper presented at the Symposium on Women and Minority Groups in American Science and Engineering, California Institute of Technology.
    1977   "Deviant Behavior and Social Control in Science." In *Deviant Behavior and Social Control in Science,* vol. 1, ed. E. Sagarin. Sage Annual Reviews of Studies in Deviance. Beverly Hills, Calif.: Sage Publications.
    1977   *Scientific Elite: Nobel Laureates in the United States.* New York: Free Press.
————, AND J. R. COLE
    1975   "Women in American Science." *Minerva* 13, no. 1:84.
————, AND R. K. MERTON
    1971   "Patterns of Evaluation in Science: Institutionalization, Structure and Functions of the Referee System." *Minerva* 9, no. 1:66–100. Reprinted in R. K. Merton, *The Sociology of Science.* Chicago: University of Chicago Press.
    1972   "Age, Aging and Age Structure in Science." In *A Sociology of Age Stratification,* vol. 3 of *Aging and Society,* ed. M. W. Riley, M. Johnson, and A. Foner. New York: Russell Sage Foundation.

SUBJECT INDEX

NAME INDEX